T0139866

Lecture Notes in Intelligent Transportation and Infrastructure

Series Editor

Janusz Kacprzyk, Systems Research Institute, Polish Academy of Sciences, Warszawa, Poland

The series "Lecture Notes in Intelligent Transportation and Infrastructure" (LNITI) publishes new developments and advances in the various areas of intelligent transportation and infrastructure. The intent is to cover the theory, applications, and perspectives on the state-of-the-art and future developments relevant to topics such as intelligent transportation systems, smart mobility, urban logistics, smart grids, critical infrastructure, smart architecture, smart citizens, intelligent governance, smart architecture and construction design, as well as green and sustainable urban structures. The series contains monographs, conference proceedings, edited volumes, lecture notes and textbooks. Of particular value to both the contributors and the readership are the short publication timeframe and the world-wide distribution, which enable wide and rapid dissemination of high-quality research output.

More information about this series at http://www.springer.com/series/15991

Antonio Sciarretta · Ardalan Vahidi

Energy-Efficient Driving of Road Vehicles

Toward Cooperative, Connected, and Automated Mobility

 Springer

Antonio Sciarretta
Rueil Malmaison, France

Ardalan Vahidi
Department of Mechanical Engineering
Clemson University
Clemson, SC, USA

ISSN 2523-3440 ISSN 2523-3459 (electronic)
Lecture Notes in Intelligent Transportation and Infrastructure
ISBN 978-3-030-24129-2 ISBN 978-3-030-24127-8 (eBook)
https://doi.org/10.1007/978-3-030-24127-8

This Springer imprint is published by the registered company Springer Nature Switzerland AG
The registered company address is: Gewerbestrasse 11, 6330 Cham, Switzerland

To Lorenzo, the boy who prefers writing narrative.
To Cyrus who is passionate to learn and to Parimah for her constant inspiration.

Preface

This book is the result of our research in the area of energy-efficient driving of road vehicles over the past decade. Having worked independently across the pond for most of these years, the idea of writing a book on energy-efficient driving came about when we got together, in Summer of 2017 at IFP Energies Nouvelles in Rueil-Malmaison, a suburb of Paris, to write a review paper on this topic. Parts of that review paper became the first chapter of this book. And the effort continued after, thanks to tools such as overleaf that allowed us to interactively work on the book despite our geographic distance.

The topic is very timely as rapid proliferation of vehicle and infrastructure connectivity extends vehicles' perception horizon which is important for efficient driving. At the same time, debut of automated vehicles can be a game changer, as automated vehicles are expected to be better than human drivers in processing information and following commands. Therefore, while many ideas in the book apply to both human-driven vehicles and Connected and Automated Vehicles (CAV), we place a particular emphasis on CAV technology.

A main purpose of the book is to highlight the significant potential that CAVs offer in increasing energy efficiency and reducing the environmental impact of the transportation sector and to introduce formal methods to achieve these benefits. The intended audience is students, researchers, and practicing engineers in the automotive sector but perhaps policy-makers in the transportation field and roboticists developing autonomous cars could find value in the book. Due to the book's reliance on formal methods in optimization and optimal control theory, those with a background in optimization and control can move along at a faster pace. Yet, we did our best to introduce the mathematics of energy-efficient driving gradually so those without such background can benefit from the book as well.

The book can be used as the main text for a one-semester graduate course on energy-efficient mobility. The first four chapters are tutorial introductions and assume no or little domain knowledge. Chapter 1 is a broad introduction to energy-efficient driving where we discuss the expected energy gains with various eco-driving strategies. Chapter 2 lays out the modeling foundations for relating energy consumption of vehicles with conventional, hybrid, or electric powertrains

to the road loads they experience. Chapter 3 is a brief introduction to communication, perception, and control systems of CAVs. In Chap. 4, the state of the art of
modeling the road network as well as microscopic and macroscopic traffic models
are covered.

Chapters 5–7 form the core of the book and describe the mathematical foundations and algorithms for eco-routing and eco-driving based on optimization and
optimal control theory. Chapter 5 focuses on energy-efficient routing; skipping it
will not disrupt the flow of the book. In Chap. 6, energy-efficient driving of gasoline,
electric, and hybrid powered vehicles is posed as optimal control problems of various degree of complexity and solved analytically as well as numerically. Chapter 7
goes more in-depth and further exposes the nature of energy-efficient profiles under
different driving and traffic scenarios.

Several practical and implementation techniques of eco-driving are covered in
Chap. 8. The final chapter of the book is dedicated to more complex and rather
complete case studies that are based on the authors' previous research and publications. Students could replicate and expand the case studies of Chap. 9 in a
project-based course.

Two Appendixes complete the book in providing detailed calculations that
would have disrupted the flow of the main text. In particular, Appendix B presents
several new results on the feasibility of analytical solutions of the eco-driving
problem for electric vehicles.

Rueil Malmaison, France   Antonio Sciarretta
Clemson, South Carolina, USA   Ardalan Vahidi
April 2019

Acknowledgements

This work would have not been possible without the contribution and input from our students and colleagues.

The first author would like to acknowledge his current and former colleagues, postdoctoral, and doctoral students at IFP Energies Nouvelles, Jihun Han, Giovanni De Nunzio, Luis León Ojeda, Caroline Ngo, Jiamin Zhu, Matěj Kubička, and Jianning Zhao. The publications they have co-authored form the basis of several sections of the book. He also gratefully acknowledges his colleagues Ibtihel Ben Gharbia and Gabriel Ducret for the useful comments on some sections.

The second author would like to acknowledge the contribution of his current and former graduate students, Behrang Asadi, Austin Dollar, Alireza Fayazi, Grant Mahler, Nianfeng Wan, and Chen Zhang. Their joint papers with the second author were the foundation for a number of the case studies in Chap. 9. The case study in Sect. 9.1 summarizes the research of Behrang Asadi and Grant Mahler and the case study in Sect 9.2 is written based on Alireza Fayazi's work. The case study in Sect. 9.3 combines the formulation and results from the work of Austin Dollar, Nianfeng Wan, and Chen Zhang. The problem formulation and the results in the case study in Sect. 9.4 is out of Austin Dollar's research. The probabilistic approach to traffic signal prediction in Sect. 8.2.4 is due to Grant Mahler. We also thank Dr. Yunyi Jia from Clemson University for his input on automated drive in Sect 8.3.3. The second author gratefully acknowledges the financial support provided by IFP Energies Nouvelles, which made possible a summer stay on their premises where the idea of writing this book took shape.

The authors gratefully acknowledge the valuable help of Austin Dollar in carefully proofreading the first draft of the book and his thoughtful comments.

Most importantly, we are indebted to our families for their patience and support in the one year that this book was being written, a lot of times after work hours.

Contents

Acronyms

3GPP	Third generation partnership project
AC	Alternating current
ACC	Adaptive cruise control
ADS	Automated driving system
AMT	Automated manual transmission
API	Application programming interface
APU	Auxiliary power unit
ATP	Adenosine triphosphate
AWS	Amazon web services
BF	Bellman-Ford (algorithm)
CACC	Cooperative adaptive cruise control
CAN	Control area network
CAV	Connected and autonomous vehicle
CC	Constant current
CC	Cruise control
CMEM	Comprehensive modal emission and energy model
CP	Critical power
CPM	Chebyshev pseudospectral method
CV	Constant voltage
CVT	Continuously-variable transmission
DBMS	Database management system
DC	Direct current
DDT	Dynamic driving task
DEM	Digital elevation model
DP	Dynamic programming
DSM	Digital surface model
DSRC	Dedicated short range communication
DTM	Digital terrain model
E85	Ethanol fuel blend of 85% ethanol
ECMS	Equivalent consumption minimization strategy

ED-OCP	Eco-driving OCP
EI	Energy intensity
EKF	Extended Kalman filtering
EMS	Energy management strategy
ER-SPP	Eco-routing shortest path problem
EV	Electric vehicle
FACE	Fully-analytical consumption estimation
FCC	Federal communication commission
FI	Fuel intensity
FP	Fastest path
GIS	Geographic information system
GNSS	Global navigation satellite system
GPM	Gauss pseudospectral method
GPS	Global positioning system
HD	High definition
HEV	Hybrid-electric vehicle
HMI	Human-machine interface
HPV	Human-powered vehicle
ICE	Internal combustion engine
ICEV	Internal combustion engine vehicle
IDM	Intelligent driver model
IEEE	Institute of electrical and electronics engineers
ITIC	International transportation innovation center
IMU	Inertial measurement unit
IP	Interior point (method)
LCO	Lithium cobalt oxide
LFP	Lithium iron phosphate
LIDAR	Light detection and ranging
LMO	Lithium ion manganese oxide
LPG	Liquid petroleum gas
LPM	Legendre pseudospectral method
LTV	Linear time-varying
MET	Maximum endurance time
MILP	Mixed-integer linear programming
MIQP	Mixed-integer quadratic programming
MOBIL	Minimizing overall braking induced by lane changes
MOOP	Multi-objective optimization problem
MPC	Model predictive control
MPG	Miles per gallon
MVC	Maximum voluntary contraction
NCA (lithium)	Nickel cobalt aluminium oxide
NHTSA	National highway traffic safety administration
NLP	Nonlinear programming
NMC (lithium)	Nickel manganese cobalt oxide
NMPC	Nonlinear MPC

OBD	On-board diagnostics
OCP	Optimal control problem
ODD	Operational design domain
OEDR	Object and event detection and response
OOL	Optimal operating line
P&G	Pulse and glide
PCC	Predictive cruise control
PCM	Phase-change material
PMP	Pontryagin minimum principle
PSL	Posted speed limit
QP	Quadratic programming
RCSPP	Resource-constrained SPP
ROS	Robot operating system
RRT	Rapidly-exploring random tree
RTK	Real-time kinematic
SAE	Society of automotive engineers
SDP	Stochastic DP
SHS	Sensible heat storage
SLAM	Simultaneous localization and mapping
sMAPE	Symmetric mean absolute percentage error
SoC	State of charge
SoF	State of fatigue
SP	Shortest path
SPaT	Signal phase and timing
SQP	Sequential QP
TMC	Traffic management center
TPBVP	Two-point boundary value problem
UDP	User datagram protocol
UI	Use intensity
UTM	Universal transverse Mercator
V2C	Vehicle to cloud
V2D	Vehicle to device
V2G	Vehicle to grid
V2I	Vehicle to infrastructure
V2N	Vehicle to network
V2P	Vehicle to pedestrian
V2V	Vehicle to vehicle
V2X	Vehicle to everything
VIL	Vehicle in the loop
VMT	Vehicle miles traveled
WAVE	Wireless access in vehicular environment
WLAN	Wireless local area network
xEVR	(hybrid-)electric vehicle with recharge

Chapter 1
Energy Saving Potentials of CAVs

1.1 Introduction

The shift that we are witnessing toward vehicle connectivity and autonomy is going to be perhaps, the most disruptive since the early days of automobiles and could revolutionize movement of people and goods. According to one estimation, the number of connected cars sold globally will grow to 152 million across the globe by 2020, a sixfold increase with respect to 2015 [1]. Another estimate puts the number of connected vehicles at 250 million vehicles by 2020 [2], a fourth of the billion cars that are in service today. In 2016 the US Department of Transportation issued a notice of proposed rule making, that if implemented would require Vehicle-to-Vehicle (V2V) connectivity on all new light-duty vehicles and was intended to reduce the number of car accidents [3]. Similar provisions and guidelines are envisioned for Vehicle-to-Infrastructure (V2I) communication [4]. With implementation of such mandates, the number of connected cars with access to information and data will rapidly increase.

On a different front, major auto manufacturers, technology firms, and startup companies have started a race toward building fully automated cars. Many automated functions such as adaptive cruise control and lane keeping assist are already available on several production vehicles. It is expected that first fully automated vehicles be available for sale in the next few years [5, 6]. A projection is that 20–40% of vehicle sales be automated by 2030 and full penetration could happen in several stages over the next few decades [7].

This level of connectivity and autonomy will transform transportation of people and goods in several dimensions with important societal and economical impacts: improved safety, increased comfort, time saving potential, and more efficient road utilization are among the most widely discussed positive impacts of CAVs. Fully automated vehicles could improve mobility for young, elderly, and people with disability who are unable to drive today. Ride sharing and on-demand mobility services could gain more popularity due to reduced labor cost, in turn influencing urban planning and land use.

© Springer Nature Switzerland AG 2020
A. Sciarretta and A. Vahidi, *Energy-Efficient Driving of Road Vehicles*,
Lecture Notes in Intelligent Transportation and Infrastructure,
https://doi.org/10.1007/978-3-030-24127-8_1

Table 1.1 Potential Impact of CAVs on **a** energy intensity or user intensity according to [8], **b** operational energy use by year 2050 according to [9]. Here EI, UI, and FI respectively represent Energy, Use, and Fuel Intensity. Reference [8] defines factors affecting vehicle miles traveled (VMT)/vehicle as Use Intensity (UI); factors affecting Energy/VMT as Energy Intensity (EI); and factors affecting Liquids Energy Fuel Intensity (FI)

Contributing factors	Ref. [8]	Ref. [9]
Platooning	(−)10% EI	(−) 2–10%
Eco-driving	(−)15–40% EI	(−) 20%
Eco-routing	(−) 5% EI	NA
Congestion mitigation	NA	(−)2–4%
De-emphasized performance	NA	(−) 5–23%
Vehicle light-weighting	(−) 50% EI	(−) 5–23%
Vehicle right-sizing	(−) 12% UI	(−) 20–45%
Changed mobility services	NA	(−) 0–20%
Infrastructure footprint	NA	(−) 2–5%
Reduced parking search	(−) 4% UI	NA
Enabling electrification	(−) 75% FI	NA
Higher highway speeds	(+) 30%	(+) 5–25%
Increased features	NA	(+) 0–10%
Travel cost reduction	(+) 50% UI	(+) 5–60%
New user groups	(+) 40% UI	(+) 2–10%

Energy use has not been the core consideration in development of connected and automated vehicles, but it could be impacted significantly. The impact could be positive or negative according to [8, 9] as summarized in Table 1.1. A careful scenario analysis in [9] shows vehicle automation could cut energy use and green house gas emissions in half in an optimistic scenario or double them in a "dystopian nightmare", depending on the effects that come to dominate. Increased opportunities for eco-driving and platooning, traffic harmonization, vehicle light-weighting enabled by lower crash risk, vehicle right-sizing for number of travelers, de-emphasized vehicle performance, car-sharing and on-demand mobility, and reduced infrastructure footprint of automated vehicles all contribute to improved energy utilization according to [9]. But according to the same study, the increase in vehicle distance traveled due to lower travel costs, addition of new user groups (young, elderly, disabled), higher highway speeds, and increased vehicle features can also dramatically increase the energy footprint of vehicle automation. The outcomes depend on which scenarios prevail and proactive policy making is essential to steer the technology toward energy efficiency as also emphasized in [6, 9, 10]. The authors of [11] speculate that the aggregate energy and environmental impact of automated and on-demand mobility could be positive, but acknowledge a big shift from historical trends that needs to be carefully watched by policy makers and planners.

This chapter provides an overview of increased opportunities for energy efficient driving with connected and automated vehicles, disregarding second order effects of connectivity and automation, such as increased vehicle distance traveled or reduced vehicle weight. Because CAVs are capable of sensing more accurately, processing more information, and can be more tightly controlled, they benefit more from information offered by connectivity and road preview. With higher penetration rate of CAVs, opportunities increase for vehicle to vehicle communication and cooperative control, which can lead to additional energy efficiency gains. Despite these prospects, connected and automated vehicle research and development have been mostly on software, sensing, and safety and there are limited results on energy efficiency potentials.

In particular in this chapter we discuss opportunities that arise for individual CAVs by choosing energy efficient routes, anticipating future road slope and geometry, macroscopic state of traffic, state of upcoming traffic signals, and motion of their neighboring vehicles. This allows CAVs to more judiciously choose their velocity and lane to minimize wasteful braking and idling and also enables predictive powertrain control due to increased certainty about future vehicle motion. With higher penetration of CAVs, more opportunities arise for collaborative driving which could further enhance energy efficiency as discussed in Sect. 1.4. In particular we discuss platooning, cooperative adaptive cruise control, cooperative lane change and merge, and cooperative intersection control for a CAV fleet. The impact on mixed traffic is discussed briefly in Sect. 1.4.4. This sets the stage for the rest of the book where algorithms for energy efficient driving and detailed case studies are presented.

1.2 Minimal-Energy Route Navigation

Modern navigation systems are becoming indispensable parts of our every day commutes. Relying on their up-to-date maps and efficient routing algorithms, onboard or mobile navigation systems can calculate the shortest or fastest route between origin-destination pairs across a continent in a blink of an eye in what would have been an unthinkable feat not very long ago. Enabled by connectivity, an increasing number of navigation systems receive the latest traffic information and road closings and adjust their route recommendation accordingly. Road elevation and slope information may be embedded in onboard maps or could be retrieved by connecting to an online Geographic Information System (GIS) server. One could envision real-time access to more rapidly changing road information such as state of traffic signals or weather related road conditions, in the near future. With access to information such as road topography and traffic conditions, it is now possible to find routes with the lowest energy cost or environmental impact in what is referred to as energy-efficient or *eco* routing. Eco-routing has been the subject of many recent publications focusing on algorithms, simulation case studies, or real-world deployments. And it is not only the environmentally concerned and energy conscious that favor eco-routes. The limited driving range of electric vehicles requires accurate estimation of each route's energy cost and determination of the minimal energy route as an option.

Eco-routing algorithms start by assigning an energy consumption (or pollutant emission) cost to each link of a road network. A routing algorithm searches for the path connecting an origin-destination pair with the lowest sum of link costs. More details of well-established optimization-based routing algorithms are described in Chap. 5. Here we provide a short overview of potential energy benefits of eco-routing based on published literature.

In one of the first published case studies on eco-routing [12] the authors used a collection of 15437 recorded commutes to compute typical consumption on streets of Lund, Sweden. Using a smaller subset of 109 real journeys, the authors estimated that fuel efficiency could be enhanced for 46% of the trips and that fuel savings would be 8.2% on the average.

Field study results reported in [13] of 39 trips taken between same origin-destination pair but some via a highway route and others via a slower arterial route show 18–23% energy saving when taking the arterial road. Such a trend is also observed in a case study conducted in Netherlands [14] showing energy advantage of local and provincial roads over highways could be as much as 45%. The network-wide effect of eco-routing algorithms is simulated for two US major cities in [15] which shows the average fuel savings of eco-routing vehicles range from 3.3% to 9.3% compared to algorithms that find the shortest-time path. By trading travel time and driving more slowly, energy loss to aerodynamic drag is lower which contributes to the reported energy savings on slower roads.

Eco-routing for electric and hybrid vehicles present new algorithmic challenges as discussed in [16]. For electric vehicles limited range, long recharge times, and the ability to regenerate energy during deceleration invalidate use of standard shortest-path search methods and require new treatments [17] and different solutions have been proposed. A model-based eco-routing strategy for electric vehicles in large urban networks is presented in [18]. A model-based eco-routing approach tailored for hybrid vehicles is presented in [19] while [20] addresses eco-routing for plug-in hybrid vehicles. We note that eco-routing for electric vehicles with consideration of in-trip charging poses new challenges as described in [21, 22].

Eco-routing may be even more attractive for fleet operators where small energy benefits could add up to significant savings. This could include operators of delivery trucks, long-haul heavy vehicles, or mobility-on-demand services. Other factors such as delivering the heavier items first could be considered in the routing algorithm as discussed in [23] and could result in additional energy savings.

Table 1.2 provides a summary of energy benefits of eco-routing as reported in the literature.

Table 1.2 Summary of selected published results on energy efficiency gain enabled by eco-routing. Simulation and Experimental results are denoted by S and E respectively

Refs.	Methods and conditions	Efficiency gain (%)
[12]	S, Lund street network, Sweden	+8.2
	Links' fuel consumption estimated based on 15437 recorded trips	
	22 street classes, peak/off-peak hours, 3 types of cars	
	Eco-route compared to original route for 109 of these trips	
	50 of 109 trips could benefit from Eco-routing	
[14]	E, Delft city network, Netherlands	+45
	Delft-Zoetermeer routes, off-peak	
	Compares highway, local, and provincial routes	
	40 trips per route	
	Ford Focus vehicles with 1.6 liter engines	
	Empirical model for fuel consumption estimation	
[13]	E, Washington DC suburb, USA	+18–23
	21 highway and 18 arterial recorded trips	
	Same origin and destination for all	
	Fuel consumption estimated using several empirical models	
	Arterial trips 17% slower but more energy efficient	
[15]	S, downtown Cleveland and Columbus, Ohio, USA	+3–9
	Network wide effects of eco-routing using micro-simulations	
	Fuel consumption estimated using an empirical model	
[18]	S, city center of Lyon network, France	+6 w.r.t
	5400 nodes and 9500 links	Shortest
	1,000 different origin-destination pairs	+10 w.r.t
	Network topology, road grade, traffic from HERE Maps [24]	Fastest
[23]	S, truck delivery routing	+4.9–6.9
	Time-constrained, multiple-stop, class 8 truck-routing	w.r.t.
	Optimize to unload heavier loads first	Shortest

1.3 Anticipation in CAV Driving

CAVs offer huge potentials for boosting road safety, capacity, and efficiency, because of their ability to process data from many more sources and their ability for more precise positioning and control than human drivers. While similar information can be processed, and provided to connected human-driven vehicles [25, 26] (e.g. as optimal speed/lane advisories), only fully automated vehicles can be made to comply with and reliably follow real-time energy-efficient commands. Even in mixed-traffic that involves other non-automated vehicles, energy-efficient automated vehicles can have a positive impact on the energy efficiency of surrounding traffic as will be illustrated later. Automated cars have the potential to uncover the "driving signature" of their neighboring vehicles and predict their most likely actions. They can also anticipate probable locations of slow-downs by systematic evaluation of historical data. Connectivity between cars and infrastructure can make much more information available to each vehicle and the vehicles can form groups and act cooperatively. All of these advances, when put into an organized framework, can help better anticipation and improve traffic flow, increased safety, and reduce energy consumption.

1.3.1 Anticipating the State of the Road

Prior knowledge of road speed limits, safe speeds on curved roads, and an estimate of average traffic speed allows for more energy efficient velocity transitions in anticipation of the change in velocity constraints. Speed limit is a standard feature on modern onboard navigation units. Road curvature may be extracted from navigation maps to calculate the likely (safe) speed on a curve. Curve speeds can also be inferred from connected vehicle data. Average traffic speeds for upcoming segments of a trip can be queried from a Traffic Management Center (TMC) that operates based on local sensors and cameras or estimated from traffic feeds that mostly rely on crowdsourced information. As of 2019, sources of such data include feeds of Google, Here, Waze, and Inrix. Dynamic spatiotemporal evolution of traffic speed can be estimated via a faster-than-real-time traffic simulation model which is initialized by current traffic speed, deterministically [27] or probabilistically [28]. In absence of real-time traffic information services, time- and location-specific historical traffic data can be used as a baseline predictor [29]. Traffic speed can be imposed as a spatio-temporally varying upper bound on the CAV speed [27]. Speed limit, curve and traffic speeds can be unified [30] into a single spatiotemporal bound on CAV velocity and used not only to optimize velocity transitions of a CAV but also inform its predictive powertrain control functions.

Another dominating factor in vehicle power demand is road grade, in particular on steep roads, and more so for heavier vehicles. Road grade influences velocity and torque constraints and gear selection. Therefore advanced knowledge of the road grade, obtained from 3D road maps, is very beneficial in predictive powertrain control as shown for instance in [36, 38]. Additionally, due to constraints on velocity, prior knowledge of road grade will allow more judicious use of available velocity band and better gear selection [31–33, 35, 39–42]; for instance a vehicle can slow down in anticipation of a steep descent or speed up in preparation for a climb. The optimal solution can be non-trivial as shown for a heavy duty vehicle in [34]. Daimler already has a predictive cruise control function in production that adjusts a heavy duty truck speed [43] and gear [44] in anticipation of upcoming road grade to increase its energy efficiency by 3% on a highway. This level of achievable improvement is consistent with results in literature as summarized in Table 1.3.

Predicted velocity transitions and road grade can also reduce energy use via predictive power split in hybrid powertrains [45], fuel cut-off [46] and cylinder deactivation [47] in combustion engines, and thermal load management [48]. While such predic-

Table 1.3 Summary of selected published results on energy efficiency gain enabled by road grade preview. Simulation and Experimental results are denoted by S and E respectively

Refs.	Methods and Conditions	Efficiency gain (%)
[31]	S, 32 ton class 8 truck	
	Constrained NLP, preview horizon: 1500 m	
	Optimized velocity, gear, and throttle input	
	Route 1: $-3.7° \leqslant \theta \leqslant +4.7°$, $\mu_\theta = 0.29°$, $\sigma_\theta = 1.32°$	+2.6
	Route 2: $-4.3° \leqslant \theta \leqslant +3.0°$, $\mu_\theta = -0.21°$, $\sigma_\theta = 1.06°$	+2.0
[32]	E 39 ton SCANIA truck	+3.5
[33]	120 km highway, Södertälje to Norrköping, Sweden	
	Dynamic programming, preview horizon: 1500 m	
	Optimized velocity; gear was preselected	
[34]	S, 29 ton class 8 Navistar truck	+11.6
	4 km single valley profile $h(s) = 30(1 - s/2000)^2$	Over a
	Pontryagin Min. Principle & numerical continuation	Single
	Horizon $= 4000$ m, optimized velocity and gear	Valley
[35]	S, 1.3 Liter gasoline engine passenger car	+4–7
	Simplified polynomial fuel consumption model	
	Model predictive control, optimized velocity	
	2.5 km Yuniba Dori Road, Fukuka City, Japan	
	$-5.0° \leqslant \theta \leqslant +6.0°$	
[36]	S, 2000 kg hybrid electric vehicle	
	Dynamic programming, preview horizon: full trip	
	Constant speed, optimized power split	
	36 and 48 km hilly roads, Contra Costa, California	
	PSAT [37] fuel economy evaluation	
	Route 1 $-4.3° \leqslant \theta \leqslant +3.0°$, $\mu_\theta = -0.21°$, $\sigma_\theta = 1.04°$	+0–3.0
	Route 2: $-8.0° \leqslant \theta \leqslant +5.3°$, $\mu_\theta = -0.17°$, $\sigma_\theta = 2.3°$	+0–6.0

tive powertrain control functions can be exercised in conventionally driven vehicles and some have been extensively studied, they will have a larger impact in CAVs. Real-time access to information due to connectivity and absence of a human driver in a CAV increases certainty of predictions and therefore effectiveness of predictive powertrain control as schematically depicted in Fig. 1.1.

1.3.2 Anticipating Signal Phase and Timing

When driving on arterial roads, repetitive stops at traffic signals result in loss of energy due to braking and idling, engine and brake wear, and cause passenger discomfort and frustration. Some of these stops are unnecessary, in particular under light to medium

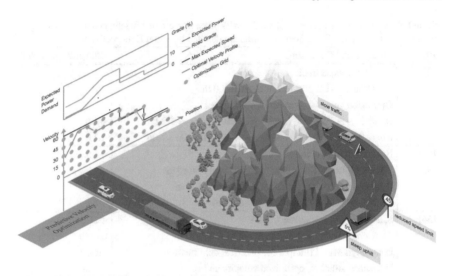

Fig. 1.1 Eco driving in anticipation of upcoming hills, changes in speed limit, and slow traffic. The white CAV solves for the fuel minimal velocity trajectory given road power demand and constraints. The image was created on https://icograms.com

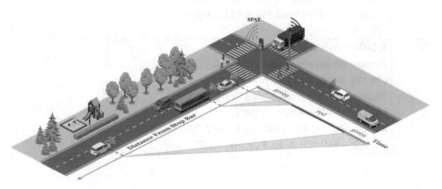

Fig. 1.2 Schematic of eco-driving with SPaT preview. Shaded triangles contain feasible paths to green intervals of the traffic light for the 3 vehicles moving from bottom left to top right. Most parts of the image were created on https://icograms.com

traffic conditions, and are due to lack of information about the state of traffic lights. In an ideal connected urban area with Vehicle-to-Infrastructure (V2I) connectivity, Signal Phase and Timing (SPaT) can be broadcast to approaching vehicles so that connected vehicles adjust their speed for a timely arrival at a green light as shown schematically in Fig. 1.2. Vehicle autonomy further facilitates this scenario by taking the burden of speed adjustments away from human drivers.

Energy efficient driving at signalized intersections and its impact on energy use has been the topic of much research and development in recent years. One of the earlier works was presented in [49] and expanded in [50] and showed potential for significant fuel savings in a simulation study. These positive results have been corroborated in [51–53] and many more publications that have followed them. Experimental results in isolated environments [54, 55] and in real-world traffic conditions [56–58] show that considerable fuel saving (5–15%) is possible with human drivers in the loop. Even more energy saving is expected in automated driving (or with automated cruise control) where vehicles can adjust their speeds more precisely and effortlessly.

The technology for transmitting traffic signal information to subscribing vehicles has been demonstrated in several research projects [54, 58, 59] and described in more detail in Chap. 3. The SPaT information may be directly transmitted to vehicles within range using Dedicated Short Range Communications (DSRC) technology [57] or may become available by the traffic control center via cellular networks as shown in [58]. A software architecture for cellular communication of SPaT from a server to subscribing connected vehicles is described in [58]. Alternative means of inferring SPaT information via on-board cameras [59] and via crowd-sourcing [60, 61] have also been proposed. There has also been commercial efforts to build a SPaT information repository to provide speed advisory to human drivers via a mobile app [62]. However, a much needed real-time server that covers large urban areas is still missing. In absence of real-time SPaT information, it is still possible to use history of observation during daily commutes and to estimate the probability of a green or red over a future horizon, conditioned on the current color of the light [53] to allow vehicles to target probable green windows. Even when SPaT is available in real-time, the future color of the light is not known with certainty, for instance when the light is actuated by the state of loop detectors. In such a scenario one can still use historical trends to predict the probability of a red or green over a future time horizon [63].

While simple logical rules, such as those in [49], can be used when approaching a traffic signal, velocity planning can benefit from more formal optimization methods and will be discussed in more detail in Chaps. 6 and 7 of this book where the goal is reducing energy consumption subject to the constraint imposed by red signals. Passenger comfort and travel time could also be considered in formulating the approach to traffic lights; for instance penalizing both travel time and acceleration results in smoother trajectories and less braking which also saves energy.

"Selfish" optimization that focuses on eco-driving of a single vehicle could be disruptive to the flow of following vehicles. In [71], while a vehicle-centric optimization is still solved, the interest of following vehicles are taken into account.

Because such traffic signal speed advisory technology is unlikely to be implemented in every vehicle in the near future, it is important to evaluate the influence of equipped vehicles on other vehicles in mixed traffic flow. It is currently prohibitively difficult to do field experiments of a large number of CAVs in mixed traffic. Therefore traffic simulation tools have been used in most studies. The impact of traffic signal advisory on mixed traffic is studied, via microsimulations in [67, 69, 70]. In [3] the impact of CAVs on mixed traffic near signalized intersections is studied in traffic microsimulations. The CAVs receive the timing of signals in advance and adjust their speed for a timely arrival at green. It is shown that CAVs not only improve their energy efficiency but as their penetration increases they reduce the energy consumption of conventional vehicles as well. With the increment of CAVs, other conventional vehicles are more likely to follow a smoother-moving CAV. By their simple car following strategy, such conventional vehicles may reduce the chance of stopping at intersections as well. Potential impact on energy efficiency is summarized in Table 1.4.

1.3.3   Anticipative Car Following

Human drivers are often reactive when following other cars as their view is often blocked by the preceding car and therefore their event horizon is very limited. In sudden slow downs, they often fail to consider the vehicles approaching from behind. This is not only disruptive to traffic flow and is unsafe, but it can result in inefficient slow-down of multiple vehicles. Balancing the position dynamically with respect to the cars in the front and back is cognitively demanding for humans. Most autonomous cars without connectivity do not necessarily do better. Many are designed to behave like human-driven vehicles and could be reactive to the perception of their immediate surrounding which results in similar short-sighted decisions. In [72] a simulation scenario depicts an automated vehicle that uses 3% more energy than a conventional vehicle baseline due to its aggressive car following strategy.

The challenge is anticipation of road events, although experienced drivers do exercise anticipation to some extent in driving [26, 73]. We pay attention to clues and drive accordingly. For example, if we observe that a lead vehicle is accelerating and decelerating erratically we increase our following distance or change lanes. If we observe that a following vehicle is tail-gating us we try to induce a larger gap or allow that vehicle to pass. But most of these precautions are practiced in an adhoc manner, are constrained by our limited sensory and cognitive limits [74], and are inconsistent across different drivers [75] and traffic scenarios. These cause poor local judgments that could lead to shock waves that slow us down to inefficient crawls. Today much more can be done: thanks to better sensing capabilities, CAVs have the potential to anticipate the motion of their preceding vehicle and finely adjust their speeds for a more steady and smooth motion. Additional information of the intent of preceding vehicles via V2V communication can enhance such anticipative car following.

While the main goal should be to robustly maintain a safe following distance to the preceding vehicle (imposed as position constraints), the inter-vehicle gap can be judiciously used as a degree of freedom to filter abrupt slow-downs and application of brakes [76] and increase energy efficiency of the ego vehicle as schematically shown in Fig. 1.3. Smoother velocity transitions of the host vehicles are expected to positively influence the motion of upstream traffic and reduce the chance of a phantom jam, in which traffic comes to a halt with no apparent reason and because of small disturbances [77–79]. This lowers fuel used by the entire queue of vehicles as experimentally shown in [80].

Because of shorter relevant time scales in car-following, a moving horizon energy optimization is a natural choice (as opposed to full trip optimization). The main challenge that arises here is dependence of the inter-vehicle constraint on the position

Table 1.4 Summary of selected published results on energy efficiency gain enabled by SPaT anticipation with respect to conventional vehicles without SPaT information. Simulation and Experimental results are denoted by S and E respectively. The efficiency gains are in vicinity of signalized intersections and not an entire trip gain

Refs.	Methods and Conditions	Efficiency Gain (%)
[49] [50]	S, lone vehicle, 10 fixed time lights	+24–29
	Real SPaT: Greenville, SC timing cards	
	1.7 L 4-cylinder gasoline engine,	
	High fidelity vehicle model in PSAT [37]	
[51]	S, 10 fixed time lights,	+12–14
	Stochastic parameter variation	
	Passenger car and SUV, CMEM models [64]	
[65]	S, 1 fixed time light	+20
	Varying road conditions, random initialization	
	Virginia Tech fuel consumption model [66]	
[53]	S, 3 fixed and variable timing lights	+16
	Probabilistic SPaT, probabilistic planning	
	Monte Carlo Evaluation (3000 scenarios)	
[54]	E, no traffic	+13
	1 fixed time signal, 4G cellular comm.	
	2011 BMW 535i	
[56] [58]	E, real city traffic,	+9
	Real-time TMC data, 4G cellular comm.	
	Mix of 10 fixed time and actuated signals	
	2011 BMW 535i, 4 complying drivers	
[57]	E, real city traffic	+2–6
	Coordinated actuated signals, DSRC comm.	
	2008 Nissan Altima, 2 complying drivers	
[59]	E, real city traffic	+25
	2 fixed time lights	
	Camera SPaT estimation- V2V comm.	
	2001 2.4 L PT Cruiser, 1 complying driver	
[67]	S, network wide effect	+25
	4 fixed time signals, multi lane	(CAV)
	Paramics [68] microsimulations, mixed traffic	+6
	50% CAV penetration, 900 veh/hour/lane	(surrounding traffic)
	Polynomial fuel consumption model	
[69]	S, network wide effect	+12.5
	11 fixed time signals, one lane	(CAV)
	Paramics microsimulations, mixed traffic	+7.5
	50% CAV penetration, 300 veh/hour/lane	(all traffic)
	CMEM [64] fuel consumption model	

(continued)

Table 1.4 (continued)

Refs.	Methods and Conditions	Efficiency Gain (%)
[70]	S, network wide effect	+26
	1 fixed time signals, single lane at grade	(100% CAV)
	INTEGRATION microsimulation package	none
	Varied CAV penetration	(≤50%
	VT-micro fuel consumption model	CAV)

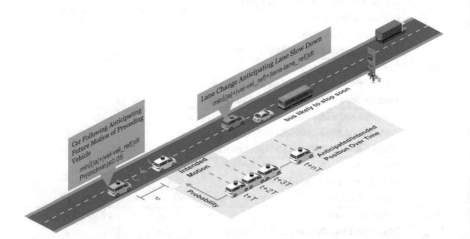

Fig. 1.3 Anticipative car following and lane selection. The yellow CAV receives the imminent intentions of its preceding white CAV or predicts it using past statistical data and plans its motion to minimize its acceleration and velocity deviation while enforcing safe gap constraints. The blue CAV which is preceded by a bus anticipates right lane traffic slow down near a bus stop and proactively starts a lane change. Its goal could be minimizing a weighted sum of its acceleration and deviations from desired velocity and lane. Most parts of the image were created on https://icograms.com

of the preceding vehicle which is typically unknown. Therefore despite a relatively simple optimization problem formulation, we are faced with a difficult prediction problem.

In absence of any information and when only instantaneous velocity or acceleration of the preceding vehicle is known, the position of the preceding vehicle can be projected over the horizon assuming that it travels with constant speed [81] or constant acceleration [82]. Or perhaps it is reasonable to assume that acceleration of the preceding vehicle decays over the horizon to zero with some time constant [30]. When information from the road and infrastructure is available one can construct a deterministic profile that the preceding vehicle is expected to follow as will be discussed later in Chaps. 4 and 8.

More complex driver modeling methods have also been proposed in the literature. For instance [83] proposes fitting a nonlinear autoregressive model to historical data to predict the motion of preceding vehicle. In [76] the future motion of a group of preceding vehicles is estimated via traffic microsimulations. Many have used probabilistic models to capture the statistics of velocity transitions [29, 84–86]. This topic is visited in more detail in Sect. 8.2.3 of this book.

In an ideal scenario when all vehicles communicate, each vehicle can pass on its intended action to the vehicles that follow it [87, 88]. This allows the ego vehicle to know, with more certainty, the position of the preceding vehicle(s) over its planning horizon and is believed to result in smoother flow and improved overall energy efficiency. Note that in this scenario, the vehicles are just sharing intentions and do not necessarily cooperate toward a common goal. Later in Sect. 1.4.1 we discuss a cooperative cruise control scenario where the vehicles could cooperate toward a "social" optimum.

A different approach is proposed in [76] where it is assumed that all vehicles in a queue communicate their immediate state (position, velocity, acceleration) but not their intentions. The ego vehicle assumes a standard car following model for the preceding vehicles to anticipate their positions over its optimization horizon. A similar approach is discussed in [89, 90]. In a less than ideal scenario, when only a portion of the vehicles in a queue communicate, the position of non-communicating vehicles is inferred in [91] at signalized intersections. Communication delay makes the problem even more complex and is discussed in [89, 92]. Packet drops result in stochastic delays and their impact is discussed in [93].

Table 1.5 highlights selected results that show the impact of anticipative car following on energy efficiency. As can be seen, the reported gains vary significantly even for vehicles of the same size. This could be due to design and parameters of the car-following algorithms and scenario setups.

1.3.4 Anticipative Lane Selection and Merging

Most existing literature on eco-driving assume the vehicle maintains a single lane, reducing planning of the vehicle motion to the choice of its velocity. In multi-lane roads, the freedom to choose a different lane provides a new dimension and many more possibilities for optimizing the motion of the vehicle to safely improve its energy efficiency and may also harmonize traffic. But every day driving experience indicates that choice of lane is a complex decision making problem, perhaps due to its combinatorial nature and typical lack of information about the average speed (or efficiency) of adjacent lanes. The same is true when merging into a highway from an on-ramp or exiting to an off-ramp. Lane selection can be a dilemma point for average drivers; aggressive lane change on the other hand can be unsafe and disruptive to the flow and efficiency of upstream traffic.

In a connected and automated vehicle environment, more information about the intention of neighboring vehicles can become available via V2V communication, speed of each lane could be broadcast from roadside sensors, and therefore automated vehicles can change lanes more judiciously and smoothly as schematically illustrated in Fig. 1.3. A rather comprehensive survey of lane change/merge for CAVs can be

Table 1.5 Summary of selected published results on energy efficiency gain enabled by anticipative car following. Simulation and Experimental results are denoted by S and E respectively

Refs.	Methods and conditions	Efficiency gain (%)
[94]	S, 1.6 ton vehicle	+13–35
	3 standard driving cycles for phantom lead vehicle	w.r.t.
	Rule-based preview car following, horizon = 50 s	No preview
[84]	S, 2 ton vehicle	
	Recorded real-data for lead vehicle	+15
	Winding road from Clemson, SC to Highland NC	w.r.t.
	Chance constrained MPC, horizon = 15 s	Lead
	Markov chain prediction of lead vehicle velocity	Vehicle
	Fuel economy evaluated in ANL's PSAT [37]	
[82]	S, 1.4 ton electric vehicle, no regeneration loss	+12–44
	3 real city driving profiles for lead vehicle	w.r.t.
	MPC, horizon = 100 s	Lead
	Assumes constant acceleration for lead vehicle	Vehicle
	Physical polynomial model for energy use	
[95]	S, 1.8 ton simulated vehicle with combustion engine	+0–32
	Following lead car with constant speed	w.r.t
	Optimal control yields pulse and glide strategy	Lead
	Efficiency gain is speed dependent	Vehicle
[83]	E, engine-in-the-loop simulations,	+6.5–22
	Microsimulation + engine test bench measurement	
	Driver prediction: nonlinear autoregressive model	
	Prediction horizon =15 s	
	Results depend on allowable inter-vehicle gap	
[86]	E, real ego vehicle, 2007 Ford Edge	+3.6
	12 rounds city/highway driving on Michigan-39	w.r.t.
	Following phantom vehicle with constant speed	Lead
	Stochastic DP policy calculated offline	Vehicle
	Restricted to ± 2 mph speed difference w.r.t. lead	
	Resulting strategy is pulse and glide	
[96]	E, real ego vehicle, 3.8 L V6 engine, 8 speed trans.	+39–50
	Hyundai-Kia proving grounds, California	w.r.t.
	Simulated lead vehicle with sinusoidal velocity	Imperfect
	MPC tracking, perfect preview, horizon = 6 s	Preview

Table 1.6 Summary of selected published results on energy efficiency gain enabled by anticipative lane selection. All results are simulation results denoted by S

Refs.	Methods and conditions	Efficiency gain (%)
[100]	S, microscopic simulations	
	MPC velocity & lane selection, horizon=15 s	
	Tested 2 cases, 2 km road, varying CAV levels	
	At 50% penetration, w.r.t. conventional vehicles:	$+14.3\ (12.9)^a$
	At 50% penetration, w.r.t. CACC vehicles:	$+7.8\ (5.1)^a$
[101]	S, micro-simulation in merging zone, 30 vehicles	$+48$
	Optimal coordinated merging into a highway	w.r.t.
	Fuel economy via polynomial metamodel in [102]	Yield
	Reported gain for merging period only	& merge
[103]	S, microscopic simulations	$+8.4$
	2.3 km two lane road, 4 CAVs, full intent communication	w.r.t.
	MPC velocity & lane selection, horizon=10 s	Rule-based
	Mixed integer quadratic program formulation	

[a]Equipped vehicles (all traffic)

found in [97, 98]. One of the original formulations in this area can be found in [99, 100] where choice of lane is an additional integer decision variable in the energy cost of the vehicle.

Merging from ramps often causes breakdown and a traffic jam in a highway. Today, solutions such as ramp metering are being used to remedy the situation [104, 105], requiring infrastructure investment and maintenance. With CAV technology the merge can be coordinated much more safely as experimentally shown in [106] resulting in smoother traffic [107, 108] and higher energy efficiency [101]. Equipped with more data and processing power, the CAV can anticipate more systematically the motion of the neighboring vehicle during a merge as shown schematically in Fig. 1.4. The impact could go beyond individual vehicles; by reducing the chance of a phantom jam, the overall energy efficiency of traffic will improve. Table 1.6 lists the limited results that authors could find on the impact of lane selection on energy efficiency.

1.4   Increased Opportunities for Cooperative Driving

In a connected vehicle world, deliberate exchange of intentions by vehicles and infrastructure reduces the need for guessing the surrounding traffic patterns and therefore enables better coordination. Automated vehicles can cooperate rather than compete for right of way in urban areas and highways, thus contributing to harmony in motion and improved mobility and efficiency of a group of vehicles. Therefore "cooperation" in what follows, refers to sharing information and coordinating movements

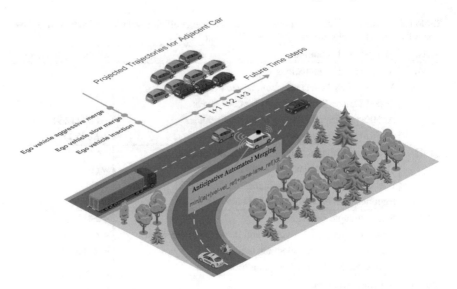

Fig. 1.4 Anticipative merging. The white CAV anticipates the imminent motion of its neighboring vehicle in response to its intended lane change. The image was created on https://icograms.com

for a "common" good. Even with the best intentions of human drivers, cooperation among conventional vehicles is rather challenging due to often unknown plans of neighboring vehicles and complexity of coordination at speed. For instance, merging from a ramp into a highway lacks a clear protocol and is often done in an "ad hoc" manner in the hope that fast approaching vehicles act with "consideration". This is not only unsafe, but the need for frequent braking in dilemma zones increases energy use and could negatively impact traffic flow. Information sharing via connectivity allows establishing more systematic coordination protocols that increase safety and efficiency. Automated vehicles can be programmed to take full advantage of such protocols that may require precise movement coordination. We describe below cooperation in car following, merging, lane changing, and intersection crossing and also discuss their potential impact on efficiency of cooperating vehicles as well as benefits to mixed traffic.

1.4.1   Cooperative Car Following

Cooperative car following in which vehicles coordinate in longitudinal formations is perhaps the most researched topic in cooperative driving, under the contexts of platooning and cooperative adaptive cruise control. A schematic is shown in Fig. 1.5.

Tight platooning gained popularity in the 1990s for its potential to increase highway throughput. In a platoon of communicating and partially automated vehicles, the gap between a group of following vehicles can be safely reduced to increase

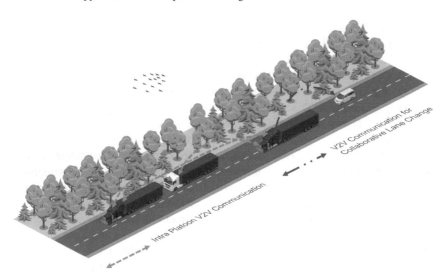

Fig. 1.5 Collaborative car following and lane selection. The two front trucks maintain a platoon formation relying on V2V communication and automated longitudinal control. A third truck communicates with the platoon to join. In a collaborative lane change maneuver enabled by V2V communication, the passenger car leaves a gap for the third truck to change lane. Most parts of the image were created on https://icograms.com

road capacity. Moreover at short following distances, the aerodynamic drag coefficient is smaller resulting in considerable energy savings, in particular for heavy duty vehicles. Recognized research programs in the USA [109, 110], Europe [111–113], and Japan [114] have demonstrated the feasibility of the technology in well documented road experiments as discussed in [115] showing potential for 5–15% energy saving. Experimental results in [112, 113] show between 4 and 7% energy saving potential for a heavy truck. Over the years, important technical challenges such as platoon string stability [116], communication needs [117, 118], control design [119, 120], and formation scheduling [121] have been addressed. Today the technology has matured to the extent that major manufacturers and startup companies plan to deliver truck platooning solutions to market in the near future with the goal of reducing energy and personnel costs [122].

Over the past few years and with increased prospects for vehicle connectivity, Cooperative Adaptive Cruise Control (CACC), has gained popularity in the research community. CACC is essentially an enhanced Adaptive Cruise Control (ACC) system that, in addition to range sensor feedback, relies on wireless communication of the acceleration of the preceding vehicle(s) for feedforward control. V2V communication is intended to increase safety and allows string-stable reduction of the inter-vehicle gap for improved road utilization [123]. With a correct design, velocity variations are much better attenuated than in ACC car following, as shown in road experiments with six equipped vehicles in [124]. The 2011 Driving Challenge in Netherlands was a successful showcase of CACC technology by multiple teams. An overview of this competition is presented in [125, 126]. CACC formations could positively or negatively impact surrounding traffic as demonstrated in a simulation study [127], for instance long formations may prevent those that intend to merge into a highway. But overall, CACC is expected to have a harmonizing impact on participating vehicles and on surrounding traffic, reducing braking events and lowering energy consumption. Despite these benefits there are few papers documenting the energy efficiency impact of CACC, for instance [83]. It appears that increasing energy efficiency has been mostly the focus of truck platooning projects.

While the platoon and CACC terminologies are sometimes interchangeably used in the literature, there are some differentiating features. The original concept of a platoon relied on a designated lead vehicle and a hierarchical control structure from the lead to the following vehicles. This hierarchy is not needed in CACC car following and each vehicle can individually switch to its CACC mode as long as it receives messages communicated by its preceding vehicles. The information flow between vehicles can vary from one implementation to the other. A vehicle can receive information from the lead vehicle only, from its preceding vehicle only, or from multiple preceding vehicles as schematically shown in [89, 128].

Depending on the information flow and content shared between vehicles, we can envision enhanced versions of current platooning and CACC practices. Ideally each vehicle will share its intended acceleration profile over a future horizon, rather than its instant acceleration, with all its following vehicles [87, 88]. This reduces the uncertainty about the movement of preceding vehicles, as was discussed in Sect. 1.3.3, aiding each vehicle to better plan its motion and reduce braking events.

Note that in this scenario, cooperation is only via information sharing, and each vehicle plans "selfishly". In a true collaborative environment, collaborating CAVs not only share information but also plan for their common good [88] or follow formation consensus rules [129]. Their common goal for instance could be reducing the energy consumption of the entire fleet [71, 130, 131], string stability [132], or collision mitigation [133]. Optimization can be distributed onboard each vehicle based on information communicated by neighboring vehicles to reach a consensus [131, 132]. Alternatively in a centralized control framework described in [131], the fleet energy useage is optimized on a central cloud server for a group of freight trucks and the decision is issued to low-level controllers of individual trucks. Table 1.7 summarizes some of the limited results on energy efficiency impact of cooperative car following, including platooning.

1.4.2 Cooperative Lane Change and Merge

In Sect. 1.3.4 we discussed that individual CAVs can benefit from connectivity and autonomy and more safely and efficiently merge and change lanes. Additional gains are expected if CAVs cooperate, not only by sharing intentions but also by being "considerate" of neighboring vehicles. In such a cooperative scenario, each vehicle considers the impact of its decision on neighboring vehicles. Lane change and merge

Table 1.7 Summary of selected published results on energy efficiency gain enabled by cooperative car following. Simulation and Experimental results are denoted by S and E respectively

Refs.	Methods and conditions	Efficiency gain (%)
[131]	S, group of five 1.2 ton electric vehicles Eco-platooning for reduced group consumption Considered drag reduction Nonlinear MPC, prediction horizon = 120 s Studied centralized and distributed solutions	+10.5
[87]	S, microsimulation, combustion engine vehicles 10 CAVs follow lead vehicle, share partial info Drag reduction is not considered Each CAV solves MPC, horizon = 12 to 20 s Fuel use evaluated using an engine map Compared against IDM car following	+50 For FTP Cycle Following
[109]	E, truck platooning 2.4 km unused runway, Crows Landing Two identical Freightliner tractors, 16 m trailers 90 km/h constant speed, 3–10 m spacing	+8–11
[113]	E, truck platooning, 45 km Swedish highway Three 18 m, 37–39 ton Scania tractor-trailers Wirelessly communicate vel., accel., parameters Time headway = 1 s	+4–6.5
[114]	E, truck platooning on a test track 3 fully-automated 25 ton trucks & 1 light truck Communicate vel., accel., brake via DSRC 80 km/h constant speed, 4.7 m gap	+15
[110]	E, truck platooning on test track 2 Peterbilt tractors, full aerodynamic packages 16 m trailers weighing 30 ton High-speed oval track, banked turns 105 km/h and 10 m following distance	+7.0

decisions can then be made in a distributed manner with each vehicle deciding (optimizing) its motion and sharing its intentions [134]. Alternatively, in a centralized framework, a single decision-making (optimization) problem is solved for a group of cooperative vehicles [135]. Cooperative lane selection and merge not only contributes to efficiency of the cooperating fleet but can also have a positive harmonizing effect on surrounding traffic.

There is a large body of literature on lane change models for traffic microsimulations, such as the widely used MOBIL lane change model which will be described later in Chap. 4. However cooperative lane selection and merging for CAVs has only been recently discussed. In [136] a cooperative lane-changing algorithm is simulated that considers follower vehicles in current and target lanes when making a lane change decision. The simulations in [136] show improvement with respect to MOBIL, in terms of merge time and rate, wait time, fuel consumption, average velocity, and flow at the cost of slightly increased travel time for main road vehicles. In [137] a merging assistant system that relies on vehicle cooperation, reduces the number of "late-merging" vehicles and subsequent likelihood of flow break-downs. Different algorithms for cooperative merging have been proposed, for instance [138] proposes a decentralized control method and [135] formulates it in a receding horizon optimization framework. A cooperative V2V "negotiation" process for lane changing is described in [139] while [140] proposes interaction protocols for cooperative lane changing. Experiments with three CAVs performing a semi-automated cooperative lane change maneuver are described in [141] and show the potential for smoother velocity trajectories. The focus of the above results has not been energy efficiency and only [136] reports energy efficiency gains. However, we expect considerable energy saving from wide deployment of cooperative lane changing and merging systems due to reduced braking events and harmonizing effect on traffic flow.

1.4.3 Cooperative Intersection Control

The coordination and optimal timing of traffic signals are by nature complex problems and backed by years of research in traffic engineering and operations research. Current signal timings are mostly scheduled offline. The optimized timings are then deployed as fix timetables for different times of the day and are further discussed in Chap. 4. Many signals are actuated by traffic and have rules to override their pre-optimized timetables based on the state of their loop-detectors to reduce idling at intersections. While these traffic responsive control strategies calculate their timing in real-time [142], they act based on the immediate state of loop-detectors. On the other hand, smart traffic signal controllers in connected vehicle environments will do more than just signaling right of ways by acting intelligently as hubs that sense, route, and harmonize the flow of arterial traffic.

The research on uni-directional signal to vehicle communication for improving efficiency by providing speed advisory to individual vehicles was discussed in Sect. 1.3.2. Another research direction has focused on improving intersection flow by optimizing timing of traditional traffic signals informed by uni-directional communication from connected vehicles [143, 144]. In addition, bi-directional vehicle-signal communication allows the geographical data of the connected vehicles to be also wirelessly transmitted in real-time to smart traffic signal controllers [145]. This increases energy efficiency and intersection flow as signals adjust their timings and vehicles their speeds [146].

Automated vehicles can further benefit from the communicated traffic signal information because they not only process the incoming information rather effortlessly but also can precisely control their speed and arrival time at a green light. The situation can get even better with 100% penetration of automated vehicles since a physical

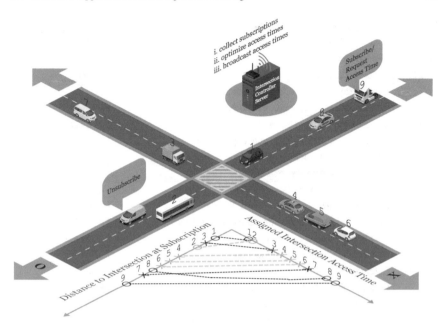

Fig. 1.6 Schematic of a cooperative intersection. CAVs subscribe to an intersection control server as they approach the intersection, the controller assigns access times to each approaching vehicle allowing only vehicles of the same movement in the intersection area at the same time. In this schematic vehicles on movements X and O are grouped together when assigning access times which reduces idling and saves energy. Most parts of the image were created on https://icograms.com

traffic light is not needed anymore as shown schematically in the cooperative intersection of Fig. 1.6 and in concept papers in [147–149]. Also because automated cars have much faster reaction times than human driven vehicles, the intersection controller can rapidly switch between phases [150]. Some of the benefits of eliminating traffic signals in an all-automated vehicle environment is discussed in [147] and demonstrated by interesting simulation results in a recent publication [151]. In [152] the authors show the potential for 50% energy efficiency gain via such reservation-based intersection control systems. In [153] increasing the intersection throughput is formalized as an optimization problem with details in a case study discussed in Chap. 9 of this book. They show significant reduction in number of stops and fuel use compared to traditional intersection control schemes. In a one hour microsimulation case study it is shown that the number of stops can be reduced 100 times [154]. Via a vehicle-in-the-loop experiment also described in Chap. 9, the authors measure 20% improvement in energy efficiency of a real-vehicle that interacts with the intersection controller and hundreds of simulated vehicles. Simulations indicate benefits of such systems greatly increase if vehicles move in platoons, in certain cases doubling the arterial network capacity with the coordination of platoons and intersections [155]. In [156] a platoon-based approach shows up to 20% energy efficiency benefit with

Table 1.8 Summary of selected published results on energy efficiency gain enabled by cooperative intersection control. Simulation and Experimental results are denoted by S and E respectively

Refs.	Methods and Conditions	Efficiency gain (%)
[152]	S, microsimulation in Paramics [68]	+50%
	Real-world road network, 3 intersections	
	CMEM [64] fuel efficiency evaluation	
[156]	S, microsimulation in Sumo	+11–21
	1 intersection with 2 single lane approaches	
	Vehicles form platoons to pass intersection	
	CMEM fuel efficiency evaluation	
[149]	S, Microsimulation in DIVERT [157]	+25
	Simulated entire network of Porto, Portugal	(fuel)
	EMIT [158] fuel consumption and emission model	+1–18
	1.3 ton combustion engine vehicle	(CO_2 emissions)
	Mostly studied emissions reduction	
[159]	E, real vehicle interacting with microsimulation	+20
	Real vehicle: 2011 Honda Accord 2.4 L engine	For
	Custom microsimulation written in Java	Real
	12 laps, 1.6 km track, single virtual intersection	Vehicle
	Scheduling via mixed integer linear programming	

respect to signalized intersections, but under simulation conditions of [156] energy efficiency was slightly sacrificed to form platoons. Table 1.8 highlights some of the key results on energy benefits of cooperative intersection control.

1.4.4   Indirect Benefits Through Traffic Harmonization

Coordinated and smoother motion of CAVs could harmonize the surrounding traffic and contribute to energy efficiency of conventional vehicles, even at low penetration levels. While it is difficult to establish the network-wide benefits experimentally, there are microsimulation case studies and isolated experiments that show such positive impacts. For instance in Sect. 1.3.2 we explained that according to [67] traffic signal speed advisory can reduce the energy consumption of conventional vehicles at moderate penetration rates. Several papers have shown the harmonizing effect of automated cruise control on the upstream traffic [76, 160] which is expected to positively influence energy efficiency of upstream traffic. CACC not only increases road utilization due to smaller gaps [123, 127, 161], but is shown to attenuate velocity variations as shown in road experiments in [124, 162]. These findings are corroborated by microsimulation studies, reported in [163], that show reduction of shock waves with increased penetration of connected and automated vehicles. In [164] a "theory for jam-absorption driving" is presented which is a method for driving a single car to attenuate a traffic shockwave, followed by experiments in [165]. An interesting experiment with a group of 22 vehicles moving on circle, showed that a single automated vehicle using relatively simple control rules, could dissipate the phantom jam waves formed by the 21 human driven vehicles. This contributed to between 20 and 40% improvement in average fuel economy of the fleet across the 3 experiments [80]. Harmonizing impact of CAVs in an open highway is demonstrated via interesting experiments in [166]. Three CAVs were driven side-by-side in real shock-wave traffic and their influence was measured by three probe vehicles that were deployed downstream and upstream. It was observed that CAVs reduced the oscillation induced by the shock-waves and harmonized the traffic with expected network-wide energy efficiency impact. Secondary effects due to reduced number of accidents, could further lower delays and loss of energy which is difficult to quantify.

*
* *

As mentioned several times in this chapter, CAVs have a huge potential in terms of energy consumption reductions of road mobility, thanks to their enhanced control capabilities. However, to exploit this potential at its best, simple rule-based control strategies are often not sufficient and a more systematic approach, based on mathematical optimization, is needed.

In the rest of this book we formalize the science of energy-efficient driving by laying out the mathematical modeling, control, and optimization foundations for it. We also describe the software and hardware technologies needed to implement energy-efficient driving functions and conclude by presenting a number of simulation and experimental case studies.

Traffic control and planning might benefit from CAV technology and help reduce overall energy consumption but are not the subject of this book.

References

1. McCarthy N (2015) Connected cars by the numbers. Forbes. Accessed 27 Jan 2015
2. Gartner (2015) Gartner says by 2020, a quarter billion connected vehicles will enable new in-vehicle services and automated driving capabilities. http://www.gartner.com/newsroom/id/2970017. Accessed 26 Jan 2015
3. NHTSA (2016) Federal motor vehicle safety standards; V2V communications. Technical Report NHTSA-2016-0126, US Department of Transportation National Highway Traffic Safety Administration
4. FHWA (2015) Vehicle-to-infrastructure deployment guidance and products. Technical Report FHWA-HOP-15-015, US Department of Transportation Federal Highway Administration
5. University of Michigan Center for Sustainable Systems (2016) Autonomous vehicles factsheet. Technical Report Pub. No. CSS16-18, Center for Sustainable Systems, University of Michigan
6. Alexander-Kearns M, Peterson M, Cassady A (2016) The impact of vehicle automation on carbon emissions. Technical report, Center for American Progress

7. Litman T (2017) Autonomous vehicle implementation predictions. Victoria transport policy institute
8. Brown A, Gonder J, Repac B (2014) An analysis of possible energy impacts of automated vehicle. In: Road vehicle automation, pp 137–153. Springer
9. Wadud Z, MacKenzie D, Leiby P (2016) Help or hindrance? The travel, energy and carbon impacts of highly automated vehicles. Transp Res Part A: Policy Pract 86:1–18
10. Simon K, Alson J, Snapp L, Hula A (2015) Can transportation emission reductions be achieved autonomously?
11. Greenblatt JB, Shaheen S (2015) Automated vehicles, on-demand mobility, and environmental impacts. Curr Sustain/Renew Energy Rep 2(3):74–81
12. Ericsson E, Larsson H, Brundell-Freij K (2006) Optimizing route choice for lowest fuel consumption-potential effects of a new driver support tool. Transp Res Part C: Emerg Technol 14(6):369–383
13. Ahn K, Rakha H (2008) The effects of route choice decisions on vehicle energy consumption and emissions. Transp Res Part D: Transp Environ 13(3):151–167
14. Minett CF, Salomons AM, Daamen W, Van Arem B, Kuijpers S (2011) Eco-routing: comparing the fuel consumption of different routes between an origin and destination using field test speed profiles and synthetic speed profiles. In: Proceedings of forum on integrated and sustainable transportation system (FISTS), pp 32–39. IEEE
15. Ahn K, Rakha HA (2013) Network-wide impacts of eco-routing strategies: a large-scale case study. Transp Res Part D: Transp Environ 25:119–130
16. Sachenbacher M, Leucker M, Artmeier A, Haselmayr J (2011) Efficient energy-optimal routing for electric vehicles. In: AAAI, pp 1402–1407
17. Artmeier A, Haselmayr J, Leucker M, Sachenbacher M (2010) The shortest path problem revisited: optimal routing for electric vehicles. In: Proceedings of annual conference on artificial intelligence, pp 309–316. Springer
18. De Nunzio G, Thibault L, Sciarretta A (2017) Model-based eco-routing strategy for electric vehicles in large urban networks. In: Comprehensive energy management–eco routing & velocity profiles, pp. 81–99. Springer
19. Jurik T, Cela A, Hamouche R, Natowicz R, Reama A, Niculescu S-I, Julien J (2014) Energy optimal real-time navigation system. IEEE Intell Transp Syst Mag 6(3):66–79
20. Houshmand A, Cassandras CG (2018) Eco-routing of plug-in hybrid electric vehicles in transportation networks. In: Proceedings international conference on intelligent transportation systems (ITSC), pp 1508–1513. IEEE
21. Baum M, Sauer J, Wagner D, Zündorf T (2017) Consumption profiles in route planning for electric vehicles: Theory and applications. In: Proceedings of international symposium on experimental algorithms (SEA 2017). Schloss Dagstuhl-Leibniz-Zentrum fuer Informatik
22. Liu C, Zhou M, Jing W, Long C, Wang Y (2017) Electric vehicles en-route charging navigation systems: joint charging and routing optimization. IEEE Trans Control Syst Technol 99:1–9
23. Suzuki Y (2011) A new truck-routing approach for reducing fuel consumption and pollutants emission. Transp Res Part D: Transp Environ 16(1):73–77
24. HERE. HERE APIs (2019) https://developer.here.com/develop/rest-apis
25. Barth M, Boriboonsomsin K (2009) Energy and emissions impacts of a freeway-based dynamic eco-driving system. Transp Res Part D: Transp Environ 14(6):400–410
26. Stahl P, Donmez B, Jamieson GA (2016) Supporting anticipation in driving through attentional and interpretational in-vehicle displays. Accid Anal Prev 91:103–113
27. Asadi B, Zhang C, Vahidi A (2010) The role of traffic flow preview for planning fuel optimal vehicle velocity. In: Proceedings of dynamic systems and control conference, pp 813–819. American Society of Mechanical Engineers
28. Wan N, Gomes G, Vahidi A, Horowitz R (2014) Prediction on travel-time distribution for freeways using online expectation maximization algorithm. In: Proceedings of transportation research board annual meeting
29. Wan N, Zhang c, Vahidi A (2019) Probabilistic anticipation and control in autonomous car following. IEEE Trans Control Syst Technol 27:30–38

30. Schepmann S, Vahidi A (2011) Heavy vehicle fuel economy improvement using ultracapacitor power assist and preview-based MPC energy management. In: Proceedings of American control conference (ACC), pp 2707–2712. IEEE
31. Huang W, Bevly DM, Schnick S, Li X (2008) Using 3D road geometry to optimize heavy truck fuel efficiency. In: Proceedings of international conference on intelligent transportation systems (ITSC), pp 334–339. IEEE
32. Hellström E, Ivarsson M, Åslund J, Nielsen L (2009) Look-ahead control for heavy trucks to minimize trip time and fuel consumption. Control Eng Pract 17(2):245–254
33. Hellström E, Åslund J, Nielsen L (2010) Design of an efficient algorithm for fuel-optimal look-ahead control. Control Eng Pract 18(11):1318–1327
34. He C, Maurer H, Orosz G (2016) Fuel consumption optimization of heavy-duty vehicles with grade, wind, and traffic information. J Comput Nonlinear Dyn 11(6):061011
35. Kamal MAS, Mukai M, Murata J, Kawabe T (2011) Ecological driving based on preceding vehicle prediction using MPC. IFAC Proc Vol 44(1):3843–3848
36. Zhang C, Vahidi A, Pisu P, Li X, Tennant K (2010) Role of terrain preview in energy management of hybrid electric vehicles. IEEE Trans Veh Technol 59(3):1139–1147
37. Argonne National laboratory (ANL) (2010) Argonne launches new tool to help auto industry reduce costs. https://www.anl.gov/article/argonne-launches-new-tool-to-help-auto-industry-reduce-costs
38. Back M, Terwen S, Krebs V (2004) Predictive powertrain control for hybrid electric vehicles. IFAC Proc Vol 37(22):439–444
39. Terwen S, Back M, Krebs V (2004) Predictive powertrain control for heavy duty trucks. IFAC Proc Vol 37(22):105–110
40. Fröberg A, Hellström E, Nielsen L (2006) Explicit fuel optimal speed profiles for heavy trucks on a set of topographic road profiles. Technical report, SAE technical paper
41. Lu J, Hong S, Sullivan J, Hu G, Dai E, Reed D, Baker R (2017) Predictive transmission shift schedule for improving fuel economy and drivability using electronic horizon. SAE Int J Engines 10(2017-01-1092):680–688
42. Danninger A, Armengaud E, Milton G, Lutzner J, Hakstege B, Zurlo G, Schoni A, Lindberg J, Krainer F (2018) IMPERIUM - IMplementation of Powertrain Control for Economic and Clean Real driving emission and fuel consUMption. In: Proceedings of transport research Arena TRA, Vienna, Austria
43. Freightliner (2009) Freightliner trucks launches RunSmart Predictve Cruise for Cascadia. https://daimler-trucksnorthamerica.com/influence/press-releases/#freightliner-trucks-launches-runsmart-predictive-cruise-2009-03-019. Accessed 19 Mar 2009
44. Barry K (2012). Trucks use GPS to anticipate hills, save fuel. Wired magazine. Accessed 21 May 2012
45. Sun C, Hu X, Moura SJ, Sun F (2015) Velocity predictors for predictive energy management in hybrid electric vehicles. IEEE Trans Control Syst Technol 23(3):1197–1204
46. Dornieden B, Junge L, Pascheka P (2012) Anticipatory energy-efficient longitudinal vehicle control. ATZ Worldw 114(3):24–29
47. Sujan VA, Frazier TR, Follen K, Moon SM (2014) System and method of cylinder deactivation for optimal engine torque-speed map operation. https://www.google.ch/patents/US8886422. US Patent 8,886,422. Accessed 11 Nov 2014
48. Braun M, Linde M, Eder A, Kozlov E (2010) Looking forward: predictive thermal management optimizes efficiency and dynamics. dSPACE Mag
49. Asadi B, Vahidi A (2009) Predictive use of traffic signal state for fuel saving. IFAC Proc Vol 42(15):484–489
50. Asadi B, Vahidi A (2011) Predictive cruise control: utilizing upcoming traffic signal information for improving fuel economy and reducing trip time. IEEE Trans Control Syst Technol 19(3):707–714
51. Mandava S, Boriboonsomsin K, Barth M (2009) Arterial velocity planning based on traffic signal information under light traffic conditions. In: Proceedings of international conference on intelligent transportation systems (ITSC). IEEE, pp 1–6

52. Rakha H, Raj Kishore Kamalanathsharma. Eco-driving at signalized intersections using V2I communication. In: Proceedings of international conference on intelligent transportation systems (ITSC), pp 341–346. IEEE
53. Mahler G, Vahidi A (2014) An optimal velocity-planning scheme for vehicle energy efficiency through probabilistic prediction of traffic-signal timing. IEEE Trans Intell Transp Syst 15(6):2516–2523
54. Xia H, Boriboonsomsin K, Schweizer F, Winckler A, Zhou K, Zhang W-B, Barth M (2012). Field operational testing of eco-approach technology at a fixed-time signalized intersection. In: Proceedings of international conference on intelligent transportation systems (ITSC), pp 188–193. IEEE
55. Jin Q, Wu G, Boriboonsomsin K, Barth MJ (2016) Power-based optimal longitudinal control for a connected eco-driving system. IEEE Trans Intell Transp Syst 17(10):2900–2910
56. Mahler G (2013) Enhancing energy efficiency in connected vehicles via access to traffic signal information. PhD thesis, Clemson University
57. Hao P, Wu G, Boriboonsomsin K, Barth M (2017) Eco-approach and departure (EAD) application for actuated signals in real-world traffic. Technical report, University of California, Riverside
58. Mahler G, Winckler A, Fayazi SA, Vahidi A, Filusch M (2017) Cellular communication of traffic signal state to connected vehicles for eco-driving on arterial roads: system architecture and experimental results. In: Proceedings of intelligent transportation systems conference, pp 1–6. IEEE
59. Koukoumidis E, Peh L-S, MR Martonosi (2011). SignalGuru: leveraging mobile phones for collaborative traffic signal schedule advisory. In: Proceedings of international conference on mobile systems, applications, and services, pp 127–140. ACM
60. Fayazi SA, Vahidi A, Mahler G, Winckler A (2015) Traffic signal phase and timing estimation from low-frequency transit bus data. IEEE Trans Intell Transp Syst 16(1):19–28
61. Fayazi SA, Vahidi A (2016) Crowdsourcing phase and timing of pre-timed traffic signals in the presence of queues: algorithms and back-end system architecture. IEEE Trans Intell Transp Syst 17(3):870–881
62. Marshall A (2016) Enlighten app uses AI to predict when lights will turn green. Wired Magazine. Accessed 30 Oct 2016
63. Bodenheimer R, Brauer A, Eckhoff D, German R (2014) Enabling GLOSA for adaptive traffic lights. In: Proceedings of vehicular networking conference (VNC), pp 167–174. IEEE
64. Scora G, Barth M (2006) Comprehensive modal emissions model (CMEM), version 3.01. User guide. Centre for environmental research and technology. University of California, Riverside
65. Kamalanathsharma R, Rakha H (2013) Multi-stage dynamic programming algorithm for eco-speed control at traffic signalized intersections. In: Proceedings of international conference on intelligent transportation systems (ITSC), pp 2094–2099. IEEE
66. Rakha HA, Ahn K, Moran K, Saerens B, Van den Bulck E (2011) Virginia tech comprehensive power-based fuel consumption model: model development and testing. Transp Res Part D: Transp Environ 16(7):492–503
67. Wan N, Vahidi A, Luckow A (2016) Optimal speed advisory for connected vehicles in arterial roads and the impact on mixed traffic. Transp Res Part C: Emerg Technol 69:548–563
68. Paramics Q (2009) The paramics manuals, version 6.6. 1. Quastone Paramics LTD, Edinburgh, Scotland, UK
69. Xia H, Boriboonsomsin K, Barth M (2013) Dynamic eco-driving for signalized arterial corridors and its indirect network-wide energy/emissions benefits. J Intell Transp Syst 17(1):31–41
70. Kamalanathsharma RK, Rakha HA, Yang H (2015. Network-wide impacts of vehicle eco-speed control in the vicinity of traffic signalized intersections. In: Proceedings of transportation research board annual meeting, pp 91–99
71. HomChaudhuri B, Vahidi A, Pisu P (2017) Fast model predictive control-based fuel efficient control strategy for a group of connected vehicles in urban road conditions. IEEE Trans Control Syst Technol 25(2):760–767

72. Mersky AC, Samaras C (2016) Fuel economy testing of autonomous vehicles. Transp Res Part C: Emerg Technol 65:31–48
73. Hoogendoorn S, Ossen S, Schreuder M (2006) Empirics of multianticipative car-following behavior. Transp Res Rec: J Transp Res Board 1965(1):112–120
74. Vanderbilt T (2009). Traffic: why we drive the way we do, 1st edn. Vintage
75. Ossen S, Hoogendoorn SP (2011) Heterogeneity in car-following behavior: theory and empirics. Transp Res Part C: Emerg Technol 19(2):182–195
76. Kamal MAS, Imura J, Hayakawa T, Ohata A, Aihara K (2014) Smart driving of a vehicle using model predictive control for improving traffic flow. IEEE Trans Intell Transp Syst 15(2):878–888
77. Dirk H (2001) Traffic and related self-driven many-particle systems. Rev Mod Phys 73:1067–1141. https://doi.org/10.1103/RevModPhys.73.1067
78. Sugiyama Y, Fukui M, Kikuchi M, Hasebe K, Nakayama A, Nishinari K, Tadaki S, Yukawa S (2008) Traffic jams without bottlenecks–experimental evidence for the physical mechanism of the formation of a jam. New J Phys 10(3):033001
79. Flynn MR, Kasimov AR, Nave J-C, Rosales RR, Seibold B (2009) Self-sustained nonlinear waves in traffic flow. Phys Rev E 79(5):056113
80. Stern RE, Cui S, Delle Monache ML, Bhadani R, Bunting M, Churchill M, Hamilton N, Haulcy R, Pohlmann H, Wu F, Piccoli B, Seibold B, Sprinkle J, Work DB (2017) Dissipation of stop-and-go waves via control of autonomous vehicles: Field experiments. http://arxiv.org/abs/1705.01693
81. McDonough K, Kolmanovsky I, Filev D, Yanakiev D, Szwabowski S, Michelini J (2013) Stochastic dynamic programming control policies for fuel efficient vehicle following. In: Proceedings of American control conference (ACC), pp 1350–1355. IEEE
82. Han J, Sciarretta A, Ojeda LL, De Nunzio G, Thibault L (2018) Safe- and eco-driving control for connected and automated electric vehicles using analytical state-constrained optimal solution. IEEE Trans Intell Veh 3(2):163–172
83. Lang D, Schmied R, Del Re L (2014) Prediction of preceding driver behavior for fuel efficient cooperative adaptive cruise control. SAE Int J Engines 7(2014-01-0298):14–20
84. Zhang C, Vahidi A (2011) Predictive cruise control with probabilistic constraints for eco driving. In: Proceedings of dynamic systems and control conference and symposium on fluid power and motion control, pp 233–238. American Society of Mechanical Engineers
85. Bichi M, Ripaccioli G, Di Cairano S, Bernardini D, Bemporad A, Kolmanovsky IV (2010) Stochastic model predictive control with driver behavior learning for improved powertrain control. In: Proceedings of conference on decision and control (CDC), pp 6077–6082. IEEE
86. McDonough K, Kolmanovsky I, Filev D, Szwabowski S, Yanakiev D, Michelini J (2014) Stochastic fuel efficient optimal control of vehicle speed. In: Optimization and optimal control in automotive systems, pp 147–162. Springer
87. Dollar RA, Vahidi A (2017) Quantifying the impact of limited information and control robustness on connected automated platoons. In: Proceedings of international conference on intelligent transportation systems (ITSC), pp 1–7. IEEE
88. Zheng Y, Li SE, Li K, Borrelli F, Hedrick JK (2017) Distributed model predictive control for heterogeneous vehicle platoons under unidirectional topologies. IEEE Trans Control Syst Technol 25(3):899–910
89. Orosz G (2016) Connected cruise control: modelling, delay effects, and nonlinear behaviour. Veh Syst Dyn 54(8):1147–1176
90. Li NI, He CR, Orosz G (2016) Sequential parametric optimization for connected cruise control with application to fuel economy optimization. In: Proceedings of conference on decision and control (CDC), pp 227–232. IEEE
91. Goodall NJ, Park B, Smith BL (2014) Microscopic estimation of arterial vehicle positions in a low-penetration-rate connected vehicle environment. J Transp Eng 140(10):04014047
92. Ge JI, Orosz G (2014) Dynamics of connected vehicle systems with delayed acceleration feedback. Transp Res Part C: Emerg Technol 46:46–64

93. Qin WB, Gomez MM, Orosz G (2017) Stability and frequency response under stochastic communication delays with applications to connected cruise control design. IEEE Trans Intell Transp Syst 18(2):388–403
94. Manzie C, Watson H, Halgamuge S (2007) Fuel economy improvements for urban driving: Hybrid vs. intelligent vehicles. Transp Res Part C: Emerg Technol 15(1):1–16
95. Li SE, Peng H (2012) Strategies to minimize the fuel consumption of passenger cars during car-following scenarios. Proc Inst Mech Eng, Part D: J Automob Eng 226(3):419–429
96. Turri V, Kim Y, Guanetti J, Johansson KH, Borrelli F (2014) A model predictive controller for non-cooperative eco-platooning. In: Proceedings of American control conference (ACC), pp 2309–2314
97. Rios-Torres J, Malikopoulos AA (2017) A survey on the coordination of connected and automated vehicles at intersections and merging at highway on-ramps. IEEE Trans Intell Transp Syst 18(5):1066–1077
98. Bevly D, Cao X, Gordon M, Ozbilgin G, Kari D, Nelson B, Woodruff J, Barth M, Murray C, Kurt A et al (2016) Lane change and merge maneuvers for connected and automated vehicles: a survey. IEEE Trans Intell Veh 1(1):105–120
99. Kamal MAS, Taguchi S, Yoshimura T (2015) Efficient vehicle driving on multi-lane roads using model predictive control under a connected vehicle environment. In: Proceedings of intelligent vehicles symposium (IV), pp 736–741. IEEE
100. Kamal MAS, Taguchi S, Yoshimura T (2016) Efficient driving on multilane roads under a connected vehicle environment. IEEE Trans Intell Transp Syst 17(9):2541–2551
101. Rios-Torres J, Malikopoulos AA (2017) Automated and cooperative vehicle merging at highway on-ramps. IEEE Trans Intell Transp Syst 18(4):780–789
102. Kamal MAS, Mukai M, Murata J, Kawabe T (2013) Model predictive control of vehicles on urban roads for improved fuel economy. IEEE Trans Control Syst Technol 21(3):831–841
103. Dollar RA, Vahidi A (2018) Predictively coordinated vehicle acceleration and lane selection using mixed integer programming. In: Proceedings of dynamic systems and control Conference, pages V001T09A006–V001T09A006. American Society of Mechanical Engineers
104. Papageorgiou M, Kotsialos A (2000) Freeway ramp metering: an overview. In: Proceedings of international conference on intelligent transportation systems (ITSC), pp 228–239. IEEE
105. Hegyi A, De Schutter B, Hellendoorn H (2005) Model predictive control for optimal coordination of ramp metering and variable speed limits. Transp Res Part C: Emerg Technol 13(3):185–209
106. Hafner MR, Cunningham D, Caminiti L, Del Vecchio D (2013) Cooperative collision avoidance at intersections: algorithms and experiments. IEEE Trans Intell Transp Syst 14(3):1162–1175
107. Letter C, Elefteriadou L (2017) Efficient control of fully automated connected vehicles at freeway merge segments. Transp Res Part C: Emerg Technol 80:190–205
108. Zhou M, Qu X, Jin S (2016) On the impact of cooperative autonomous vehicles in improving freeway merging: a modified intelligent driver model-based approach. IEEE Trans Intell Transp Syst
109. Browand F, McArthur J, Radovich C (2004) Fuel saving achieved in the field test of two tandem trucks. California partners for advanced transit and highways (PATH)
110. Bishop R, Bevly D, Humphreys L, Boyd S, Murray D (2017) Evaluation and testing of driver-assistive truck platooning: phase 2 final results. Transp Res Rec: J Transp Res Board 2615(1):11–18
111. Kunze R, Ramakers R, Henning K, Jeschke S (2011) Organization and operation of electronically coupled truck platoons on German motorways. In: Automation, communication and cybernetics in science and engineering 2009/2010, pp 427–439. Springer
112. Alam A, Gattami A, Johansson KH (2010) An experimental study on the fuel reduction potential of heavy duty vehicle platooning. In: Proceedings of international conference on intelligent transportation systems (ITSC), pp 306–311. IEEE
113. Alam A (2014) Fuel-efficient heavy-duty vehicle platooning. PhD thesis, KTH Royal Institute of Technology

114. Tsugawa S (2014) Results and issues of an automated truck platoon within the energy its project. In: Proceedings of intelligent vehicles symposium, pp 642–647. IEEE
115. Tsugawa S, Jeschke S, Shladover SE (2016) A review of truck platooning projects for energy savings. IEEE Trans Intell Veh 1(1):68–77
116. Swaroop D, Hedrick JK (1996) String stability of interconnected systems. IEEE Trans Autom Control 41(3):349–357
117. Segata M, Bloessl B, Joerer S, Sommer C, Gerla M, Cigno RL, Dressler F (2015) Toward communication strategies for platooning: simulative and experimental evaluation. IEEE Trans Veh Technol 64(12):5411–5423
118. Willke TL, Tientrakool P, Maxemchuk NF (2009) A survey of inter-vehicle communication protocols and their applications. IEEE Commun Surv Tutor 11(2)
119. Swaroop DVAHG, Hedrick JK (1999) Constant spacing strategies for platooning in automated highway systems. J Dyn Syst Meas Control 121(3):462–470
120. Horowitz R, Varaiya P (2000) Control design of an automated highway system. Proc IEEE 88(7):913–925
121. Larson J, Liang K-Y, Johansson KH (2015) A distributed framework for coordinated heavy-duty vehicle platooning. IEEE Trans Intell Transp Syst 16(1):419–429
122. Muoio D (2017) Here's how Tesla, Uber, and Google are trying to revolutionize the trucking industry. Business Insider. Accessed 20 June 2017
123. Naus GJL, Vugts RPA, Ploeg J, van de Molengraft MJG, Steinbuch M (2010) String-stable CACC design and experimental validation: a frequency-domain approach. IEEE Trans Veh Technol 59(9):4268–4279
124. Ploeg J, Scheepers BTM, Van Nunen E, Van de Wouw N, Nijmeijer H (2011) Design and experimental evaluation of cooperative adaptive cruise control. In: Proceedings of international conference on intelligent transportation systems (ITSC), pp 260–265. IEEE
125. Ploeg J, Shladover S, Nijmeijer H, van de Wouw N (2012) Introduction to the special issue on the 2011 grand cooperative driving challenge. IEEE Trans Intell Transp Syst 13(3):989–993
126. Van Nunen E, Kwakkernaat RJAE, Ploeg J, Netten BD (2012) Cooperative competition for future mobility. IEEE Trans Intell Transp Syst 13(3):1018–1025
127. Van Arem B, Van Driel CJG, Visser R (2006) The impact of cooperative adaptive cruise control on traffic-flow characteristics. IEEE Trans Intell Transp Syst 7(4):429–436
128. Zheng Y, Li SE, Wang J, Wang LY, Li K (2014) Influence of information flow topology on closed-loop stability of vehicle platoon with rigid formation. In: Proceedings of international conference on intelligent transportation systems (ITSC), pp 2094–2100. IEEE
129. Di Bernardo M, Salvi A, Santini S (2015) Distributed consensus strategy for platooning of vehicles in the presence of time-varying heterogeneous communication delays. IEEE Trans Intell Transp Syst 16(1):102–112
130. Besselink B, Turri V, van de Hoef SH, Liang K-Y, Alam A, Mårtensson J, Johansson KH (2016) Cyber–physical control of road freight transport. Proc IEEE 104(5):1128–1141
131. Lelouvier A, Guanetti J, Borrelli F (2017) Eco-platooning of autonomous electrical vehicles using distributed model predictive control. In: Proceedings of conference on intelligent transportation systems, pp 464–469. IEEE
132. Dunbar WB, Caveney DS (2012) Distributed receding horizon control of vehicle platoons: stability and string stability. IEEE Trans Autom Control 57(3):620–633
133. Wang J-Q, Li SE, Zheng Y, Lu X-Y (2015) Longitudinal collision mitigation via coordinated braking of multiple vehicles using model predictive control. Integr Comput-Aided Eng 22(2):171–185
134. Nie J, Zhang J, Ding W, Wan X, Chen X, Ran B (2016) Decentralized cooperative lane-changing decision-making for connected autonomous vehicles. IEEE Access 4:9413–9420
135. Cao W, Mukai M, Kawabe T, Nishira H, Fujiki N (2015) Cooperative vehicle path generation during merging using model predictive control with real-time optimization. Control Eng Pract 34:98–105
136. Awal T, Murshed M, Ali M (2015) An efficient cooperative lane-changing algorithm for sensor-and communication-enabled automated vehicles. In: Proceedings intelligent vehicles symposium (IV), pp 1328–1333. IEEE

137. Scarinci R, Heydecker B, Hegyi A (2015) Analysis of traffic performance of a merging assistant strategy using cooperative vehicles. IEEE Trans Intell Transp Syst 16(4):2094–2103
138. Mosebach A, Röchner S, Lunze J (2016) Merging control of cooperative vehicles. IFAC-PapersOnLine 49(11):168–174
139. Lombard A, Perronnet F, Abbas-Turki A, El Moudni A (2017) On the cooperative automatic lane change: Speed synchronization and automatic "courtesy". In: Proceedings of design, automation & test in Europe conference & exhibition (DATE), pp 1655–1658. IEEE
140. Kazerooni ES, Ploeg J (2015) Interaction protocols for cooperative merging and lane reduction scenarios. In: Proceedings of international conference on intelligent transportation systems (ITSC), pp 1964–1970. IEEE
141. Raboy K, Ma J, Stark J, Zhou F, Rush K, Leslie E (2017) Cooperative control for lane change maneuvers with connected automated vehicles: a field experiment. Technical report, Transportation Research Board
142. Diakaki C, Papageorgiou M, Aboudolas K (2002) A multivariable regulator approach to traffic-responsive network-wide signal control. Control Eng Pract 10(2):183–195
143. He Q, Head KL, Ding J (2012) PAMSCOD: platoon-based arterial multi-modal signal control with online data. Transp Res Part C: Emerg Technol 20(1):164–184
144. Kamal MAS, Imura J, Hayakawa T, Ohata A, Aihara K (2015a) Traffic signal control of a road network using MILP in the MPC framework. Int J Intell Transp Syst Res 13(2):107–118
145. Goodall N, Smith B, Park B (2013) Traffic signal control with connected vehicles. Transp Res Rec: J Transp Res Board 2381(1):65–72
146. De Nunzio G, Gomes G, Canudas-de Wit C, Horowitz R, Moulin P (2017). Speed advisory and signal offsets control for arterial bandwidth maximization and energy consumption reduction. IEEE Trans Control Syst Technol 25(3):875–887
147. Dresner K, Stone P (2008) A multiagent approach to autonomous intersection management. J Artif Intell Res, pp 591–656
148. Ferreira M, Fernandes R, Conceição H, Viriyasitavat W, Tonguz OK (2010) Self-organized traffic control. In: Proceedings of international workshop on VehiculAr InterNETworking, pp 85–90. ACM
149. Ferreira M, d'Orey PM (2012) On the impact of virtual traffic lights on carbon emissions mitigation. IEEE Trans Intell Transp Syst 13(1):284–295
150. Guler SI, Menendez M, Meier L (2014) Using connected vehicle technology to improve the efficiency of intersections. Transp Res Part C: Emerg Technol 46:121–131
151. Tachet R, Santi P, Sobolevsky S, Reyes-Castro LI, Frazzoli E, Helbing D, Ratti C (2016) Revisiting street intersections using slot-based systems. PloS one 11(3):e0149607
152. Huang S, Sadek AW, Zhao Y (2012) Assessing the mobility and environmental benefits of reservation-based intelligent intersections using an integrated simulator. IEEE Trans Intell Transp Sys 13(3):1201–1214
153. Fayazi SA, Vahidi A, Luckow A (2017) Optimal scheduling of autonomous vehicle arrivals at intelligent intersections via MILP. In: Proceedings of American control conference (ACC), pp 4920–4925. IEEE
154. Fayazi SA, Vahidi A (2018) Mixed-integer linear programming for optimal scheduling of autonomous vehicle intersection crossing. IEEE Trans Intell Veh 3(3):287–299
155. Lioris J, Pedarsani R, Tascikaraoglu FY, Varaiya P (2016) Doubling throughput in urban roads by platooning. IFAC Symp Control Transp Syst 49(3):49–54
156. Jin Q, Wu G, Boriboonsomsin K, Barth M (2013) Platoon-based multi-agent intersection management for connected vehicle. In: Proceedings of international conference on intelligent transportation systems (ITSC 2013), pp 1462–1467. IEEE, Oct 2013. https://doi.org/10.1109/ITSC.2013.6728436
157. Fernandes R, d'Orey PM, Ferreira M (2010) Divert for realistic simulation of heterogeneous vehicular networks. In: Proceedings of international conference on mobile adhoc and sensor systems (MASS), pp 721–726. IEEE
158. Cappiello A, Chabini I, Nam EK, Lue A, Zeid MA (2002) A statistical model of vehicle emissions and fuel consumption. In: Proceedings of international conference on intelligent transportation systems, pp 801–809. IEEE

159. Fayazi SA, Vahidi A (2017) Vehicle-In-the-Loop (VIL) verification of a smart city intersection control scheme for autonomous vehicles. In: Proceedings of conference on control technology and applications (CCTA), pp 1575–1580. IEEE
160. Koshizen T, Kamal MAS, Koike H (2015) Traffic congestion mitigation using intelligent driver model (IDM) combined with lane changes-why congestion detection is so needed? Technical report, SAE technical paper
161. Shladover S, Dongyan S, Xiao-Yun L (2012) Impacts of cooperative adaptive cruise control on freeway traffic flow. Transp Res Rec: J Transp Res Board 2324(1):63–70
162. Milanés V, Shladover SE, Spring J, Nowakowski C, Kawazoe H, Nakamura M (2014) Cooperative adaptive cruise control in real traffic situations. IEEE Trans Intell Transp Syst 15(1):296–305
163. Talebpour A, Mahmassani HS (2016) Influence of connected and autonomous vehicles on traffic flow stability and throughput. Transp Res Part C: Emerg Technol 71:143–163
164. Nishi R, Tomoeda A, Shimura K, Nishinari K (2013) Theory of jam-absorption driving. Transp Res Part B: Methodol 50:116–129
165. Taniguchi Y, Nishi R, Tomoeda A, Shimura K, Ezaki T, Nishinari K (2015) A demonstration experiment of a theory of jam-absorption driving. In: Traffic and Granular Flow' 13, pp 479–483. Springer
166. Ma J, Li X, Shladover S, Rakha HA, Lu X-Y, Jagannathan R, Dailey DJ (2016) Freeway speed harmonization. IEEE Trans Intell Veh 1(1):78–89

Chapter 2
Fundamentals of Vehicle Modeling

At least three energy conversion steps are relevant for a comprehensive analysis of energy efficiency of passenger cars and other road vehicles. As illustrated in Fig. 2.1, in a first step ("*grid-to-tank*"), energy carriers that are available at stationary distribution networks, such as gasoline, electricity, etc., are transferred to an on-board storage system. This energy is then converted by the propulsion system to mechanical energy aimed at propelling the vehicle ("*"tank-to-wheels"*"). In the third energy conversion step ("*wheel-to-distance*"), this mechanical energy is ultimately converted into the kinetic and potential energy required by the displacement. Unfortunately, all of these conversion processes may cause substantial energy losses.

Tank-to-wheels efficiency may be improved by several approaches, both at the component level and at the system control level [1]. Methods to improve grid-to-tank efficiency by choosing the appropriate charging time slots and profiles are currently being studied for electric vehicles. The maximization of wheel-to-distance efficiency is one of the main topics of this book. Although minimizing the energy at wheels E_W for a given amount of the useful energy E_M is theoretically possible and will be actually treated in the following chapters, it is often more interesting to try to minimize the energy consumed by the on-board sources ("tanks"), E_T. To analyze such a problem, we shall introduce in the following section vehicle-specific models of powertrain components.

The standard tool to evaluate tank-to-distance efficiency and predict energy consumption of road vehicles as a function of how they are driven is the use of modular modeling of the vehicle and its propulsion system. As Figs. 2.2, 2.3, 2.4, 2.5 and 2.6 suggest, such models are composed of several sub-models representing each relevant energy conversion step from E_T to E_M. To evaluate the tank energy for a given driving profile, these models are typically solved "backwards", i.e., the model inputs are at the road level (speed, acceleration), and the energy conversion chain is followed in the opposite direction of the physical energy flow, to eventually evalu-

© Springer Nature Switzerland AG 2020
A. Sciarretta and A. Vahidi, *Energy-Efficient Driving of Road Vehicles*,
Lecture Notes in Intelligent Transportation and Infrastructure,
https://doi.org/10.1007/978-3-030-24127-8_2

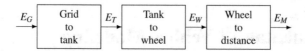

Fig. 2.1 Macroscopic energy conversion steps in road transport

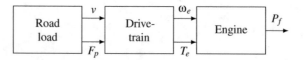

Fig. 2.2 Backward calculation of energy consumption rate in an engine-powered vehicle. Nomenclature defined in Sects. 2.1–2.2

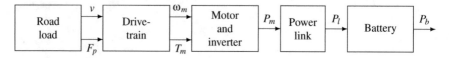

Fig. 2.3 Backward calculation of energy consumption rate in an electric vehicle. Nomenclature defined in Sect. 2.3

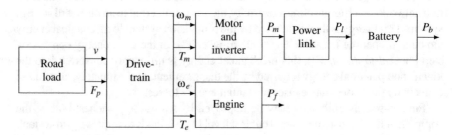

Fig. 2.4 Backward calculation of energy consumption rate in a parallel hybrid-electric vehicle. Nomenclature defined in Sect. 2.4

ate the power drained from on-board sources. In contrast, the physical energy flow (from the sources to the road) is followed to predict the effects of powertrain control strategies on driving and, consequently, on energy consumption. For this purpose, the same modular approach of Figs. 2.2, 2.3, 2.4, 2.5 and 2.6 can be used, albeit with physical causality ("forward" modeling).

In Sect. 2.1, a model for the energy required at the wheels that is common to all types of road vehicles is presented. Then, each propulsion system is modeled separately, and the main equations that are useful to evaluate the tank energy consumption are summarized (Sect. 2.2–2.5).

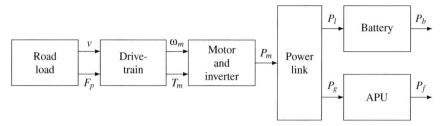

Fig. 2.5 Backward calculation of energy consumption rate in a series hybrid-electric vehicle. Nomenclature defined in Sect. 2.4

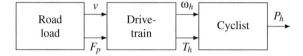

Fig. 2.6 Backward calculation of energy consumption rate in an human-powered vehicle. Nomenclature defined in Sect. 2.5

2.1 Road Load

From a dynamic viewpoint, road vehicles are generally treated as semi-rigid bodies and described by their linear and angular position along three coordinated dimensions. For our purposes, however, the longitudinal motion often suffices.[1] We shall therefore consider the dynamics of longitudinal position, $s(t)$, and speed, $v(t) \triangleq \dot{s}(t)$.

2.1.1 Forces Acting on Road Vehicles

The vehicle's longitudinal dynamics is governed by Newton's second law of motion, see Fig. 2.7,

$$m_t \frac{dv(t)}{dt} = F_p(t) - F_{\text{res}}(t) - F_b(t) . \tag{2.1}$$

In this equation, $m_t = m + m_r$ is *total effective mass*, sum of the overall mass m of the vehicle (curb mass) and of its occupants, plus a term m_r that describes the effect of the inertia of rotating parts (engine, motor, etc.) transferred to the wheels. The latter term is usually varying with time as transmission ratio changes, but this variation is often neglected. Of course, vehicle mass has a wide range, from less than 100 kg for bicycles (where the cyclist's mass is prominent over the vehicle) to several thousand kg for heavy-duty trucks.

---

[1]Even in the presence of lateral maneuvers such as lane changes (see, e.g., Sect. 9.4), the energy consumption associated with such transient maneuvers will be neglected.

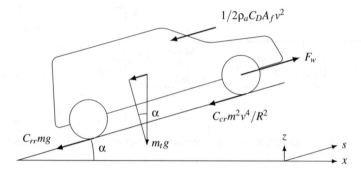

Fig. 2.7 Schematic representation of the forces acting on a vehicle in motion

The force F_p is the sum of forces applied by the powertrain at the wheels, as explained below. The term F_b is the force applied by frictional brakes only, while "regenerative" brake possibly provided by electric, hydraulic, pneumatic, or kinetic means coupled with accumulators is considered here as part of the propulsion system.

The term representing road load, F_{res}, may include several contributions. Limiting our analysis to the most relevant ones, we shall write the road load as

$$F_{res}(t) = \frac{1}{2}\rho_a C_D(t)A_f\,(v(t) - w)^2 + C_{rr}mg\cos(\alpha(s(t))) +$$
$$+ C_{cr}\frac{m^2 v(t)^4}{R(s(t))^2} + mg\sin(\alpha(s(t)))\,, \tag{2.2}$$

where C_{rr} is the coefficient of rolling resistance, α and R are the road slope and the out-of-the-pictured-plane radius of curvature, which are generally functions of the vehicle's position s, ρ_a is air density, w is longitudinal wind speed, A_f is vehicle's frontal area, C_{cr} the coefficient of cornering resistance [2, 3], and C_D is the aerodynamic drag coefficient.

The first three terms in the right-hand side of (2.2) must be regarded as approximations of physically complex phenomena [1]. Basically, these terms define the coefficients C_{rr}, C_{cr}, and C_D to represent rolling, cornering, and aerodynamic resistance forces. Typical values for C_{rr} on dry roads range from 0.002 (high-quality bicycle racing tires) to 0.02 (car tires at low pressure). Typical values for C_D range from 0.15 (low-drag concept cars) to 1.2 (utility bicycles with cyclist). The order of magnitude of C_{cr} is of 10^{-5}.

The main factors that influence such parameters are described in standard books on vehicle systems [1]. They comprise of vehicle, road, and weather factors, such that for a given trip, they can be generally considered as constants.[2] Perhaps the only relevant exception is the variation of C_D as a function of the inter-distance with a leading vehicle. Since the reduction of this coefficient when the inter-distance approaches to

---

[2]From here on, we shall always make use of this assumption, unless explicitly stated.

zero can be very relevant, both for the follower and, albeit less markedly, the leader [4, 5], this property is the basis of energy-saving techniques such as platooning (for heavy-duty trucks) or drafting (for bicycles).

Often F_{res} is identified as a whole by letting the vehicle coast ($F_p = F_b = 0$) on a flat and straight road ($\alpha = 0$, $R \to \infty$) with no wind ($w = 0$), and observing the speed variation, then expressed as a polynomial function of $v(t)$,

$$F_{res}(t) = C_0 + C_1 v(t) + C_2 v(t)^2 , \qquad (2.3)$$

where the C's are called the *road load coefficients*.

Note that, in the "backward" approach presented at the beginning of this chapter, (2.1)–(2.3) are used to evaluate $F_p(t)$ as a function of $v(t)$, its derivative dv/dt, and $s(t)$.

2.1.2 Energy Requirement at the Wheels

Based on the longitudinal dynamics introduced in the previous section, in this section we shall derive equations for the energy needed at the wheels, E_W, to follow a prescribed speed profile $v(t)$ covering a distance s_f in t_f units of time. For such a "trip", the useful energy E_M will be defined as

$$E_M = \frac{1}{2} m_t (v_f^2 - v_i^2) + mg\Delta z , \qquad (2.4)$$

where v_i and v_f are velocities at origin and destination respectively, and Δz is total elevation change during the trip.

Let us first define the net force at the wheels as $F_w(t) \triangleq F_p(t) - F_b(t)$. The instantaneous power needed at the wheels is thus $F_w(t)v(t)$. From (2.1–2.2), assuming a straight road with no wind,[3] the net energy E_W can then be calculated as

$$E_W = \int_0^{t_f} F_w(t)v(t)dt =$$
$$\int_0^{t_f} \left(m_t \frac{dv(t)}{dt} + mg(sin\alpha(s(t)) + C_{rr}cos\alpha(s(t)) + \frac{1}{2}\rho_a A_f C_D v^2(t) \right) v(t)dt .$$
$$(2.5)$$

Using (2.4), integration yields

$$E_W = E_M + mg C_{rr} \Delta x + \frac{1}{2}\rho_a A_f C_D \int_0^{t_f} v^3(t)dt , \qquad (2.6)$$

---

[3]From here on, we shall always make use of this assumption, unless explicitly stated.

where Δx is the horizontal distance covered. Note that road grade does not appear after integration. However, it shall be explained later that, because of constraints on velocity and powertrain output, the elevation profile along a trip can have a significant effect on energy use and prior knowledge of it can help save fuel via better constraint management.

The analysis of (2.6) shows that the term E_M does not offer opportunities for reducing E_W, since it is dictated by initial and terminal conditions only. The term $mgC_{rr}\Delta x$ represents the irreversible frictional loss and could be reduced by choosing shorter routes with lower C_{rr}. The last term, the energy lost to aerodynamic drag, is the only term that can be influenced by the decisions along the route and therefore should be a core consideration in eco-driving.

In this regard, lower speeds obviously result in lower losses to drag. More specifically, the drag term can be explicitly evaluated as in [6]:

$$\frac{1}{2}\rho_a A_f C_D \int_0^{t_f} v^3(t)dt = \frac{1}{2}\rho_a A_f C_D \left(\frac{b_v \sigma_v^3}{\bar{v}} + 3\sigma_v^2 + \bar{v}^2 \right) s_f , \tag{2.7}$$

where

$$\bar{v} \triangleq \frac{\int_0^{t_f} v(s)dt}{t_f} = \frac{s_f}{t_f} , \tag{2.8}$$

is the average speed (first raw moment of speed),

$$\sigma_v^2 \triangleq \frac{\int_0^{t_f} (v(t) - \bar{v})^2 dt}{t_f} , \tag{2.9}$$

is its variance (second central moment), and

$$b_v \triangleq \frac{\int_0^{t_f} (v(t) - \bar{v})^3 dt}{t_f \sigma_v^3} , \tag{2.10}$$

is speed skewness (third standardized moment). A generalization of (2.6) with the parametrization (2.3) for the flat road load thus reads

$$E_W = E_M + \left(C_0 + C_1\bar{v} + C_2\bar{v}^2 + C_1\frac{\sigma_v^2}{\bar{v}} + 3C_2\sigma_v^2 + \frac{C_2 b_v \sigma_v^3}{\bar{v}} \right) s_f . \tag{2.11}$$

In summary, the wheel-to-distance energy loss $E_W - E_M$ depends on four driving parameters (s_f, $\bar{v}$ or, alternatively, t_f, σ_v^2, and b_v) and three vehicle parameters (C_0, C_1, C_2). Their relative influence can be evaluated with a sensitivity analysis, defining the sensitivity with respect to the generic parameter π as

$$S_\pi \triangleq \frac{\partial(E_W - E_M)}{\partial \pi} \cdot \frac{\pi}{E_W - E_M} . \tag{2.12}$$

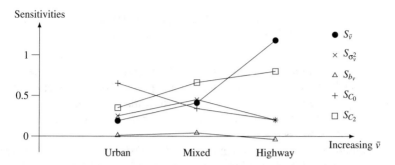

Fig. 2.8 Relative influence of driving profile parameters on the wheel-to-distance losses of a typical full-size passenger car, for urban, mixed, and highway type profiles

Typical values of sensitivities to driving and vehicle parameters are shown in Fig. 2.8 for urban, mixed, and highway driving profiles. While the influence of C_0 (largely representative of rolling resistance) is larger than that of C_2 (aerodynamic drag) at low average speeds, the situation is the opposite as $\bar{v}$ increases. For high average speeds, the relative influence of $\bar{v}$ also becomes dominant over that of the other driving parameters. The sensitivity $\otimes_{\sigma_v^2}$ is only relevant for urban and mixed driving conditions. Finally, the skewness b_v has generally a little influence.

The previous analysis shows that, while vehicle parameters play an important role in determining the energy demand, the approach of improving wheel-to-meters efficiency by "controlling" the driving profile reveals all its potential when considering that it does not require structural or material changes to the system.

2.1.3 Energy Required from the Powertrain

The wheel energy evaluated with (2.6) or (2.11) does not take into account the energy dissipated in friction brakes that are acting on the wheels. To include this term, *powertrain energy* can be defined as

$$E_p = \int_0^{t_f} F_p(t)v(t)dt = E_W + \int_0^{t_f} F_b(t)v(t)dt , \qquad (2.13)$$

where the braking force $F_b \geq 0$ explicitly appears.

The amount of this force (total braking effort) depends on the braking strategy and is usually split between the wheel axles. Of course, brakes are normally activated only for slowing a moving vehicle, thus to provide a required deceleration in addition to that induced by resistive forces. For our purposes, it is reasonable to assume that

$$F_b(t) = \begin{cases} -(1 - k_b(t))F_w(t), & \text{if } F_w(t) < 0 \\ 0, & \text{otherwise} \end{cases}, \qquad (2.14)$$

where k_b is the *ratio of regenerative braking* force (recuperated by the powertrain and stored in the onboard tank) to the total braking force. For engine-based vehicles without recuperation devices, k_b is close to zero, while for electrified vehicles $0 < k_b < 1$. An ideal vehicle would have $k_b = 1$ (perfect recuperation) and, in this case, E_p would be the same as E_W.

The condition $F_w < 0$ means that the vehicle transfers power to the powertrain or brakes and defines the *braking mode* $\mathcal{B}$. Contrarily, in the *traction mode* $\mathcal{T}$ (when $F_w > 0$), the vehicle is receiving power from the powertrain. The separating situation is called *coasting*. In such a mode, the vehicle is moving solely due to the resistive forces ($m_t dv/dt = -F_{res}$, $F_w = 0$), thus both F_p and F_b are zero. In the *stop mode* $\mathcal{S}$, if present, the speed is also equal to zero.

Inserting (2.14) into (2.13), the powertrain energy can be evaluated as

$$E_p = \int_{\mathcal{T}} \left(m_t \frac{dv(t)}{dt} + F_{res}(t) \right) v(t)dt + \int_{\mathcal{B}} k_b(t) \left(m_t \frac{dv(t)}{dt} + F_{res}(t) \right) v(t)dt .$$
$$(2.15)$$

The first term in the right-hand side of (2.15) can be explicitly computed with the methods introduced in Sect. 2.1.2, but with the driving parameters (distance, average speed, higher moments) now evaluated for the traction phase only. The second term is negative by definition. Its explicit evaluation is made complex by the variability of k_b. However, if k_b is considered as a constant parameter, the methods of Sect. 2.1.2 still apply.

*

* *

We switch now our attention to the evaluation of the tank energy, that is, the energy consumed from the onboard source(s), differentiating this evaluation by the type of powertrain.

2.2 Internal Combustion Engine Vehicles

Among fuel-powered vehicles, the vast majority are composed of vehicles (cars, trucks, buses, and others) propelled by a reciprocating internal combustion engine (ICE). Such powertrains consists of an engine that burns fuel stored in a tank, delivering mechanical power to a rotating shaft, from which it is transmitted to the wheels by a drivetrain (see Fig. 2.2).

In ICE-powered vehicles (ICEVs), the tank energy consumption corresponds to the chemical energy burned with the fuel,

$$E_T^{(ICEV)} = \int_0^{t_f} P_f(t)dt , \qquad (2.16)$$

where P_f is the fuel power, the product of fuel mass flow rate and its lower heating value. How P_f is related to the driving profile will be now described by following the approach of Fig. 2.2 and analyzing separately the drivetrain and the engine.

2.2.1 Drivetrain (Gearbox)

In ICE-powered vehicles, the output of the engine is transmitted to the driving wheels by a *drivetrain*. The components in the drivetrain vary according to the type of drive (front-wheel, rear-wheel, four-wheel) but generally include a clutch, a transmission, a drive shaft, and a final drive with differential.

Apart from rarely used continuously-variable transmissions (CVT), usually transmissions with a finite number of gear ratios (*gearboxes*) are used, with the ability to switch between them. The drivetrain transmission ratio γ_e in this case is the product of the transmission ratios of the gearbox and the final drive, and can take the G discrete values $\gamma_e \in \{\gamma_{e,g}\}$, $g = 1, \ldots, G$. The gear selected at a particular time t is determined either by the driver (manual transmissions, automated manual transmissions (AMT)) or by the transmission controller (automatic transmissions). In the latter case, gear ratio is the result of a gear shift law that can be expressed as a map

$$\gamma_e(t) = \Gamma(v(t), F_p(t)) . \tag{2.17}$$

A common speed-only-dependent form for such a map is[4]

$$\Gamma(v) = \gamma_{e,1} + \sum_{g=1}^{G-1} \frac{1}{2}(\gamma_{e,g+1} - \gamma_{e,g})(1 + \sin(\arctan(\alpha_g(v - v_{sh,g})))) , \tag{2.18}$$

where $v_{sh,g}$ are the gear-shift speeds and the coefficients α_g are chosen to make the gear transition sufficiently smooth, see Fig. 2.9.

Following the backward calculation approach illustrated in Fig. 2.2, the engine torque T_e is related to the powertrain force F_p through the overall transmission ratio,

$$T_e(t) = \frac{F_p(t)r_w}{\gamma_e(t)\eta_t^{\text{sign}(F_p(t))}} , \tag{2.19}$$

where r_w is the wheel radius and η_t is the transmission efficiency. This quantity depends on the gear ratio used, although this dependency is often neglected.

Conversely, engine rotational speed ω_e is related to the vehicle's speed v,

---

[4]This approximation is usually valid only for small and large force values. For intermediate values, the gear-shift speeds would increase with force. Also, the gear-shift speeds are usually different for upshifts than for downshifts.

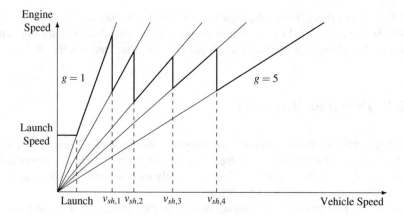

Fig. 2.9 Qualitative speed-only-dependent gear shift law described by (2.18)

$$\omega_e(t) = \frac{\gamma_e(t)}{r_w} v(t) .$$ (2.20)

Note that (2.19) implies a discontinuous derivative $\partial T_e / \partial F_p$ at $F_p = 0$ (coasting operation). A particular case of coasting is when the transmission is in the neutral state, with a clutch opening to disconnect the engine from the wheels. Commonly reserved for the stopped vehicle state, this operation is nowadays used also in *sailing* maneuvers, aimed at prolonging coasting by suppressing the engine brake effect to save fuel. The neutral cannot be described by setting $\gamma_e = 0$ in (2.19) and (2.20). In this case the engine is either stopped or idling (see below).

Note also that, during clutching maneuvers (in manual transmissions) or torque converter maneuvers (in automated transmissions), the relation between engine and vehicle speeds and torques cannot be described by (2.19)–(2.20).

2.2.2 Engine

Although engines can differ substantially concerning the type of fuel (gasoline, diesel, alternative fuels such as LPG, natural gas, E85), the thermodynamic cycle followed (four-strokes, two-strokes), the aspiration method (naturally aspirated, supercharged), and generally the technology used, they share a similar representation for our purposes.

The fuel power P_f can be modeled under a steady-state approximation using tabulated data ("engine map") as a function of engine torque and rotational speed,

$$P_f(t) = f(T_e(t), \omega_e(t)) .$$ (2.21)

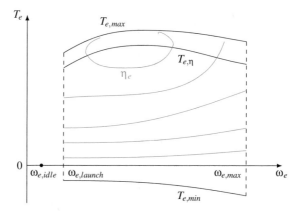

Fig. 2.10 Example efficiency map of a naturally-aspirated spark-ignition engine

Equivalently, an engine map can be expressed in terms of efficiency

$$\eta_e \triangleq \frac{T_e \omega_e}{P_f}, \qquad (2.22)$$

as depicted by the contour lines in Fig. 2.10.

This figure also shows that the engine operation is restricted by a certain number of limitations, namely: (i) the full-load torque $T_{e,max}$, which depends on the rotational speed; (ii) the fuel-cutoff torque $T_{e,min}$, which is negative due to friction and also dependent on speed; (iii) the maximum speed $\omega_{e,max}$; and (iv) the launch speed at which the engine starts producing stable torque, $\omega_{e,launch}$.

When the vehicle is stopped or coasting, the engine might be disconnected by putting the transmission in neutral, and run at the idle speed $\omega_{e,idle}$. To cover this situation, the idle consumption must be considered alongside the engine map. However, modern engines are often equipped with a stop-start device that turns the engine off during these operations.

The curve $T_{e,\eta}$ is the locus of the engine operating points for which the efficiency—defined by (2.22)—is the highest for those at the same output power. It is referred to as optimal operating line (OOL) of the engine and does not in general coincide with the maximal torque curve.

For the online applications described later in this book, e.g., in Chap. 8, approximated polynomial expressions are preferred to (2.21). The most used of such expressions is the affine-in-torque Willans model [1]

$$P_f(t) = \frac{T_e(t)\omega_e(t) + P_{e,min}(\omega_e(t))}{e(\omega_e(t))}, \qquad (2.23)$$

defined by the speed-dependent parameters e (efficiency of the thermodynamic energy conversion from fuel to cylinder pressure) and $P_{e,min} = \omega_e T_{e,min}$ (mechan-

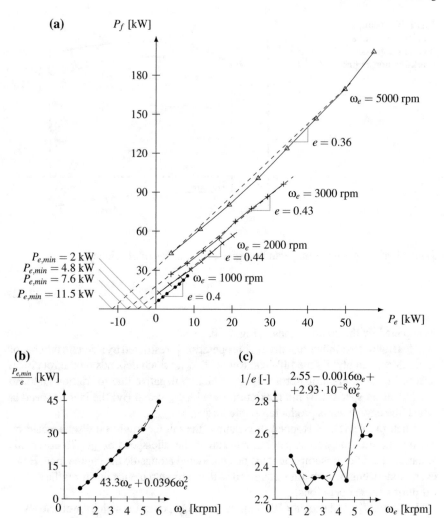

Fig. 2.11 Willans lines (dashed) and actual map data (solid marked) of a naturally-aspirated spark-ignition engine, with fitted values of e and $P_{e,min}$ for various speeds (**a**). Parametrization of $P_{e,min}/e$ (**b**) and $1/e$ (**c**) leading to (2.24) (ω_e in rad/s)

ical friction and pumping losses). An example Willans representation of an engine map is shown in Fig. 2.11a.

The further parametrization of the terms $1/e$ and $P_{e,min}/e$ as a function of speed leads to the closed-form expression [7]

$$P_f(t) = (k_{e,0}\omega_e(t) + k_{e,1}\omega_e^2(t)) + (k_{e,2} + k_{e,3}\omega_e(t) + k_{e,4}\omega_e^2(t))T_e(t)\omega_e(t) ,$$

$$(2.24)$$

where the k_{ei}'s, $i = 0, \dots, 4$ are design-dependent coefficients. An illustrative example of the data fitting leading to (2.24) is shown in Fig. 2.11b, c. Note that (2.23) or (2.24) also capture the idle speed consumption, obtained by setting $T_e = 0$ and $\omega_e = \omega_{e,idle}$.

In the example of Fig. 2.11, the affine dependency of the fuel power on torque or mechanical power is reasonably valid for all speeds shown, and up to the maximal torque allowed at each speed. In more general terms, however, the affine range extends only up to an intermediate torque, often coincident with or close to $T_{e,\eta}$, after which the curve $P_f - P_e$ becomes visibly convex.[5]

The engine brake torque curve $T_{e,min}(\omega_e)$ is obtained from (2.24) by setting $P_f = 0$ (fuel cutoff). For the maximum torque curve $T_{e,max}$, convenient parametrizations [7] are the quadratic relation

$$T_{e,max}(t) = k_{e,5} + k_{e,6}\omega_e(t) + k_{e,7}\omega_e^2(t) \tag{2.25}$$

for SI, naturally-aspirated engines, and a piecewise-affine relation of the type

$$T_{e,max}(t) = \min\left(k_{e,8} + k_{e,9}\omega_e(t), k_{e,10}, k_{e,11} + k_{e,12}\omega_e(t)\right) \tag{2.26}$$

for turbocharged engines.

2.2.3 Fuel Energy Consumption of ICEVs

With the models introduced in the previous sections, it is possible to evaluate the fuel energy consumption of an ICEV defined by (2.16). It is further assumed that in the braking phase $\mathcal{B}$ ($F_w < 0$) the fuel injection is disabled (*fuel cut-off*), thus the engine absorbs the power $P_{e,min}$ (engine brake). With the notation of Sect. 2.1.3, $k_b = -P_{e,min}\eta_t/(vF_w)$. Consequently, the integral in (2.16) is limited to the traction phase and, inserting (2.23), is evaluated as

$$E_T^{(ICEV)} = \int_T \frac{v(t)F_w(t)/\eta_t + P_{e,min}(\omega_e(t))}{e(\omega_e(t))}\, dt, \tag{2.27}$$

With the parametrization of (2.24), the integral in (2.27) can be written as a sum of terms

$$E_T^{(ICEV)} = \sum_g E_{T,g}^{(ICEV)} = \sum_g \int_{T_g} \left(k_{e,0}\omega_{e,g}(t) + k_{e,1}\omega_{e,g}^2 + \right.$$
$$\left. + \left(k_{e,2} + k_{e,3}\omega_{e,g}(t) + k_{e,4}\omega_{e,g}^2(t)\right) v(t)F_w(t)/\eta_t\right) dt, \tag{2.28}$$

---

[5]A behavior that is seen for the last point at 5000 rpm in Fig. 2.11a. For some engine technologies, e.g., downsized-supercharged engines that require spark retard to avoid knock at high torque, these curves might become more convex at high engine power, especially at low speed.

where each term $E_{T,g}^{(ICEV)}$ corresponds to the gear g and is evaluated by integrating over $\mathcal{T}_g$, the particular portion of $\mathcal{T}$ where gear g is engaged. Developing all terms, an explicit formula is obtained, which reads

$$
\begin{aligned}
E_{T,g}^{(ICEV)} = & \left(\frac{k_{e,0}\gamma_{e,g}}{r_w} + \frac{k_{e,2}C_0}{\eta_t} \right) \int_{\mathcal{T}_g} v(t)dt + \\
& \left(\frac{k_{e,1}\gamma_{e,g}^2}{r_w^2} + \frac{k_{e,2}C_1}{\eta_t} + \frac{k_{e,3}C_0\gamma_{e,g}}{r_w\eta_t} \right) \int_{\mathcal{T}_g} v^2(t)dt + \\
& \left(\frac{k_{e,2}C_2}{\eta_t} + \frac{k_{e,3}\gamma_{e,g}C_1}{r_w\eta_t} + \frac{k_{e,4}\gamma_{e,g}^2 C_0}{r_w^2\eta_t} \right) \int_{\mathcal{T}_g} v^3(t)dt + \\
& \left(\frac{k_{e,3}\gamma_{e,g}C_2}{r_w\eta_t} + \frac{k_{e,4}\gamma_{e,g}^2 C_1}{r_w^2\eta_t} \right) \int_{\mathcal{T}_g} v^4(t)dt + \left(\frac{k_{e,4}\gamma_{e,g}^2 C_2}{r_w^2\eta_t} \right) \int_{\mathcal{T}_g} v^5(t)dt + \\
& \left(\frac{m_t k_{e,2}}{\eta_t} \right) \int_{\mathcal{T}_g} v(t)dv + \left(\frac{m_t k_{e,3}\gamma_{e,g}}{r_w\eta_t} \right) \int_{\mathcal{T}_g} v^2(t)dv + \\
& \left(\frac{m_t k_{e,4}\gamma_{e,g}^2}{r_w^2\eta_t} \right) \int_{\mathcal{T}_g} v^3(t)dv
\end{aligned}
\tag{2.29}
$$

In this formulation, the fuel energy consumption is expressed as a sum of many terms, each being the product of a vehicle-dependent factor and a driving-dependent factor under the form of an integral. Each of these integrals can be further expressed as a function of the average speed over the partial phase $\mathcal{T}_g$ and higher moments, using the procedure illustrated in Sect. 2.1.2.

2.3 Electric Vehicles

Electric vehicles (EV) are powered by a battery that accumulates electrochemical energy and delivers electric power at its terminals. Batteries are reversible storage systems, thus they can be recharged during their operation if electricity is provided. Otherwise, they are filled back by external chargers when the vehicle is stopped at charging stations. The EV powertrain is completed by one or more reversible electric machines, that can be operated as motors or generators. Usually, three-phase alternating current machines are used. The direct-current electricity at the battery terminals is transformed to alternating current and back by a power electronic device, an *inverter*. The mechanical power at the machine shaft is linked to the wheels by the drivetrain.

In EVs, the tank energy corresponds to the electrochemical energy drained from or supplied to the battery,

$$
E_T^{(EV)} = \int_0^{t_f} P_b(t)dt \,,
\tag{2.30}
$$

where P_b is the electrochemical power. How this quantity is related to the driving profile, will be now described by following the approach of Fig. 2.3 and analyzing separately the drivetrain, the motor with its inverter, the power link, and the battery.

2.3.1 Drivetrain

In electric vehicles the drivetrain is usually equipped with a transmission with a fixed configuration (single-gear reductor). Defining the drivetrain transmission ratio γ_m as the product of the final drive and reductor ratios, (2.19)–(2.20) are replaced by

$$T_m(t) = \frac{F_p(t)r_w}{\gamma_m \eta_t^{sign(F_p(t))}} \tag{2.31}$$

and

$$\omega_m(t) = \frac{\gamma_m}{r_w} v(t) , \tag{2.32}$$

respectively, where T_m and ω_m are motor torque and speed.

2.3.2 Motor and Inverter

Traction motors adopted in electric vehicles are usually permanent-magnet, synchronous AC machines or, to a lesser extent, induction (asynchronous) AC machines, although newer technologies are emerging. For our purposes, all these motor types share a similar representation.

Electric power supplied to or generated by the motor, including its inverter, P_m, is usually tabulated ("motor map") as a function of motor torque and rotational speed,

$$P_m(t) = f(T_m(t), \omega_m(t)) , \tag{2.33}$$

Equivalently, a motor map can be expressed in terms of efficiency

$$\eta_m \triangleq \left(\frac{T_m \omega_m}{P_m} \right)^{sign(T_m)} , \tag{2.34}$$

as depicted by the contour lines in Fig. 2.12. Note that the motor map extends to two quadrants of the torque-speed plane, reflecting the reversible nature of these machines, that are able to operate as a motor ($T_m > 0$) as well as a generator ($T_m < 0$).

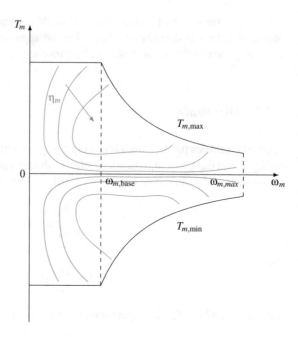

Fig. 2.12 Example efficiency map of an electric machine

   The operation of electric machines is restricted by a number of limitations, namely: (i) the maximum torque $T_{m,max}$ that can be provided continuously,[6] a speed-dependent quantity that is constant from zero speed up to a certain value known as base speed, $\omega_{m,base}$, then decreases with speed roughly hyperbolically; (ii) the maximum torque in the generator range, $T_{m,min}$ that is a similar function of speed; and (iii) the maximum speed $\omega_{m,max}$. Contrarily to engines, there is no minimum speed and the motor can produce torque at rest.

   For the online implementation described in Chap. 8, approximated closed-form expressions are used instead of (2.33). For instance, the physics of DC motor inspires the quadratic model [8]

$$P_m(t) = b_2(\omega_m(t))T_m^2(t) + b_1(\omega_m(t))T_m(t) + b_0(\omega_m(t)) , \qquad (2.35)$$

where the b's are tunable parameters. Further parametrization of the b's coefficients as a function of the motor speed leads to the closed-form expression

$$
\begin{aligned}
P_m(t) = & \left(k_{m,4} + k_{m,5}\omega_m(t) + k_{m,6}\omega_m^2(t)\right) T_m^2(t) + \\
& + k_{m,3}\omega_m(t)T_m(t) + k_{m,2}\omega_m^2(t) + k_{m,1}\omega_m(t) + k_{m,0}
\end{aligned}
\qquad (2.36)
$$

---

[6]Higher torque levels can be delivered for short times; correspondingly, motor maps often present *peak torque* curves for various delivery times.

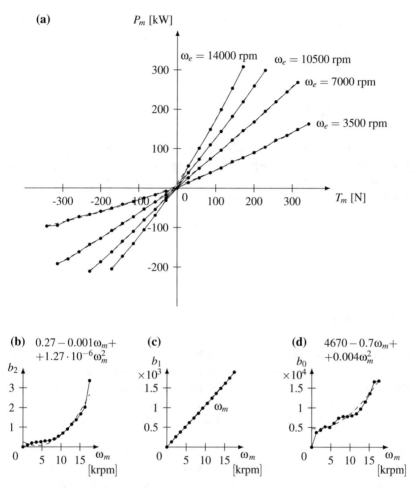

Fig. 2.13 Quadratic model (dashed) and actual map data (solid marked) of a synchronous permanent-magnet motor (**a**). Parametrization of b_2 (**b**), b_1 (**c**), and b_0 (**d**) leading to (2.36) (ω_e in rad/s)

that, as Fig. 2.13 illustrates, is often sufficiently accurate, at least in the operating regions not too far above the base speed. Even simpler representations, for instance, constant motoring and generating efficiency, have also been adopted in energy efficiency studies.

As shown in Fig. 2.12, the maximum torque curve can be effectively approximated as

$$T_{m,max}(t) = k_{m,7} \frac{\min(\omega_{m,base}, \omega_m(t))}{\omega_m(t)} \ , \tag{2.37}$$

where the coefficient $k_{m,7}$ corresponds to the nominal torque and the base speed is the ratio of nominal power to nominal torque. As for the minimum torque curve $T_{m,min}$, it is often equaled to $-T_{m,max}$.

2.3.3 Power Link

The power link electrically connects the battery terminals to the inverter input, allowing for bi-directional flow of electricity. Although losses inevitably occur in power links, usually these contributions are neglected, such that, following the backward calculation flow in Fig. 2.3, it is assumed that

$$P_l(t) = P_m(t) \, . \tag{2.38}$$

where P_l is the electric power out of the battery.

2.3.4 Battery

As of 2017, most of automotive traction batteries embody one of the variants of the lithium-ion technology, which has superseded older technologies such as nickel-metal hydride or lead-acid. Although lithium-ion batteries can differ substantially among each other concerning the specific chemistry of the cathode (NMC, LMO, LFP, LCO, NCA, etc.), the anode, and their sizes, for our purposes they share a similar representation.

The battery power is calculated by modeling the battery electro-chemistry as an equivalent electric circuit, with an ideal voltage source V_{b0} and an internal resistance R_b in series, see Fig. 2.14. With this representation, the electrochemical power to be used in (2.30) is $P_b = V_{b0}I_b$ and it is further evaluated as a function of the terminal power $P_l = V_bI_b$ as

Fig. 2.14 Equivalent circuit of a battery

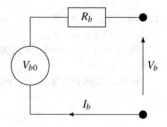

Fig. 2.15 Constant-
efficiency model (dashed)
and actual tabulated data
(solid marked: * 50% SoC,
△ 90% SoC, □ 10% SoC) of
a lithium-ion battery

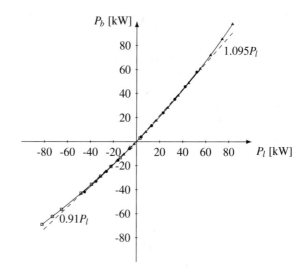

$$P_b(t) = \frac{V_{b0}^2}{2R_b} - V_{b0} \sqrt{\frac{V_{b0}^2 - 4P_l(t)R_b}{4R_b^2}}. \tag{2.39}$$

Note that P_b and P_l are defined as positive for discharging and negative for charging.

The difference between P_b and P_l is due to inner battery losses. For the online use described in Chap. 7, these effects are sometimes further simplified to a constant efficiency model

$$P_b(t) = P_l(t)\eta_b^{-\text{sign}(P_l(t))}, \tag{2.40}$$

see Fig. 2.15, or even neglected, $P_b(t) = P_l(t)$.

The open-circuit voltage V_{b0} and internal resistance R_b may actually vary with time as they depend on the level of energy accumulated in the battery. The usual measure of such energy is the State of Charge (SoC), defined as $\xi_b = q_b/Q_b$, where q_b is the remaining charge and Q_b is the nominal battery capacity, a quantity that is usually expressed in ampere-hour units (Ah). Assuming a simple "Coulomb-counting" model for the battery charge depletion, $dq_b(t)/dt = -I_b(t)$, the SoC dynamics reads

$$\dot{\xi}_b(t) = -\frac{P_b(t)}{V_{b0}Q_b}. \tag{2.41}$$

The SoC is useful to describe the limitations imposed on the battery operation. Of course, in principle $0 \le \xi_b \le 1$, however, automotive batteries are often managed in such a way that $\xi_{b,min} \le \xi_b \le \xi_{b,max}$, where the admissible range tends to be wider for batteries equipping EVs and plug-in HEVs. Additional operating limits are imposed on battery voltage and, consequently, on output power P_l.

We define for later use the energy stored in the battery as $\varepsilon_b \triangleq q_b V_{b0}$.

2.3.5 Electric Energy Consumption of EVs

With the models introduced in the previous sections, it is possible to evaluate the electric energy consumption of an EV. It is further assumed that friction brakes are not used so that the recuperation potential is exploited at its maximum ($k_b = 1$ in the nomenclature of Sect. 2.1.3).[7]

Since the drivetrain model (2.31) and the battery model (2.40) have a different form for traction/discharge ($T_m > 0, P_m > 0$) and braking/charge ($T_m < 0, P_m < 0$), the integral of (2.30) is conveniently split in two parts,

$$E_T^{(EV)} = \sum_\sigma E_{T,\sigma}^{(EV)} , \qquad (2.42)$$

with $\sigma \in \{T, B\}$ spanning the two phases defined in Sect. 2.1.3. Using the parametric models (2.36) and (2.40), and developing all terms, an involved but explicit formula is obtained, of the type

$$E_{T,\sigma}^{(EV)} = \sum_{i,j,\sigma} \mathsf{E}_{ij}^\sigma \mathsf{C}_{ij}^\sigma . \qquad (2.43)$$

Each term in the summation is the product of a vehicle-dependent factor (E) and a driving-dependent factor (C), defined as

$$\mathsf{C}_{ij}^\sigma \triangleq \int_\sigma v^i \dot{v}^j dt , \qquad (2.44)$$

with the integral extended only to times where the mode σ is active. For instance, when $k_{m,5} = k_{m,6} = 0$ and $\eta_b = 1$, one obtains

$$
\begin{aligned}
E_{T,\sigma}^{(EV)} = {}& \left(k_{m,0} + \frac{C_0^2 k_{m,4} r_w^2}{\eta_t^{2\sigma} \gamma_m^2} \right) \int_\sigma dt + \left(\frac{C_0 k_{m,3}}{\eta_t^\sigma} + \frac{\gamma_m k_{m,1}}{r_w} + \frac{2 C_0 C_1 k_{m,4} r_w^2}{\eta_t^{2\sigma} \gamma_m^2} \right) \cdot \\
& \int_\sigma v(t) dt + \left(\frac{C_1 k_{m,3}}{\eta_t^\sigma} + \frac{\gamma_m^2 k_{m,2}}{r_w^2} + \frac{C_1^2 k_{m,4} r_w^2}{\eta_t^{2\sigma} \gamma_m^2} + \frac{2 C_0 C_2 k_{m,4} r_w^2}{\eta_t^{2\sigma} \gamma_m^2} \right) \int_\sigma v^2(t) dt + \\
& + \left(\frac{C_2 k_{m,3}}{\eta_t^\sigma} + \frac{2 C_1 C_2 k_{m,4} r_w^2}{\eta_t^{2\sigma} \gamma_m^2} \right) \int_\sigma v^3(t) dt + \left(\frac{C_2^2 k_{m,4} r_w^2}{\eta_t^{2\sigma} \gamma_m^2} \right) \int_\sigma v^4(t) dt + \\
& + \left(\frac{2 C_0 k_{m,4} m r_w^2}{\eta_t^{2\sigma} \gamma_m^2} \right) \int_\sigma a(t) dt + \left(\frac{k_{m,3} m}{\eta_t^\sigma} + \frac{2 C_1 k_{m,4} m r_w^2}{\eta_t^{2\sigma} \gamma_m^2} \right) \int_\sigma v(t) a(t) dt + \\
& + \frac{2 C_2 k_{m,4} m r_w^2}{\eta_t^{2\sigma} \gamma_m^2} \int_\sigma a(t) v^2(t) dt + \left(\frac{k_{m,4} m^2 r_w^2}{\eta_t^{2\sigma} \gamma_m^2} \right) \int_\sigma a^2(t) dt ,
\end{aligned}
$$
$$(2.45)$$

where the exponents σ's stand here for $+1$ (traction) and -1 (braking), respectively.

---

[7]In practice the situation $k_b < 1$ is common, since in most braking maneuvers both vehicle axles must brake due to stability issues, and often recuperation is available only on one axle.

Each of these integrals can be further expressed as a function of the average speed and higher moments over the phases $\mathcal{T}$ and $\mathcal{B}$, using the procedure illustrated in Sect. 2.1.2.

2.4 Hybrid-Electric Vehicles

Hybrid-electric vehicles (HEV) are a combination of an engine-based vehicle and an electric vehicle. According to how the power is combined, HEVs are classified as *parallel* (coupling of the mechanical power of the engine with that of the motor), *series* (coupling of the electric power of the battery and of an engine-based electrical generation unit), or *series-parallel* (both couplings are present, often with a power-split device such as a planetary gear-set). According to the position of the coupling, parallel hybrids are further classified to several types, typically labeled from P0, if the motor is coupled to the engine belt, to P4, if the motor is mounted on one of the axles. According to the battery operation strategy, HEVs are classified as charge-sustaining hybrids, where the battery cannot be recharged from an external source, or plug-in hybrids.

In HEVs, energy is generally drained both from the fuel tank and the battery. The measure of energy consumption adopted depends on the type of battery operation. However, a common definition of an HEV energy consumption can be written as

$$E_T^{(HEV)} = \int_0^{t_f} (P_f(t) + s \cdot P_b(t))dt , \qquad (2.46)$$

where s is an "equivalence factor" that weights the electricity consumption with respect to the fuel consumption.

For charge-sustaining HEVs, s is vehicle- and cycle-dependent and must be determined by interpolation of two or more tests, e.g., using the procedure [9]. For plug-in hybrids, standards such as [9] recommend fuel economy evaluation rules that can be transcribed as (2.46), where s is a prescribed coefficient.

How the quantities P_f and P_b are related to the driving profile, will be now described by following the approach of Figs. 2.4 and 2.5. Models of the engine, the electric machine, and the battery presented in the previous sections still apply to HEVs. However, the nature and role of drivetrain and power link depends on the hybrid architecture, as will be now discussed.

2.4.1 Drivetrain and Power Link

In parallel HEVs the powertrain force results from the combination of the engine torque and the motor torque,

$$T_e(t) = u(t) \frac{F_p(t)r_w}{\gamma_e(t)\eta_t^{\mathrm{sign}(F_p(t))}} \ ,$$

$$T_m(t) = (1 - u(t)) \frac{F_p(t)r_w}{\gamma_m \eta_t^{\mathrm{sign}(F_p(t))}} \ ,$$

(2.47)

where $u(t)$, the *torque split ratio*, is the degree of freedom offered by the parallel hybrid architecture. In other terms, one between T_e and T_m can be chosen freely to satisfy a given F_p. In contrast, the speed levels are unambiguously related to the vehicle speed,

$$\omega_m(t) = \frac{\gamma_m}{r_w} v(t) \ ,$$

$$\omega_e(t) = \frac{\gamma_e(t)}{r_w} v(t) \ .$$

(2.48)

As for the power link, in parallel HEVs it links a single electric source (battery) to a single electric load (motor), therefore (2.38) applies.

In series hybrids, the drivetrain is described by (2.31) and (2.32), while the power balance at the electric link reads

$$P_g(t) = u(t)P_m(t) \ ,$$

$$P_l(t) = (1 - u(t))P_m(t) \ ,$$

(2.49)

where $u(t)$ is now the *power split ratio*, a degree of freedom offered by the series architecture, and P_g is the electric power generated by the auxiliary power unit (APU).

This electric generation unit is composed of an engine and an electric machine, the latter exclusively operated as a generator, mechanically connected such that T_e is rigidly tied to T_g and so are ω_e and ω_g. For such a system, the relation between $P_f(t)$ and $P_g(t)$ is not unambiguous, as it depends on the rotational speed at which the two machines are operated. The latter is a second degree of freedom offered by the series architecture and is usually chosen such as to maximize the APU efficiency, following the optimal operating line of the APU (Fig. 2.16). This curve ultimately provides the speeds and torques as a function of the generated electric power P_g, while the engine models of Sect. 2.2.2 eventually provide P_f.

In series-parallel HEVs, coupling relations are more complex than (2.47)–(2.49), as they involve coupling of rotational speeds. The reader is therefore referred to standard vehicle modeling books [1].

The role of determining the degree of freedom $u(t)$, either torque or power split, or even more complex combination for architectures not detailed here, is played by the hybrid energy management strategy (EMS). A description of such a strategy is therefore necessary in order to predict the fuel and electric consumption of a HEV for a given driving profile.

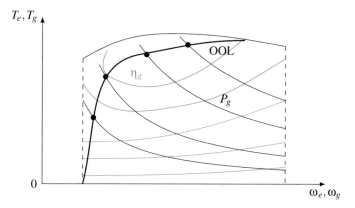

Fig. 2.16 Example optimal operating line (OOL) of an APU. Also shown are contour lines of P_g: the OOL is the locus of points that maximize η_g for different values of P_g

2.4.2 Energy Management Strategy

Several control designs are used for the EMS, including heuristic strategies and optimization-based strategies.

Heuristic EMS are based on predefined rules of the type

$$u(t) = f(F_p(t), v(t), \xi_b(t), \gamma_e(t), \theta_e(t), \ldots) , \qquad (2.50)$$

where the arguments in the right-hand side of (2.50) describe the vehicle state and the driver's request. The function $f(\cdot)$ is implemented in the onboard control unit under the form of look-up tables, algorithms, or finite-state machines. As an example, in a parallel HEV these rules might prescribe the use of the purely electric mode ($u = 0$) only under certain driving situations, typically low speed and acceleration, and for sufficiently high SoC. Heuristic strategies are heavily dependent on threshold values or maps, which are to be tuned point by point with a relevant calibration effort.

Alternatively, optimal EMS are inspired by the solution of an optimal control problem that aims to minimize the fuel consumption over the horizon t_f, under a constraint over the battery state of charge at the end of the horizon. In mathematical terms,

$$\min_{u(t)} \int_0^{t_f} P_f(u, v(t), \ldots)dt , \qquad (2.51)$$

with $u(t)$ subject to all physical limits described in the previous sections, and such that

$$\int_0^{t_f} P_b(u(t), v(t), \ldots)dt = \Delta E_b , \qquad (2.52)$$

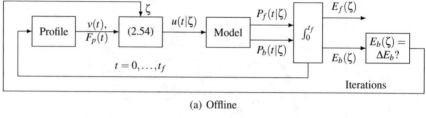

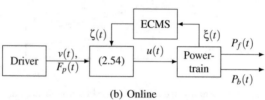

(a) Offline

(b) Online

Fig. 2.17 Schematic flowcharts of offline and online optimal EMS

where ΔE_b is the target electric consumption. Generally, $\Delta E_b = 0$ for charge-sustaining HEVs, while it is often taken as the remaining useful energy in the battery for plug-in hybrids, in such a way that the EMS provides full discharge at the end of the trip with minimum fuel consumption.

Using the theory of optimal control, this problem is solved by forming a Hamiltonian function H and then finding $u(t)$ such that

$$u(t) = \arg\min H(u, v(t), \ldots) \tag{2.53}$$

$$H(u, v(t), \ldots) = P_f(u, v(t), \ldots) + \zeta^* P_b(u, v(t), \ldots). \tag{2.54}$$

The quantity ζ^* is determined as the particular value that, when applied in (2.54), gives place to trajectories $u(t|\zeta^*)$ and consequently $P_f(t|\zeta^*)$, $P_b(t|\zeta^*)$, that fulfill (2.52). As such, ζ^* is a vehicle-dependent constant that is additionally a function of the whole profile $v(t)$, $t \in [0, t_f]$. It can be found by iterations for a prescribed driving profile known in advance ("offline optimal EMS"), see Fig. 2.17a.

However, the optimal ζ^* is generally not predictable during a real operation, since the future driving profiles are generally not known in advance. Therefore, online EMS can only be sub-optimal by adopting some kind of rule to provide estimations $\zeta(t)$ of ζ as time advances. Often the *Equivalent Consumption Minimization Strategy* (ECMS) approach is used, where $\zeta(t)$ is regulated as a function of deviations of SoC from its target value, see Fig. 2.17b.

The use of connectivity in CAVs and conventional vehicles can improve the estimation of ζ^* and thus the optimality of the EMS, by anticipating the fact that future road slopes and traffic conditions will lead to an over- or under-usage of the battery. The resulting *predictive energy management strategies* are currently a subject of research [10] but won't be treated further in this book.

2.4.3 Energy Consumption of HEVs

A closed-form expression of the energy consumed by an HEV along a given driving profile is harder to obtain than for ICEVs or EVs, even with suitable approximations for P_f and P_b, because of the fundamental role played by the energy management strategy in defining such consumption. In particular, if the minimal fuel consumption is to be predicted, that would imply an iterative process to find the optimal ζ^*.

The method called *fully-analytical consumption estimation* (FACE) [7] is based on the assumption, confirmed by observation, that the overall consumption (2.46) is only slightly dependent on the particular choice of ζ, if an optimal EMS (2.54) based on this value is applied. Under this assumption, any "reasonable" value of ζ would provide a good estimate of the overall energy consumption (2.46).[8] A reference value can, for instance, be evaluated "offline" for a known driving profile, then applied in (2.46) to evaluate the energy consumption of any other profile, even if ζ does not satisfy condition (2.52).

According to this method, an estimation of the energy consumed is thus

$$E_T^{(HEV)} = \int_0^{t_f} P_f(t|s)dt + \zeta \int_0^{t_f} P_b(t|s)dt . \qquad (2.55)$$

Using the models and methods already illustrated for ICEVs and EVs, (2.55) can be reorganized as

$$E_T^{(HEV)} = \sum_{i,j,g,\sigma} \mathsf{F}_{ij}^{g\sigma} \mathsf{C}_{ij}^{g\sigma} + \zeta \sum_{i,j,g,\sigma} \mathsf{E}_{ij}^{g\sigma} \mathsf{C}_{ij}^{g\sigma} , \qquad (2.56)$$

that is, a sum of many terms, each being the product of a vehicle-dependent factor (F or E) and a driving-dependent factor (C). The latter are defined as

$$\mathsf{C}_{ij}^{g\sigma} \triangleq \int_{g,\sigma} v^i \dot{v}^j dt , \qquad (2.57)$$

with the integral extended only to times when the gear g and the mode σ are active.

Note that (2.29) and (2.43) are particular cases of (2.56). However, in HEVs the set of possible operating modes is larger than in ICEVs or EVs, where $\sigma \in \{\mathcal{T}, \mathcal{B}\}$, since it may include a fully-electric mode ($u = 0$), a fully-ICE mode ($u = 1$), a "boost" mode ($0 < u < 1$), a recharge mode ($u > 1$), etc. Additionally, how a given profile is split into the various modes also depends on ζ.

---

[8]Only the overall, or "equivalent", consumption can be estimated with this method; it does not allow to find separately the two energy contributions, i.e., the minimal fuel consumption for a given battery consumption.

2.5 Human-Powered Vehicles (Bicycles)

The most common human-powered vehicles (HPV) are bicycles. We shall thus limit
our analysis to such systems, where the powertrain is usually composed of the cyclist,
pedals, and a chain transmission. Besides purely-human-powered bicycles, motor-
ized bicycles also exist. While this concept has been historically dominated by engine-
based designs, today an increasingly greater role is played by *electric bicycles*, which
assist the cyclist with integrated electric motor and battery. Electric bicycles are further classified as *pedal-assist*, where the motor augments the efforts of cyclists when
a sensor detects that they are pedaling, and *power-on-demand* systems (sensorless),
where the motor is activated by a throttle. The European legislation limits the electric
assistance at 250 W and up to 25 km/h speed. Those pedal-assist systems that exceed
these limits (e-bikes or s-pedelecs) are usually legally classed as motorcycles.

The human body is (also) a power converter that takes energy through food and
drink and produces useful energy in the form of muscular movements. The "tank"
energy of a bicycle is thus defined as the metabolic energy spent by the cyclist,

$$E_T^{(HPV)} = \int_0^{t_f} P_h(t)dt , \qquad (2.58)$$

where P_h is the *metabolic power*, usually defined with respect to the oxygen uptake,
that is, as the product of the volumetric consumption rate and the energy density of
oxygen.

How this quantity is related to the cycling profile will be now described by following the approach of Fig. 2.6 and analyzing separately the drivetrain and the cyclist's
physiology.

2.5.1 Drivetrain

For a discrete-gear chain transmission, the force exerted by cyclist on pedals is
evaluated as a function of the needed force at wheels as

$$F_c(t) = \frac{1}{\gamma_t(t)\eta_t} \frac{r_w}{l_c} F_p(t) , \qquad (2.59)$$

where γ_t is the chosen transmission ratio, η_t its efficiency (supposed constant), r_w is
the wheel radius, l_c the crank arm length. The *cadence*, or pedal rotational speed, is

$$\omega_c(t) = \frac{\gamma_t(t)v(t)}{r_w} . \qquad (2.60)$$

Consequently, the mechanical power is $P_c = T_c\omega_c = F_c l_c \omega_c$.

For electric bicycles, (2.59) is replaced by

$$F_c(t) = u(t) \frac{1}{\gamma_t(t)\eta_t} \frac{r_w}{l_c} F_p(t) , \qquad (2.61)$$

where $u(t)$ is the power split ratio, while the motor power is

$$P_m(t) = (1 - u(t)) \frac{v(t)F_p(t)}{\eta_t} . \qquad (2.62)$$

For pedal-assist electric bicycles (pedelecs), $u(t)$ is constant and determined by design. For power-on-demand electric bicycles, the determination of $u(t)$ is the role of an energy management strategy, similarly to HEVs. However, in this case, u is usually not a continuous variable but can be chosen only among a discrete set of values.

2.5.2 Cyclist

The evaluation of the oxygen consumption as a function of the cycling conditions is a complex subject that lacks a general and widely accepted model. Some simple equations could be nevertheless found in the specialized physiology literature [11–15].

The conversion from the metabolic power to the mechanical power is the result of two pathways to produce ATP (adenosine triphosphate), the energy carrier to the muscles: *aerobic* and *anaerobic*. The former is a process that takes place in the presence of oxygen and is thus associated with oxygen uptake. In contrast, the anaerobic pathway depletes stored resources, leading to lactate formation whose accumulation in the muscles generates feelings of fatigue and exhaustion. While the aerobic pathway is characterized by unlimited available energy but finite power capability, the anaerobic pathway is characterized by larger power but finite energy, thus it is available only for short times.

Under quasi-stationary conditions, the anaerobic pathway is not active and the oxygen uptake is a function of the power exerted and the pedaling cadence. This dependency can be expressed with a Willans-type model,

$$P_h(t) = \frac{P_c}{e_h(\omega_h(t))} + P_{h,0}(\omega_h(t)) , \qquad (2.63)$$

where $P_{h,0}$ is the metabolic power with cyclist freewheeling (zero-work) and e_h the net efficiency [11]. These parameters are subject-dependent but experimental tests suggest that on average they both increase with the pedaling cadence. Consequently, the (gross) cycling efficiency

Fig. 2.18 Illustration of the various mechanisms leading to the maximum force exerted by the cyclist. Area between the curves $F_{c,max}$ and $F_{c,max,aer}$ is proportional to the anaerobic work capacity AWC

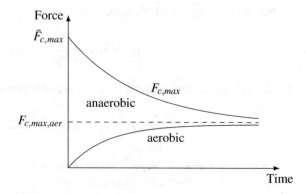

$$\eta_h \triangleq \frac{P_c}{P_h} \qquad (2.64)$$

is a function that increases with power but decreases with cadence. Typical values of η_h do not exceed 20%.

The maximum power that can be generated under stationary conditions, $P_{h,max}$, is proportional to the maximum oxygen uptake (also known as $\dot{V}_{O_2,max}$) and often called anaerobic threshold or *critical power* (CP). However, as Fig. 2.18 suggests, this limit power is not instantaneously available from the aerobic pathway, while, on the other hand, it can be exceeded for a limited amount of time by consuming the reserves in the anaerobic pathway. This limited energy reservoir is often called anaerobic work capacity (AWC).

A general result of these processes is that, contrarily to engines or electric machines,[9] the maximal force that can be generated, $F_{c,max}$, varies with time as the effort above the CP progresses, see Fig. 2.18. This phenomenon is generally called *fatigue*. If the effort falls below the CP or ceases, the stored energy is replenished by the aerobic pathway. A simple model that describes such effects, adapted from [16], is

$$\frac{dF_{c,max}(t)}{dt} = \begin{cases} -k_{fat}\left(F_{c,max}(t) - F_{c,max,aer}\right), & \text{if } F_c \geq F_{c,max,aer} \\ k_{rec}\left(\bar{F}_{c,max}(t) - F_{c,max}\right), & \text{otherwise} \end{cases}, \qquad (2.65)$$

where k_{fat} and k_{rec} are two subject-dependent coefficients, while $\bar{F}_{c,max}$ is the maximum force that an individual can develop from rest, also known as *maximum voluntary contraction* (MVC). The model (2.65) clearly captures the experimental observations that the drop of $F_{c,max}$ below MVC is quasi-exponential in time, and that the recovery mechanism tends to bring $F_{c,max}$ back to the MVC.

---

[9]As it was mentioned in Sect. 2.3.2, electric motors can actually deliver extra torque for short periods of time before thermal limitations are reached, and this is also true for some engines, particularly turbocharged ones. Although their time constants is much smaller, these processes could be described similarly to (2.65).

The coefficients k_{fat} and k_{rec} are physically related to the conservation of the available anaerobic energy AWC. Experimental observation suggests that k_{fat} increases with the relative force ratio $(F_c - F_{c,max,aer})/\bar{F}_{c,max}$. Correspondingly, the *maximum endurance time* (MET), i.e., the time at which $F_{c,max} = F_c$ after having applied a constant F_c, which is roughly proportional to the time constant of (2.65) and thus to $1/k_{fat}$, decreases with the relative force ratio [17].

Since the drop of $F_{c,max}$ described by (2.65) is an effect of fatigue accumulation, a measure of the fatigue level has been naturally proposed in [16] as

$$\xi_h(t) = \frac{\bar{F}_{c,max} - F_{c,max}(t)}{\bar{F}_{c,max} - F_{c,max,aer}} \ . \tag{2.66}$$

In analogy to battery SoC, the ratio ξ_h is often called *state of fatigue* (SoF).

2.5.3 Cycling Profiles

There is limited literature on behavioral models of cyclists. The model proposed in [18] consists of a parametrization of the cyclist power as a function of speed, $P_c(t) = f(v(t))$. This model is based on the assumption that cyclists preferably ride at constant nominal power, rather than at constant speed. The nominal power equals the power needed to cruise (zero acceleration) at a certain "comfort" speed on a flat terrain. Above the comfort speed, cyclists stop pedaling, while above a higher "feared" speed (that might be reached on steep downhills), they brake. In contrast, as the speed falls below a minimum speed (typically, at steep uphills), cyclists exert a power higher than the nominal.

The cycling speed and power profile could in principle be optimized along a trip, in particular for electric bikes. The optimization objective could be related to the cyclist's and the electric consumption. However, this concept is not adequately documented in the literature and won't be treated further in this book.

References

1. Guzzella L, Sciarretta A (2013) Vehicle propulsion system
2. Rill G (2011) Road vehicle dynamics: fundamentals and modeling. CRC Press
3. Kobayashi T, Sugiura H, Ono E, Katsuyama E, Yamamoto M (2016) Efficient direct yaw moment control of in-wheel motor vehicle. In: Proceedings of international symposium on advanced vehicle control
4. Wolf-Heinrich H (2013) Aerodynamics of road vehicles: from fluid mechanics to vehicle engineering. Elsevier
5. Bonnet C, Fritz H (2000) Fuel consumption reduction in a platoon: experimental results with two electronically coupled trucks at close spacing. Technical report, SAE technical paper

6. Kubička M, Klusáček J, Sciarretta A, Cela A, Mounier H, Thibault L, Niculescu S-I (2016) Performance of current eco-routing methods. In: Proceeding of intelligent vehicles symposium (IV), pp 472–477. IEEE
7. Zhao J, Sciarretta A (2017) A fully-analytical fuel consumption estimation for the optimal design of light-and heavy-duty series hybrid electric powertrains. Technical report, SAE technical paper
8. Dib W, Serrao L, Sciarretta A (2011) Optimal control to minimize trip time and energy consumption in electric vehicles. In: Proceedings of vehicle power and propulsion conference (VPPC), pp 1–8. IEEE
9. Hybrid-EV Committee et al (2010) Recommended practice for measuring the exhaust emissions and fuel economy of hybrid-electric vehicles, including plug-in hybrid vehicles. SAE International
10. Kermani S, Delprat S, Guerra T-M, Trigui R, Jeanneret B (2012) Predictive energy management for hybrid vehicle. Control Eng Pract 20(4):408–420
11. Chavarren J, Calbet JAL (1999) Cycling efficiency and pedalling frequency in road cyclists. Eur J Appl Physiol Occup Physiol 80(6):555–563
12. Morton RH (2006) The critical power and related whole-body bioenergetic models. Eur J Appl Physiol 96(4):339–354
13. Olds TS, Norton KI, Craig NP (1993) Mathematical model of cycling performance. J Appl Physiol 75(2):730–737
14. Hettinga FJ et al (2008) Optimal pacing strategy in competitive athletic performance. PhD thesis, PrintPartners Ipskamp
15. Rosero N, Martinez JJ, Leon H (2017) A bio-energetic model of cyclist for enhancing pedelec systems. IFAC-PapersOnLine 50(1):4418–4423
16. Fayazi SA, Wan N, Lucich S, Vahidi A, Mocko G (2013) Optimal pacing in a cycling time-trial considering cyclist's fatigue dynamics. In: Proceedings of American control conference (ACC), pp 6442–6447. IEEE
17. Ma L, Chablat D, Bennis F, Zhang W (2009) Dynamic muscle fatigue evaluation in virtual working environment. arXiv:0901.0222
18. Goussault R, Chasse A, Lippens F (2017) Model based cyclist energy prediction. In: Proceedings of international conference on intelligent transportation systems (ITSC), pp 1–6. IEEE

Chapter 3
Perception and Control for Connected and Automated Vehicles

In this book, by connected vehicles we are referring to vehicles that use communication technologies such as DSRC and cellular for vehicle-to-everything (V2X) communication. The U.S. Department of Transportation's National Highway Traffic Safety Administration (NHTSA) defines fully automated vehicles as those in which operation of the vehicle occurs without direct driver input to control the steering, acceleration, and braking and are designed so that the driver is not expected to constantly monitor the roadway while operating in self-driving mode [1]. In categorizing partial automation, NHTSA's federal automated vehicles policy adopts that of Society of Automotive Engineers (SAE) definitions for levels of vehicle automation as shown in the reproduced Table 3.1. Automation levels range from no automation with full driver control (Level 0) to full automation with no driver control (Level 5). Many of the benefits discussed in this book are realizable with partial level 2 or 3 automation as they mostly rely on automated speed and steering control which can be overseen and overridden by a human driver.

In this chapter, after a review of V2X technologies for connected vehicles in Sect. 3.1, we provide a brief overview of automated vehicle localization and perception in Sect. 3.2 and planning and control in Sect. 3.3. A schematic overview is shown in Fig. 3.1.

3.1 V2X Communication

Connected vehicles could ideally benefit from Vehicle-to-everything (V2X) communication channels and protocols to exchange data and information with a wide variety of entities. Some of the main benefits are increased road safety, harmonized traffic flow, and energy savings. For instance Vehicle-to-Vehicle (V2V) communication allows equipped vehicles to exchange their coordinates and intentions to prevent

© Springer Nature Switzerland AG 2020   63
A. Sciarretta and A. Vahidi, *Energy-Efficient Driving of Road Vehicles*,
Lecture Notes in Intelligent Transportation and Infrastructure,
https://doi.org/10.1007/978-3-030-24127-8_3

Table 3.1 Society of Automotive Engineering Vehicle automation levels reproduced from SAE Standard J3016 [2]. Copyright © 2018 SAE International. The following abbreviations are carried from J3016: Dynamic Driving Task (DDT), Object and Event Detection and Response (OEDR), Operational Design Domain (ODD), and Automated Driving System (ADS)

Level	Name	Narrative definition	DDT		DDT fallback	ODD
			Sustained lateral and longitudinal vehicle motion control	OEDR		
Driver performs part or all of the DDT						
0	No driving automation	The performance by the driver of the entire DDT, even when enhanced by active safety systems	Driver	Driver	Driver	n/a
1	Driver assistance	The sustained and ODD-specific execution by a driving automation system of either the lateral or the longitudinal vehicle motion control subtask of the DDT (but not both simultaneously) with the expectation that the driver performs the remainder of the DDT	Driver and system	Driver	Driver	Limited
2	Partial driving automation	The sustained and ODD-specific execution by a driving automation system of both the lateral and longitudinal vehicle motion control subtasks of the DDT with the expectation that the driver completes the OEDR subtask and supervises the driving automation system	System	Driver	Driver	Limited
ADS ("System") performs the entire DDT (while engaged)						
3	Conditional driving automation	The sustained and ODD-specific performance by an ADS of the entire DDT with the expectation that the DDT fallback-ready user is receptive to ADS-issued requests to intervene, as well as to DDT performance- relevant system failures in other vehicle systems, and will respond appropriately	System	System	Fallback-ready user (becomes the driver during fallback)	Limited
4	High driving automation	The sustained and ODD-specific performance by an ADS of the entire DDT and DDT fallback without any expectation that a user will respond to a request to intervene	System	System	System	Limited
5	Full driving automation	The sustained and unconditional (i.e., not ODD- specific) performance by an ADS of the entire DDT and DDT fallback without any expectation that a user will respond to a request to intervene	System	System	System	Unlimited

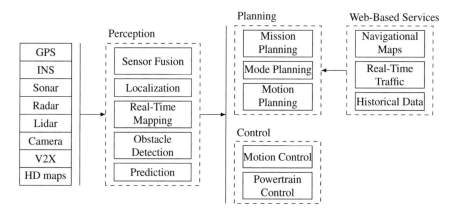

Fig. 3.1 Sensing, perception, planning, and control in CAVs

collision or to move in coordination. Vehicle-to-Infrastructure (V2I) communication allows vehicles to communicate with roadside units and infrastructure such as traffic signals enabling better coordination between them. A few other communication modes to name are Vehicle-to-Pedestrian (V2P), Vehicle-to-Device (V2D), Vehicle-to-Network (V2N), Vehicle-to-Cloud (V2C) and Vehicle-to-Grid (V2G) communication. In this book our main results only require V2V or V2I.

Today there exists two main communication technologies for V2X: (i) Wireless Local Area Network (WLAN) and (ii) Cellular Network.

WLAN technology allows vehicles moving at high speeds to establish ad-hoc and direct communication channels with neighboring vehicles and roadside traffic units without the need for additional communication infrastructure. Several countries have allocated a spectrum for Intelligent Transportation Systems communication that enables WLAN V2X. For instance in the United States, a 75 MHz band in the spectrum of 5.850–5.925 GHz has been set by the US Federal Communication Commission (FCC) since 1999. In Europe 30 MHz has been assigned for the same purpose. Currently the IEEE 1609 family, IEEE 802.11p, and the Society of Automotive Engineers (SAE) J2735 [3] form the key parts of the currently proposed Wireless Access in Vehicular Environments (WAVE) protocols. [4]. The architecture, communications model, management structure, security mechanisms and physical access for high speed (up to 27 Mb/s) short range (up to 1000 m) low latency wireless communications in the vehicular environment is defined by the IEEE 1609 Family of Standards [5]. Society of Automotive Engineers (SAE) uses the term Dedicated Short Range Radar Communication (DSRC) for the WAVE technology with J2735 set of standards which define the message payload at the physical layer. The SAE J2735 [6] supports interoperability among DSRC applications through the use of standardized message sets, data frames, and data elements.

Cellular V2X or in short C-V2X technology was initially defined as LTE in the Third Generation Partnership Project (3GPP) Release 14 [7] and is designed to operate in several modes: (1) Device-to-device and (2) Device-to-cell-tower. The device-to-device mode allows direct communication without necessarily relying on cellular network involvement. On the other hand, device-to-cell-tower relies on existing cell towers, network resources, and scheduling. Direct device-to-device communication improves latency and supports operation in areas without cellular network coverage.

3.2   Localization and Perception for Automated Driving

A key to successful automated driving is effective localization, obstacle detection, and perception. The vehicle must not only determine with high precision its location in the world and on the road but it should perceive accurately its surrounding environment such as neighboring vehicles, pedestrians, animals crossings, lane markings, traffic signs and signals, street signs, curbs and shoulders, buildings and trees, etc, and measure their relative distance and speed. These are perhaps the hardest technical challenges to overcome for highly automated driving. Here we present a brief overview of sensors and algorithms that are currently used for localization and perception.

3.2.1   Sensors for Perception and Localization

An overview of sensors for perception and localization is provided in Fig. 3.2. Self- or proprioceptive-sensors measure the ego vehicle internal states such as its velocity,

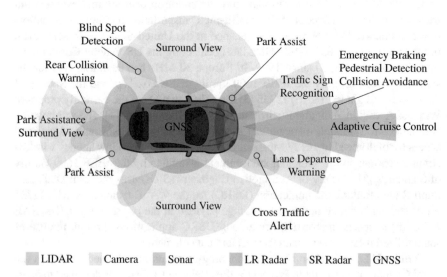

Fig. 3.2 Schematic of vehicle sensors for perception and localization. Adapted from [8]

acceleration, wheel speed, yaw, steering angle, engine speed, and engine torque. Odometers, accelerometers, Inertial Measurement Units (IMU), and information from Control Area Network (CAN) bus are used for proprioceptive-sensing and are not limited only to automated vehicles; many modern human driven vehicles rely on them for state estimation and advanced control functions. For example, IMUs contain gyroscopes, accelerometers, and sometimes magnetometers along each axis that provide dead reckoning capability in combination with the vehicle's wheel speed sensors. Since an IMU relies on integrating acceleration to determine positions, they are prone to drift and may require GPS-fusion (or camera-fusion when indoor) for more accurate localization.

Global Navigation Satellite Systems (GNSS) sensors also known commonly as Global Positioning System (GPS) are becoming standard on modern vehicles for navigation and localization. Other regional GNSS systems are Russia's GLONASS, Europe's Galileo, and China's Beidou. While current GNSS may not provide the needed sub-meter precision for localization of automated vehicles, filtering algorithms that fuse GNSS and IMU readings could offer more precise localization. Still centimeter precision levels needed for some automated vehicle functions, such as lane determination, may benefit from more precise positioning systems. Reduction in cost of highly accurate GNSS is expected in near future making it available to the mass market [9]. Today Real-Time Kinematic (RTK) GPS technology is available and relies on a roadside base station to correct GPS readings to within centimeter accuracy. Simultaneous Localization And Mapping (SLAM), which we discuss in more detail later under localization algorithms, are used by many autonomous vehicle developers to localize the vehicle with respect to the surrounding environment.

Extroceptive sensors such as sonar, radar, LIght Detection And Ranging (LIDAR) and cameras are used for sensing the surrounding environment and objects as summerized in Table 3.2. Sonar, radar, and Lidar are called active sensors because they emit energy in the form of sound and electromagnetic waves and measure the return to map the surrounding environment, e.g. distance to nearby objects. On the other hand light and infrared cameras are called passive sensors since they do not emit energy and only measure the light/electromagnetic waves in the environment [10].

Sonar can measure distance to nearby objects but only has a very limited range (<2m) and has low angular resolution. Radars rely on reflection of radio waves that they emit to measure distance to and velocity of moving objects and have a much higher range than sonar but are weak in classification, pedestrian detection, and in detecting static objects. Also radars may suffer from interference from other radars and create false alarms. Lidar works similarly to radar but relies on infrared light (laser) instead of radio waves. Lidars emit laser at wavelengths beyond the visual light spectrum at typical scan frequency of 10–15 Hz. They emit millions of pulses per second giving them high resolution, a large field of view, and the capability to create a 3D point cloud of surrounding environment. This has made them an essential sensor for most automated vehicle developers. Nevertheless, Lidar cannot directly measure velocity, may have difficulty with detecting highly reflective objects and has degraded performance in fog, rain, or snow. Segmentation, classification, and sometimes time integration algorithms are still needed to convert the 3D raw data

Table 3.2 Comparison of different extroceptive sensor technologies for automated driving. The results are compiled from [10, 12–14]

	Sonar	Radar	Lidar	Single vision camera
Perceived Energy	Sound waves	Millimeter wave radio signal	600–1000 nanometer wave laser signal	Visible light
Range [m]	2–5	0.15–250	2–100	250
Vehicle recognition versus other objects	Tracking	Tracking	Spatial segmentation, motion	Appearance, motion
Resolution	–	o	+	++
Field of view	–	+ (short range radar)- (long range radar)	++	++
Distance measurement	+	++	++	–
Velocity measurement	– –	++	+	–
Operation in poor weather	++	++	o	–
Poor lighting performance	++	++	++	– –
Other challenges	Poor classification low range low resolution	Poor classification, poor pedestrian detection, poor static object detection, prone to interference	Poor classification compared to vision, difficulty with highly reflective objects	high computational cost
Cost (US $)	50	50–200	7,000–70,000	100–200

++Very good   +good   o average   – poor   ––very poor

to classified objects [11]. While laser emitting detection technology is not new, it was not till 2005 that Velodyne put 64 rotating lasers in one compact package for 360° detection needed in automated driving. Since then Lidar technology has been adopted by almost all autonomous vehicle teams. Still current Lidars are not designed to withstand many years of harsh conditions in open road driving. Both radar and Lidar also are weak in detecting very near objects (<2m) where sonar performs well [10].

Cameras provide high field of vision and high resolution, and capture information that Lidar cannot such as color and texture which helps object classification. However with monocular camera vision it is more difficult to measure depth; this can be overcome with stereo vision provided by two cameras. Computationally, camera vision is more demanding than Lidar. Converting 2D images to 3D understanding of the environment requires computationally demanding software and machine learning algorithms. Camera vision is sensitive to lighting conditions and its performance degrades in bad weather [12].

The algorithmic aspects of perception and localization are briefly discussed next.

3.2.2 Algorithms for Perception and Localization

Given the pros and cons of extroceptive sensors, in particular camera and Lidar, it is common to use both and rely on filtering and data fusion algorithms to increase accuracy and robustness. Measurement error covariance when using two sensors is always smaller than the error covariance achieved by each individual sensor. So it often makes sense to fuse data from two inexpensive sensors and achieve similar accuracy of a single high end sensor [11]. V2X communication can provide additional information from other vehicles and roadside units for higher accuracy perception and localization.

3.2.2.1 Perception Algorithms

Perception algorithms could be vision-based relying on camera data, or rely on active sensors which capture objects by a large number of points on their surface, also called point clouds. Camera and active sensors can be employed together to detect and perceive the surrounding environment and objects (such as vehicles, pedestrians, animals, curbs) more precisely. While there are mature machine vision and statistical learning and classification algorithms for parsing information embedded in an image or point cloud, recent advances in deep learning and artificial intelligence provide new supervised learning methods for real-time object detection. Rapidly growing training datasets, increased computing power, cheaper storage, and widely available open-source algorithms seem to be bringing about revolutionary advances. For instance an open source real-time object detection algorithm presented recently in [15] based on convolutional neural networks has the ability to process 45–150 frames per second, label objects in it with a bounding box, and assign a confidence score to each as illustrated in Fig. 3.3.

In automated driving three paradigms have been proposed for perception: (i) mediated perception, (ii) behavior reflex perception, and (iii) direct perception [16]. In the more common mediated perception, a detailed map and distance to relevant objects around the ego vehicle including other vehicles, pedestrians, trees, and road markings are extracted first using standard machine vision or deep learning algorithms. Planning and control algorithms will then use this map to plan the motion of the vehicle considering the constraints imposed by the road and stationary and dynamic obstacles. Quite differently, behavior reflex perception algorithms use artificial intelligence to construct a direct mapping from the sensory input to a driving action thus bypassing intermediate layers such as localization, path planning, decision making

Fig. 3.3 An example of application of YOLO real-time object detection [15] to a driving scene. The numbers next to each label show the confidence in that label. Picture courtesy of Austin Dollar and Tyler Ard of Clemson University

and control [17]. While they reduce complexity, such end-to-end solutions, lack transparency, are too low-level missing the big picture, and sometimes may be ill-posed in training. For instance in [18] it is shown that stability can be lost when applying supervised learning to a training set of locally exponentially stable controllers. Direct perception methods proposed in [16] aim to strike a balance between the former two approaches. They abstract an image to a selected and meaningful set of indicators of the road situation, such as the angle of the car relative to the road, the distance to the lane markings, and the distance to cars in the current and adjacent lanes. The outcome is much more compact than what a mediated perception approach would generate and only contains the most relevant information to the the planning and control layers which could now be simplified according to [16].

3.2.2.2 Prediction

While perception by itself is an important and challenging step, predicting the motion of neighboring vehicles or pedestrians based on perceived current and historical information may be as important for CAV planning purposes. This is a difficult topic and still an open problem. In Sect. 1.3.3 we discussed relevant prediction literature in the context of anticipative car following where probabilistic prediction was a common theme for predicting the longitudinal motion of a preceding vehicle. Other examples are assuming a constant speed in [19], a speed-dependent acceleration in [20], probabilistic trajectory prediction in horizontal plane using a variational mixture model in [21], a Gaussian mixture model in [22], or classification and particle filtering in [23]. Most of these prediction methods target a 1–3 second prediction window which may be limited. With V2X connectivity the opportunity exists for receiving future intentions of neighboring vehicles and nearby traffic controllers which should enable predicting with more accuracy over longer horizons. We will come back to this topic in Sect. 8.2.3 in the context of energy efficient driving.

3.2.2.3 Localization and Mapping

CAVs require rather precise localization not only for navigational purposes but also to situate themselves within the road and the lane, with respect to other (connected) vehicles, and for use of mapped information such as location of traffic signals, upcoming hills, curves, dynamic congestion tail, etc. While localization and mapping is a well established topic in indoor robot navigation and mature algorithms exist [24], outdoor, dynamically changing, and high speed road environments present extra challenges for CAV localization.

Fusing GPS, IMU, and wheel odometer readings could provide meter-level precision in determining the position of the vehicle on the road. The raw coordinates determined by GPS may not match a logical model of the world where vehicles are expected to be on a road. Established map matching methods [25] are commonly used to correct the raw GPS recordings to a logical position on a road. The (corrected) GPS data can be fused with IMU and odometry readings via Extended Kalman Filtering (EKF) methods that rely on a model of vehicle kinematics or dynamics. Velocity of the vehicle could be determined as a by-product. Accurately determining vehicle heading is more difficult due to reliance on IMU readings which are subject to drift.

Algorithms relying on GPS inertial navigation could be challenged in urban canyons with tall buildings due to loss of GPS signals [26]. Also autonomous vehicle control may require centimeter level position accuracy not provided by conventional GPS/IMU fusion. While RTK GPS provides high level of position accuracy, its reliance on additional roadside stations makes it impractical on today's roads. To overcome this challenge many automated driving vehicles such as Waymo's and Uber's rely on a priori mapped roads. Instrumented mapping vehicles drive roads of interest and collect detailed 3D image or Lidar data linked to highly accurate GPS information, process and store them in large databases. Subsequent CAVs can localize by comparing their sensor readings against these a priori maps and triangulating their position with the aid of fixed objects. Moreover they can more easily distinguish dynamic objects absent from a priori maps. An early successful implementation can be found in [27]. Such a method works as long as the mapped roads remain unchanged. Construction zones, changes in lane markings or road geometry could render parts of these maps obsolete.

This problem can be overcome by High Definition (HD) mapping where a priori maps are dynamically updated in the cloud based on latest sensory information communicated from CAVs traversing these roads [28]. For instance a consortium formed by BMW, HERE, and Mobileye aims to crowdsource HD maps relying on accurate prior maps from HERE, BMW connected fleet, and Mobileye REM technology that transmits changes detected with respect to prior map to cloud servers to update the maps. The dynamically updated maps become then accessible to the connected fleet in real-time via HERE servers.

In this context Simultaneous Localization And Mapping (SLAM) arises when the vehicle has to simultaneously localize and map the environment and obviously are more difficult than only localization or only mapping. SLAM is well established in indoor robotic navigation [24] often in well-structured and well-lit environments. SLAM is more challenging for automated vehicles due to variable lighting, less structured road environments and higher speeds that require faster computations [10].

3.2.3   Web Services

Connected vehicles can query web-based Application Programming Interfaces (API) to retrieve map, traffic congestion, and weather information in real-time. For instance the cloud based Google Map Platform [29] provides several APIs for retrieving maps, elevation, traffic, directions, travel times and distances, and places in real-time. Similar services are provided by HERE APIs [30]. Inrix offers a traffic and a parking API [31]. There are several weather information APIs such as Yahoo Weather API [32]. Today computational clouds such as Amazon Web Services (AWS) offer their computing and machine learning tools to connected [33] and automated [34] vehicle developers. The idea is to offload the onboard computations and data analytics partially to the cloud.

3.3   Planning and Control

Once an automated vehicle localizes itself with respect to a 3D map of the environment and identifies constraints imposed by the surrounding stationary and moving objects, traffic rules, traffic control infrastructure, and road geometry, it can plan its long- and short-term moves. This plan is then executed by a hierarchy of motion planners and controllers in the longitudinal and lateral directions. Both planning and control layers can benefit from the extended preview of the upcoming road and traffic scene provided by V2X connectivity to make longer term judicious decisions. Here we provide a brief overview of the planning and control layers as shown schematically in Fig. 3.4.

Fig. 3.4 Logical scheme of
planning and control layers
in CAVs

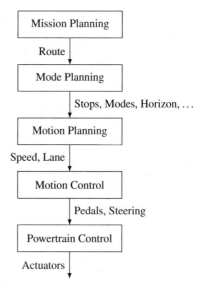

3.3.1 Mission Planning

At the highest planning layer, the route is decided, for instance to minimize trip distance, time, delay, or energy. The road network is often modeled as a directed graph with its edge weights reflecting the relevant cost of travel on that link. The minimum cost path can then be found via optimization which can be executed very efficiently today as explained in [35]. For electric vehicles, visits to charging stations may also be planned at this stage. The mission planning layer can then set waypoints along the chosen route as targets for the lower level motion planning layer. More details of algorithms employed in the mission planning layer in the context of eco-routing are described in Chap. 5.

3.3.2 Mode Planning

Another distinct planning layer may exist that chooses between a finite set of driving modes in consideration of mission waypoints, road rules, and traffic conditions. For instance the vehicle may choose lane keeping, a lane change, (adaptive) cruise control, stopping at a stop sign, or emergency braking. This will be a finite set of modes that can be handled in a finite state machine framework or via decision trees. We refer to this layer as Mode Planning, but in the literature other terms such as driving strategy [36], maneuver planning [37] and behavioral decision making [35] are also used.

We will show in Chaps. 6 and 7 that optimal eco-driving in a trip could consist of several modes for example maximum acceleration, constant speed cruising, coasting, and maximal braking between two stopping intervals.

3.3.3 Motion Planning

After a driving mode is selected, the motion planning layer generates legal, collision-free, smooth, comfortable, and efficient paths or trajectories for longitudinal and lateral motion of the vehicle. The literature distinguishes a trajectory from a path in that a path is in the spatial configuration space of the vehicle while a trajectory has a temporal component as well [38]. For instance in the longitudinal direction s, usually the velocity trajectory $\dot{s}(t)$ is planned with safety, ride comfort, travel time, and energy efficiency considerations while respecting constraints imposed by speed limits, traffic lights and stop signs, surrounding vehicles, road curvature, and longitudinal vehicle dynamics.

For example in Cruise Control (CC) mode the vehicle tracks a constant reference speed while Adaptive Cruise Control (ACC) adjusts the velocity to maintain a safe time or distance headway to the preceding vehicle. More details on that are discussed in Sect. 4.2.2. In Predictive Cruise Control (PCC) mode, the velocity is adjusted relying on V2I communication and in anticipation of future events such as changes in road slope or traffic signal phase and timing. Cooperative Adaptive Cruise Control (CACC) mode relies on V2V communication to allow vehicles cruise in coordination with neighboring vehicles. In emergency braking mode, the vehicle could apply maximal braking to avoid a collision.

Lane change, merge, and collision avoidance need to determine a feasible path in the 2D x-y plane which is by itself complex due to many choices in a 2 dimensional space and the non-convex drivable regions. Furthermore due to velocity and time dependent constraints arising from vehicle dynamics and movement of surrounding vehicles, the motion planning algorithms should also determine safe and comfortable acceleration and velocity profiles on these paths; thus a trajectory planning problem.

In Sect. 3.3.6 we discuss optimal planning algorithms applicable to motion planning.

3.3.4 Motion Control

The trajectory or path planned at the motion planning layer is issued as a reference to the vehicle longitudinal and lateral controllers for feedforward and feedback tracking. In the longitudinal direction, throttle and braking control adjusts acceleration and velocity. Lateral control relies mainly on steering and sometimes on differential braking to control lateral acceleration, velocity, and vehicle yaw rate.

3.3.4.1 Longitudinal Control

When the reference speed is determined at the planning layer, well-established classical or modern control techniques can be used at the motion control layer, to follow the planned reference by accelerator or brake actuation. For instance standard fixed gain or gain-scheduled PID type controllers can actuate accelerator and brakes for velocity reference tracking [39]. An integrator anti-windup mechanism [40] must be added to properly handle actuator saturation. Logical checks should be in place to ensure safe operations under all perceivable circumstances. Switching between accelerating and braking modes needs to be handled with care for smooth performance [41].

For instance [42] proposed the following PID type controller with an added nonlinear term shown below:

$$u(s) = -k_p e_v - k_i \frac{1}{s} \left(e_v - \frac{1}{T_t}[u - \text{sat}(u)] \right) - k_d \frac{\tau_d s}{\frac{1}{N}\tau_d s + 1} e_v - k_q e_v |e_v| , \quad (3.1)$$

where the equation should be read in the Laplace domain with s denoting the Laplace variable. Here u commands accelerator or braking, and e_v is the velocity tracking error. Tunable proportional, integral, and derivative gains are denoted by k_p, k_i, k_d respectively, while k_q is a tunable gain for the last nonlinear term. The term $\frac{1}{T_t}[u - \text{sat}(u)]$ prevents integrator windup where the $\text{sat}(u)$ function saturates at actuator limits and T_t is a time constant that determines how fast the integrator is reset. Because a pure derivative term would be non-causal and prone to noise, a pseudo-derivative term is employed by augmenting a first order lag, wherein parameter N determines the amount of filtering on the derivative term. The last nonlinear term $k_q e_v |e_v|$ is termed the quadratic component in [42] and is intended to achieve fast tracking while limiting the overshoot. Asymptotic convergence of tracking error to zero is established in [42] via a Lyapunov analysis.

Feedforward control along with feedback control can enhance the responsiveness of the longitudinal control loop. For instance when the planning layer commands an acceleration profile, a feedforward pedal/braking input can be issued [43] based on pedal-to-acceleration and braking-to-deceleration response mappings along with a feedback controller.

Input saturation, vehicle state constraints, and toggling between accelerator and braking actuators can be more systematically handled in a constrained control framework. For heavy vehicles sensitivity to often unknown mass of the truck can also be handled by adaptive control techniques as shown in [44].

3.3.4.2 Lateral Control

Lateral control engages steering and sometimes differential braking to control the vehicle in scenarios such as lane changing, merging, turning, and parallel parking. The assumption is that an appropriate reference path or trajectory is already deter-

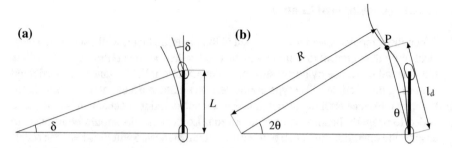

Fig. 3.5 A simplified bicycle model of a 4 wheeled vehicle: geometric bicycle model (**a**) and pure pursuit geometry (**b**). Adapted from [46]

mined in the motion planning layer. A widely used approach for path tracking with mobile robots and autonomous vehicles is pure pursuit control that was first introduced in [45] and is relatively simple to implement. The pure pursuit algorithm has a simple formula for choice of the steering angle that steers the rear axle on a circular arc to the center of the path. If a bicycle model, as shown in Fig. 3.5, is used to relate the steering angle of the front wheels $\delta(t)$ to vehicle heading $\theta(t)$, the pure pursuit algorithm formula is

$$\delta(t) = \tan^{-1}\left(\frac{2L\sin(\theta(t))}{l_d}\right), \tag{3.2}$$

where L is the wheelbase, l_d is the the distance from the rear wheel to a look-ahead point on the center of the path, and θ is angle between the heading vector and the look-ahead vector pointing to the center of the path l_d units ahead. In practice the look-ahead distance is chosen as a function of vehicle speed [46].

Another method [47], adjusts the steering as a function of vehicle heading misalignment with the path and a nonlinear function of cross track error,

$$\delta(t) = \theta(t) - \theta_p(t) + \tan^{-1}\left(\frac{ke_y}{v_x(t)}\right), \tag{3.3}$$

where θ is heading angle of the vehicle, θ_p is path heading at the point nearest to the front wheel, e_y is the cross-track error measured from the center of the front wheels to the nearest point on the track, v_x is vehicle's forward velocity, and k is a gain parameter. Using an idealized bicycle model, the cross-tracking error is shown to be monotonically convergent to zero.

The above lateral control methods are easy to implement, but rely on feedback from a single point of the lane at each time. For smoother performance, the lane tracking problem can be formulated as a finite horizon optimal control problem with full horizon preview of lane reference trajectory. The optimal steering control action will not only be a function of instantaneous vehicle state but also will include a feedforward term that integrates the entire lane preview. An analytical solution to this preview optimal control problem exists when the vehicle model is linear, the tracking cost is quadratic, and input and states are unconstrained as shown in [48]. Input and state constraints must be considered for aggressive or emergency maneuvers or when driving on slippery roads with the tires at their traction limit. In such scenarios the trajectory tracking problem can be formulated in a model predictive control framework with higher fidelity vehicle models and with explicit consideration of traction constraints. Successive model linearizion results in a Linear Time-Varying (LTV) MPC problem as shown in [49] along with experimental results that demonstrates the feasibility of real-time implementation.

Planning and control algorithms that can handle more sophisticated conditions than the relatively simple longitudinal and lateral control methods described above, are described later in Sect. 3.3.6.

3.3.5 Powertrain Control

The powertrain control modules of a CAV can be programmed to take advantage of extra information that is available to them due to connectivity and increased certainty in that information due to driving automation as highlighted in Sect. 1.3.1. Depending on the powertrain type as described in Chap. 2, we have several actuators to coordinate such as throttle, braking, ignition, injection, cam phasing, wastegate, valve lift, cylinder deactivation, and transmission for SI ICEVs, battery utilization in HEVs, and vehicle-level actuators for accessory loads. Anticipated future velocity and road grade profile provide an estimate of the future power demands. This anticipated power demand profile can be used to better schedule choice of gears, battery utilization in hybrid vehicles, thermal load management, and handling of the powertrain auxiliary loads such as air-conditioning load.

The powertrain controllers can benefit from longer term plans of the mission planning and mode planning layers as well as more imminent intentions of motion planning and motion control layers. For example scheduling a hybrid vehicle's battery utilization can benefit from the long term mission plan due to the slow dynamics associated with the battery state of charge; so is thermal management due to relatively slow thermal dynamics as discussed respectively in Sects. 2.4.2 and 4.4.4. On the other hand, shorter term decisions at motion planning and control layers could be beneficial to functions with faster dynamics such as anticipative gear shift, fuel cut-off, engine start/stop, and cylinder deactivation.

3.3.6 Algorithms for Planning and Control

Two main schools of thoughts dominate the planning and control literature and practice. One approach guided by the robotics and computer science community employs (model-free) learning methods that aim to emulate human drivers, leveraging abundant training data and advances in deep learning and reinforcement learning algorithms. The second approach spearheaded by the automatic control community casts planning in a (model-based) optimal control framework aiming to minimize a mathematical cost of the motion (be it time, discomfort, energy, risk, etc.) while respecting all motion constraints. For instance, in a reinforcement learning approach to lane changing, the motion planning layer gradually learns a lane change policy that maximizes a cumulative reward function. The policy defines what action to take given the state of the road and neighboring vehicles and associates a reward to a successful lane change and gets penalized for a collision. The algorithm goes through a systematic trial and error process in a realistic simulation or real-world environment until it is "sufficiently" trained. It can then employ its learned policy in real-world driving.

An alternative to learning from training scenarios is optimal control that relies on models of the vehicle and its surrounding environment, a carefully designed objective function, and well characterized motion constraints. The plan can then be determined by solving a dynamic constrained optimization problem. For example in an optimal control approach to lane selection, the objective could be to balance a trade off between deviation from a desired velocity and a deviation from a desired lane. The predicted path of the surrounding vehicles can be imposed as motion constraints and a bicycle model of the vehicle can approximate the ego vehicle motion under candidate input sequences [50].

Closed-form analytic solutions for optimal control and planning problems rarely exist. Exact numerical solutions are often NP hard and not solvable in polynomial time. But one can often find approximations that simplify the problem. For example discretization, linearizing the models and constraints, and using a quadratic cost are common and reduce an optimal control problem to a quadratic program for which computationally efficient solution methods exist and enable its real-time implementation. The planning problem can be solved over a receding temporal or spatial horizon using feedback from the current state of the vehicle to update its plan at each optimization stage in what is referred to as Model Predictive Control (MPC) [51]. More details of numerically solving an planning problem in an MPC framework are described later in Sect. 8.2.5.

Numerical methods for optimal planning problems can be categorized to variational, graph search, and incremental search sample methods [38]. Under this categorization, Pontryagin Minimum Principle (PMP) is a variational approach that reduces the optimal control problem to a two point boundary value problem using variational calculus, the more details of which is described in Sect. 6.2.2.1. PMP is considered an *indirect* method because it is based on analytical construction of the necessary and sufficient conditions for optimality, and then discretizing these conditions and solving them numerically. *Direct* methods on the other hand discretize state and control trajectories and convert the optimal control problem to a nonlinear program [52], which is then solved using well-known optimization techniques. Pseudospectral optimal control methods [53] are among direct variational methods.

In graph search methods, the configuration space is discretized and represented by a graph consisting of vertices and edges. The graph is then explored to find the mini-

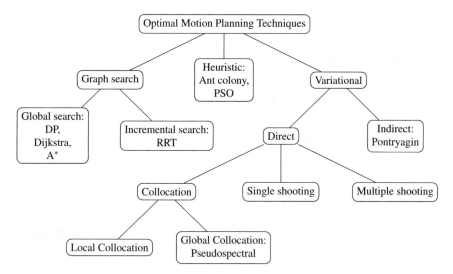

Fig. 3.6 Numerical methods for optimal motion planning

mum cost motion. Dijsktra [54], A* [55] and its variants, and Dynamic Programming (DP) [56] are among graph search methods. We will describe Dijkstra's algorithm in more detail in Sect. 5.1.2.1 in the context of eco-routing (mission planning) and DP in Sect. 6.2.2.2 in the context of eco-driving (motion planning).

A popular incremental search method is the Rapidly-exploring Random Tree (RRT) algorithm [57] designed to efficiently search nonconvex, high-dimensional spaces by randomly growing a space-filling tree in the reachable set of the vehicle. RRT algorithm is suited to problems with obstacles and differential constraints and is therefore widely used in robotic motion planning.

Heuristic methods such as ant colony optimization [58] and particle swarm optimization [59] have also been employed for path planning of autonomous agents and robots. A schematic of these categorizations is shown in Fig. 3.6.

$$*$$
$$* \quad *$$

In the rest of this book, the main focus in on higher level decisions at Mission Planning, Mode Planning, and Motion Planning layers. Readers interested in Motion Control and Powertrain Control may refer to many articles and books that exist on vehicle control such as [60].

References

1. USDOT. U.S. (2013) Department of Transportation releases policy on automated vehicle development. https://www.transportation.gov/briefing-room/us-department-transportation-releases-policy-automated-vehicle-development. Accessed 30 May 2013
2. SAE International (2018) Taxonomy and definitions for terms related to on-road motor vehicle automated driving systems. Surface vehicle information report J3016. Technical report, SAE International
3. Committee DSRC et al (2009) Dedicated short range communications (DSRC) message set dictionary. SAE Standard J 2735:2015
4. Li YJ (2010) An overview of the DSRC/WAVE technology. In: Proceedings of international conference on heterogeneous networking for quality, reliability, security and robustness, pp 544–558. Springer
5. USDOT. IEEE 1609 - family of standards for wireless access in vehicular environments (WAVE) (2009). https://www.standards.its.dot.gov/Factsheets/Factsheet/80
6. USDOT. SAE J2735 - Dedicated Short Range Communications (DSRC) message set dictionary (2009). https://www.standards.its.dot.gov/Factsheets/Factsheet/71
7. Araniti G, Campolo C, Condoluci M, Iera A, Molinaro A (2013) Lte for vehicular networking: a survey. IEEE Commun Mag 51(5):148–157
8. Energetics Incorporated and INC Z (2017) Study of the potential energy consumption impacts of connected and automated vehicles. https://www.eia.gov/analysis/studies/transportation/automated/pdf/automated_vehicles.pdf
9. Galileo GNSS. The path to high GNSS accuracy (2018). https://galileognss.eu/the-path-to-high-gnss-accuracy/
10. Van Brummelen J, O'Brien M, Gruyer D, Najjaran H (2018). Autonomous vehicle perception: the technology of today and tomorrow. Transp Res Part C: Emerg Technol 89:384–406
11. Pendleton SD, Andersen H, Du X, Shen X, Meghjani M, Eng YH, Rus D, Ang MH (2017) Perception, planning, control, and coordination for autonomous vehicles. Machines 5(1):6
12. Sivaraman S, Trivedi MM (2013) Looking at vehicles on the road: a survey of vision-based vehicle detection, tracking, and behavior analysis. IEEE Trans Intell Transp Syst 14(4):1773–1795
13. Rasshofer RH, Gresser K (2005) Automotive radar and lidar systems for next generation driver assistance functions. Adv Radio Sci 3(B. 4):205–209
14. Dhar SP (2017) From google to tesla, it's a war of LiDAR or RADAR. https://www.unitedlex.com/news-and-insights/blog/2017/google-tesla-it%E2%80%99s-war-lidar-or-radar
15. Redmon J, Divvala S, Girshick R, Farhadi A (2016) You only look once: unified, real-time object detection. In: Proceedings of conference on computer vision and pattern recognition, pp 779–788. IEEE
16. Chen C, Seff A, Kornhauser A, Xiao (2015) Deepdriving: learning affordance for direct perception in autonomous driving. In: Proceedings of international conference on computer vision, pp 2722–2730. IEEE
17. Bojarski M, Del Testa D, Dworakowski D, Firner B, Flepp B, Goyal P, Jackel LD, Monfort M, Muller U, Zhang J et al (2016) End to end learning for self-driving cars. arXiv:1604.07316
18. Da X, Hartley R, Grizzle JW (2017) Supervised learning for stabilizing underactuated bipedal robot locomotion, with outdoor experiments on the wave field. In: Proceedings of international conference on robotics and automation (ICRA), pp 3476–3483. IEEE
19. Kamal MAS , Mukai M, Murata J, Kawabe T (2011) Ecological driving based on preceding vehicle prediction using MPC. IFAC Proc Vol 44(1):3843–3848
20. Kamal MAS, Mukai M, Murata J, Kawabe T (2013) Model predictive control of vehicles on urban roads for improved fuel economy. IEEE Trans Control Syst Technol 21(3):831–841
21. Wiest J, Höffken M, Kreßel U, Dietmayer K (2012). Probabilistic trajectory prediction with gaussian mixture models. In: Proceedings of intelligent vehicles symposium (IV), pp 141–146. IEEE

22. Havlak F, Campbell M (2014) Discrete and continuous, probabilistic anticipation for autonomous robots in urban environments. IEEE Trans Robot 30(2):461–474

23. Hermes C, Wohler C, Schenk K, Kummert F (2009) Long-term vehicle motion prediction. In: Proceedings of intelligent vehicles symposium, pp 652–657. IEEE

24. Thrun S, Burgard W, Fox D (2005) Probabilistic robotics. MIT Press

25. Quddus MA, Ochieng WY, Noland RB (2007) Current map-matching algorithms for transport applications: state-of-the art and future research directions. Transp Res Part C: Emerg Technol 15(5):312–328

26. Cui Y, Ge SS (2003) Autonomous vehicle positioning with GPS in urban canyon environments. IEEE Trans Robot Autom 19(1):15–25

27. Levinson J, Montemerlo M, Thrun S (2007) Map-based precision vehicle localization in urban environments. In: Proceedings of robotics: science and systems conference, vol 4, p 1

28. Seif HG, Hu X (2016) Autonomous driving in the iCity—HD maps as a key challenge of the automotive industry. Engineering 2(2):159–162

29. Google (2018) Google maps platform. https://cloud.google.com/maps-platform/

30. HERE (2019) HERE APIs. https://developer.here.com/develop/rest-apis

31. HERE (2018) INRIX API documentation. http://docs.inrix.com/

32. Yahoo (2018) Yahoo weather API. https://developer.yahoo.com/weather/?guccounter=1

33. Amazon (2018) Amazon Web Services for connected vehicles and mobility. https://aws.amazon.com/automotive/connected-vehicles/

34. Amazon (2018). Amazon Web Services for ADAS and autonomous driving. https://aws.amazon.com/automotive/autonomous-driving/

35. Bast H, Delling D, Goldberg A, Müller-Hannemann M, Pajor T, Sanders P, Wagner D, Werneck RF (2016) Route planning in transportation networks. In: Algorithm engineering, pp 19–80. Springer

36. Aeberhard M, Rauch S, Bahram M, Tanzmeister G, Thomas J, Pilat Y, Homm F, Huber W, Kaempchen N (2015) Experience, results and lessons learned from automated driving on germany's highways. IEEE Intell Transp Syst Mag 7(1):42–57

37. Katrakazas C, Quddus M, Chen W-H, Deka L (2015) Real-time motion planning methods for autonomous on-road driving: state-of-the-art and future research directions. Transp Res Part C: Emerg Technol 60:416–442

38. Paden B, Čáp M, Yong SZ, Yershov D, Frazzoli E (2016) A survey of motion planning and control techniques for self-driving urban vehicles. IEEE Trans Intell Veh 1(1):33–55

39. Ioannou P, Xu Z (1994) Throttle and brake control systems for automatic vehicle following. IVHS J 1(4):345–377

40. Astrom KJ, Rundqwist L (1989) Integrator windup and how to avoid it. In: Proceedings of American control conference, pp 1693–1698. IEEE

41. Huang S, Ren W (1999) Vehicle longitudinal control using throttles and brakes. Robot Auton Syst 26(4):241–253

42. Yanakiev D, Kanellakopoulos I (1996) Speed tracking and vehicle follower control design for heavy-duty vehicles. Veh Syst Dyn 25(4):251–276

43. Kato S, Tsugawa S, Tokuda K, Matsui T, Fujii H (2002) Vehicle control algorithms for cooperative driving with automated vehicles and intervehicle communications. IEEE Trans Intell Transp Syst 3(3):155–161

44. Vahidi A, Stefanopoulou A, Peng H (2006) Adaptive model predictive control for co-ordination of compression and friction brakes in heavy duty vehicles. Int J Adapt Control Signal Process 20(10):581–598

45. Wallace RS, Stentz A, Thorpe CE, Moravec HP, Whittaker W, Kanade T (1985) First results in robot road-following. In: Proceedings IJCAI workshop, pp 1089–1095

46. Snider JM et al (2009) Automatic steering methods for autonomous automobile path tracking. Robotics Institute, Pittsburgh, PA, Technical Report CMU-RITR-09-08

47. Thrun S, Montemerlo M, Dahlkamp H, Stavens D, Aron A, Diebel J, Fong P, Gale J, Halpenny M, Hoffmann G et al (2006) Stanley: the robot that won the DARPA grand challenge. J Field Robot 23(9):661–692

48. Peng H, Tomizuka M (1993) Preview control for vehicle lateral guidance in highway automation. J Dyn Syst Meas Control 115(4):679–686
49. Falcone P, Borrelli F, Asgari J, Tseng HE, Hrovat D (2007) Predictive active steering control for autonomous vehicle systems. IEEE Trans Control Syst Technol 15(3):566–580
50. Dollar RA, Vahidi A (2018) Predictively coordinated vehicle acceleration and lane selection using mixed integer programming. In: Proceedings of dynamic systems and control conference, pp V001T09A006–V001T09A006. American Society of Mechanical Engineers
51. Hrovat D, Di Cairano S, Tseng HE, Kolmanovsky IV (2012) The development of model predictive control in automotive industry: survey. In: Proceeding of international conference on control applications, pp 295–302. IEEE
52. Kelly M (2017) An introduction to trajectory optimization: how to do your own direct collocation. SIAM Rev 59(4):849–904
53. Ross IM, Karpenko M (2012) A review of pseudospectral optimal control: From theory to flight. Ann Rev Control 36(2):182–197
54. Dijkstra EW (1959) A note on two problems in connexion with graphs. Numer Math 1(1):269–271
55. Hart PE, Nilsson NJ, Raphael B (1968) A formal basis for the heuristic determination of minimum cost paths. IEEE Trans Syst Sci Cybern 4(2):100–107
56. Bellman R (2013) Dynamic programming. Courier Corporation
57. LaValle SM (1998) Rapidly-exploring random trees: a new tool for path planning. Technical report
58. Porta Garcia MA, Montiel O, Castillo O, Sepúlveda R, Melin P (2009) Path planning for autonomous mobile robot navigation with ant colony optimization and fuzzy cost function evaluation. Appl Soft Comput 9(3):1102–1110
59. Zhang Y, Gong D-W, Zhang J-H (2013) Robot path planning in uncertain environment using multi-objective particle swarm optimization. Neurocomputing 103:172–185
60. Guzzella L, Sciarretta A et al (2007) Vehicle propulsion systems, vol 1. Springer

Chapter 4
Route and Traffic Description

In order to predict and minimize the energy consumption of road vehicles, modeling the vehicle and its powertrain is not sufficient. Several quantities introduced in the previous chapter (e.g., time horizon, grade, curvature, constraints to speed, etc.) are in fact functions of the road followed, its infrastructure, and the vehicle's traffic environment. This chapter aims at providing some information about state-of-the-art road network modeling (Sect. 4.1), microscopic (Sect. 4.2), and macroscopic (4.3) modeling of traffic. The chapter ends with the illustration on how to combine such information with the models of Chap. 2 to predict the energy consumption on road networks (Sect. 4.4).

4.1 Road Network Modeling

Geographic Information Systems (GIS) are systems designed to collect, store, analyze, and present geographical data. We focus here on those GIS where the particular type of information treated is about transportation networks, sometimes called GIS-T.

The heart of any GIS is its data model, i.e., the abstract representation of geographical features in terms of data and their organization in a database. GIS technology utilizes two basic types of data. These are: (i) spatial data that describe the location of geographic features, and (ii) attribute data that describe their characteristics. Attribute data are usually maintained using a Database Management System (DBMS). Spatial data is usually encoded and maintained in a proprietary file format.

Two basic types of *spatial data models* have evolved for storing geographic data digitally. These are referred to as: vector data models and raster/image data models, see Fig. 4.1. Usually a map consists of several layers of both vector and raster data.

© Springer Nature Switzerland AG 2020   83
A. Sciarretta and A. Vahidi, *Energy-Efficient Driving of Road Vehicles*,
Lecture Notes in Intelligent Transportation and Infrastructure,
https://doi.org/10.1007/978-3-030-24127-8_4

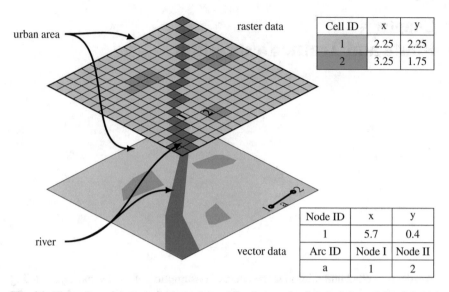

urban area

raster data

Cell ID	x	y
1	2.25	2.25
2	3.25	1.75

river

vector data

Node ID	x	y
1	5.7	0.4
Arc ID	Node I	Node II
a	1	2

Fig. 4.1 Illustration of raster and vector data, with an example of their data models. Adapted from Wikimedia Commons (Raster_vector_tikz.png by user Wegmann) under Creative Commons license

Raster data format is based on a grid-cell structure where the geographic area is divided into cells identified by their spatial coordinates. Any geographic feature is thus identified by some cell identifiers. Both regularly spaced grids and other tessellated data structures do exist in GIS systems. Probably the most important— for energy-related applications—information that can be encoded as raster data is the elevation, although vector-based representations exist, too. Digital Elevation Models (DEM) are often classified as terrain (DTM) and surface (DSM) models, where the former represent the bare ground surface without any objects like plants and buildings.

In contrast, *vector data format* is based on discrete elements, namely, vertices or shape points defined by their spatial coordinates. Several vector data models exist, however the most popular method of retaining spatial relationships among features is to explicitly record adjacency information in what is known as the topological data model. The most common topological data structure is the *arc/node data model*. This model contains two basic entities, the arc and the node. The arc is a string of points, linked by straight line segments, that start and end at a node. The node can be an intersection point where two or more arcs meet, while isolated nodes, not connected to arcs, represent point features. An area feature is comprised of a closed chain of arcs (polygons). Most GIS software record the topological relationship between these elements in tables. For instance, a node table stores information about each node and the arcs that are connected to it. An arc table contains the nodes of each arc, and the polygons that the arc is an element of. A polygon table lists the arcs that compose each polygon.

In most cases, a separate data model is used to store and maintain *attribute data*. A variety of different data models exist for this purpose. The tabular model is the manner in which most of early GIS software packages stored their attribute data. The relational models are those most commonly implemented in DBMS. The object oriented approach is newer but rapidly gaining in popularity for some applications.

The *relational data model* organizes data in tables. Each table is identified by a unique name, and is organized by rows and columns. Each column within a table also has a unique name. Columns (fields) store the values for a specific attribute. Rows represent one record in the table. In vector-based GIS, each row is usually linked to a separate spatial feature through its identifier and each column contains a specific value for that geographic feature.

As of today, many GIS deliver geographic data (maps) on the world wide web. Among the most popular web mapping services are Google maps [1], Bing maps [2], HERE maps [3], MapQuest [4], OpenStreetMap [5], etc.

4.1.1 Road Network Topology

Geometric *networks* are commonly used in GIS to digitally represent real transportation networks, for example in the so-called arc-node data model introduced above. A network is a type of graph, a mathematical structure that represents relationships among entities as a set of *nodes* interconnected with a set of *links* (also called arcs or edges). In transportation networks, nodes are point locations, while links are the road segments that connect nodes. Since transportation systems typically have important directional flow properties (e.g., one-way streets), directed networks are usually employed, where all the links have directions associated with them.

The relationship between the nodes and the links is referred to as the network *topology*. Of particular relevance are the representations of location, direction, and connectivity. The network topology's connectivity is encoded in rectangular tables, see Fig. 4.2, where the entries are the unique identifiers of links and the data accessed (fields) are pointers to connected nodes. These nodes are somehow ordered (e.g., called reference and non-reference nodes according to some geographical preference). In addition to connectivity, these tables can store attributes of links and nodes, see Sect. 4.1.2.

Including detailed information on the ability to connect from one link to another, e.g., restrictions to turn right at an intersection, can be done in several ways. A first option is to have additional link-wise table fields for the links accessible from the connected nodes (as shown in the example of Fig. 4.2). Another possibility consists of adding a turn table with node entries and fields representing the links concurring in that node. A special attention must be paid in representing connectivity of non-planar networks, where not all the intersections of two links correspond geographically to a node due to the presence of overpasses and underpasses.

Data on transportation-related events and facilities (often termed feature data) are typically located by means of a linear rather than coordinate-based system. In

Fig. 4.2 Simple road
network, with directions on
links numbered from 1 to 16
and nodes number from 1 to
8 in italic (**a**), and table
encoding its topology (**b**)

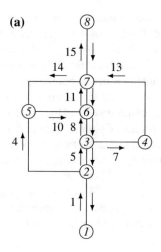

(b)

Link ID	Node I	Node II	Links through Node I	Links through Node II
1	1	2	–	4,5
4	2	5	–	10
5	2	3	1,4	7,8
7	3	4	–	13
8	3	6	5,7	8,11
10	5	6	–	8
11	6	7	8	14,15
13	4	7	–	11,14,15,
14	7	5	–	10
15	7	8	11,14	–

order to use linear-referenced attributes in conjunction with a spatially referenced
transportation network, there must be some means of linking the two referencing
systems together. There exist different linear referencing methods, e.g., distance
from a reference point (datum), control section in the link, or distance from one of
the link nodes.

4.1.2 Road Network Attributes

Besides connectivity-related attributes, the most significant attributes that can be
attached to network elements are related to basic geometry, road category, and speed
limits. More advanced functions and services consume detailed geometry attributes
and traffic-related attributes. Based on this classification, a non-exhaustive list of link
attributes that is most relevant for energy-efficient driving is shown below.

1. Geometry

   - Length.
   - Coordinates (latitude and longitude) of connected nodes.
   - Direction (positive, negative, both). To avoid unnecessary, and often unrealistic duplication of links, especially at the street level, a directional attribute can be included in the attribute table.
   - Number of lanes n_l.
   - Lane width w.

2. Categorical

   - Route type, depending on country or region (local, national, motorway, etc.).
   - Speed category depending on posted or legal speed limit.
   - Grade category (up, down, level).
   - Nature of the link, particularly for what concerns the type of vehicles allowed and the type of access (controlled access roads, low mobility roads, ramp, presence of a legal or physical divider, paved, private/public, urban, express, four-wheel-drive only, parking, etc.).

3. Speed limits

   - Speed limit v_{lim}, the maximum allowed speed established by law (also known as posted speed limit, PSL). Usually posted by increments of 10 km/h (5 mph in US) and indicated on a traffic sign. Minimum speed limits sometimes are posted where slow speeds can impede traffic flow or be dangerous. In some countries, speed limits may be variable, e.g., apply to certain classes of vehicles or depend on special conditions (daytime, weather, special zones). Advisory speed limits may provide a safe suggested speed below the legal speed in an area.

4. Detailed geometry

   - Horizontal radius of curvature R ([m]) at coordinate points along the link. This quantity is used to estimate the safe speed limit while turning, from the equation
     $$v_{turn} = \sqrt{\mu g R} \, ,$$
     where μ is the grip coefficient and g the gravity acceleration.[1]
   - Road elevation or slope α ([deg] or [%]) at coordinate points along the link.
   - Horizontal heading θ at coordinate points along the link.

5. Traffic

   - Free-flow speed v_{FF}, defined as the theoretical speed of a vehicle when traffic density and flow rate on the link both are equal to zero. In practice, operational

---

[1] This equation is valid for zero bank angle. A more general equation can be easily obtained from a free-body diagram of the turning vehicle. Also lift/downforce that are relevant, e.g., for racing cars are neglected.

definitions are used to measure free-flow speeds, such as speeds of vehicles having time headways greater than 4 s from preceding vehicles.

- Average traffic flow speed V, depending on the season, day of the week, hour (for example, it can be specified with a value each day and every 15 min). It is the result of statistics over a sufficiently large number of vehicles crossing that link. Usually the mean speed or the 85th percentile speed are used to represent such statistics. The 85th percentile speed of traffic on a road is often used in the traffic engineering literature as a guideline in setting speed limits and assessing whether such a limit is too high or low.

Some of these attributes, in particular the detailed geometry parameters α, R, the speed limit v_{lim}, and the average traffic speed V will play a key role in the models described in the next sections. Sometimes, however, detailed geometry is not available or not sufficiently accurate. In Sect. 8.2.1 we will discuss how to reconstruct instantaneous speed, curvature, and slope from sequences of coordinate points measurements.

4.1.3 Intersections

Intersections are classified by traffic control type and topology. Uncontrolled intersections do not generally present signs, although sometimes they present a warning sign. Priority (right-of-way) rules may vary by country and by the number of road segments that are involved. Yield-controlled intersections may or may not have specific signs. Stop-controlled intersections have one or more "STOP" signs. Two-way stops are common, while some countries also employ four-way stops. Signal-controlled intersections depend on traffic signals, usually electric, which indicate which movement is allowed to proceed at any particular time. Circular intersections, particularly roundabouts, have their own design and rules.

While the attributes of the other types of intersection is usually fixed and can be retrieved from GIS as described above, information on how traffic lights are operated, though essential to predict and optimize energy consumption, is harder to obtain and often must be assumed or simulated. SPaT (Signal Phasing and Timing) describes the way in which a traffic signal accommodates various users at an intersection in a safe and efficient manner. A movement reflects the user perspective and is defined by the user type and the action that is taken (turning movement for a vehicle or pedestrian crossing). Two different types of movements include those that have the right of way and those that must yield consistently with the rules of the road. These movements are regulated by the signal controller through their allocation to one or more signal phases. Signal *phase* is the right-of-way, yellow change, and red clearance intervals in a cycle that are assigned to an independent traffic movement or combination of traffic movements. An interval is a duration of time during which the signal indications do not change. Each phase at an intersection has a set of *timings*, possibly containing vehicle and pedestrian timing. A phase may control both a through movement and

Fig. 4.3 An example four-leg intersection (**a**), with movements 1–8 and concurrent pedestrian movements; left-turn movements are assigned odd number phases, while through movements are assigned to even number phases and right-turn movements share phase with through movements. In this example, the southbound left-turn movement is protected and is associated with phase 5; the westbound right-turn movement is compatible with the westbound through movement and thus shares phase 4; pedestrians crossing the northern leg of the intersection are assigned the concurrent westbound vehicular phase (phase 4), which conflicts with the eastbound left turn (phase 3). Corresponding ring-and-barrier diagram (**b**), with the two barriers as thicker vertical lines. The sequence of phases is shown as they occur in time, proceeding from left to right, with left-turn movements leading the opposing through movements [6]

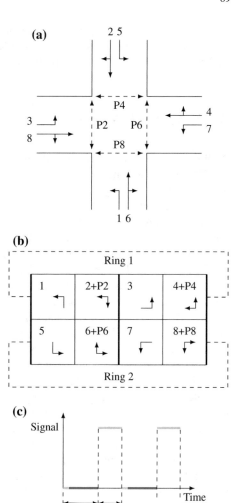

a right turn movement on an approach. In an actuated controller unit, the *cycle* is a complete sequence of all signal indications [6].

In a typical assignment for four-leg intersections, see Fig. 4.3a, left-turn movements are assigned to even number phases. The right-turn movements are not typically assigned to separate phases. Typically, a pedestrian movement is associated with the concurrent vehicular phase running parallel and adjacent to it.

Modern U.S. practice for signal control organizes phases by grouping them in a continuous loop (or ring) and separating the crossing or conflicting traffic streams with time in between when they are allowed to operate, either by making the movements sequential or adding a barrier between the movements. An example ring-and-barrier diagram is shown in Fig. 4.3b.

For a given movement, the resulting timing is a periodic alternation of green, yellow, and red signals as shown in Fig. 4.3c. We define a duty-cycle parameter r_x as the ratio of the red time t_r (including yellow) to the period $t_g + t_r$.

4.1.4 Recharge Stations

Another type of localized feature that is relevant for energy-efficient driving strategies is battery recharge stations. In a road network, different charging stations generally have different technologies.

Slow charging is mainly made at domestic socket outlets with AC output. That requires an internal AC/DC converter (on-board charger) in the vehicle. With a typical domestic rate of 3–7 kW, recharging of an EV may take several hours.

Fast charging is mostly done using DC charging stations. The two main standards are CHAdeMO and COMBO 2, offering charging powers of 20–50 kW. Also an AC technology (Type 2, available in Europe) exists, with charging at 22 kW AC. Furthermore, there exists an ad-hoc standard used only by Tesla and charging at 120 kW. As of 2018, announced plans are to bring the charging power to 100 kW for CHAdeMO and COMBO, and the latter even to 350 kW which would allow treating buses or heavy-duty vehicles. Due to the lack of a single international standard, multi-standard stations are nowadays rather popular, typically offering one or both DC options and the AC option.

Battery swap, consisting of replacing the battery at stations instead of recharging them, was once considered as a promising technology, but it seems disfavored as of 2018 . In contrast, *inductive charging* (or, more generally, wireless) concepts are currently gaining popularity. These systems use an electromagnetic field to transfer energy between a primary coil at the recharge station and a secondary coil on the vehicle. In dynamic wireless charging systems, vehicles draw power from coils buried underneath the surface of the road. Although promising in terms of charging time and comfort, these systems are costly and technically not mature yet [7].

Regardless of the technology used, charging stations can be characterized by their *charging function*, that is, the function $\Delta\tau_c(\Delta E_c, \varepsilon_{b,0})$ relating the quantity of electricity charged ΔE_c and the energy stored in the battery when the charging process begins, $\varepsilon_{b,0} \triangleq q_{b,0}V_{b0}$, to the time $\Delta\tau_c$ that is necessary. In addition to this time, the waiting time at the station should also be taken into consideration, e.g., when optimizing the recharge schedules, see Chap. 5.

In general, the charging functions are nonlinear because of the terminal voltage and current change during the charging process. As schematically shown in Fig. 4.4, this process is typically divided into two phases, a constant current (CC) phase and a constant voltage (CV) phase. The CC phase continues until the battery's terminal voltage reaches a specific value (cut-off voltage), which corresponds to, e.g., a SoC of 80%. In the second phase, the current decreases exponentially to avoid damaging the battery, and the battery SoC increases less than linearly with time. Note that charging schemes other than the CC-CV strategy exist, for example constant power-constant

Fig. 4.4 Typical charging profiles of an automotive battery

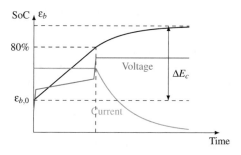

voltage, or optimal charging profiles that explicitly minimize, e.g., the battery aging or the charging losses [8].

In most cases, the energy-time relation can be approximated by a piece-wise linear function $\varepsilon_b = g(t)$ [9]. Consequently, the charging function is obtained by inverting this relation,

$$\Delta \tau_c = g^{-1}(\varepsilon_{b,0} + \Delta E_c) - g^{-1}(\varepsilon_{b,0}) . \tag{4.1}$$

Of course, each charging technology can be characterized by a specific piece-wise-linear charging function.

4.2 Microscopic Modeling of Traffic

Microscopic models describe traffic flow dynamics in terms of individual vehicles and their trajectories. These models aim to represent the behavior of human drivers or automated vehicles as they reach a target speed in free flow, adjust it in the presence of interactions with other vehicles, and make other discrete decisions such as lane change, giving priority, etc.

This section discusses in particular car-following models (Sect. 4.2.1), advanced cruise controllers (Sect. 4.2.2), and lane-changing models (Sect. 4.2.3). It is worthy to remark that a complete microscopic simulation requires to define the origins (sources) and destinations of all simulated vehicles. Vehicles can have their source at the boundaries of the area of interest, where vehicle inflows must be defined, or inside the area. Instead of specific destinations, inside or outside the area simulated, percentages of directions (turning ratios) at each intersection can be assigned [10]. However, a discussion of how these data are generated is beyond the scopes of this book.

4.2.1 Car-Following Models

Car-following models try to describe the behavior of typical drivers, that is, the vehicle's acceleration as it is influenced by the driver desires, the road infrastructure, and the surrounding traffic. These models are expressed either under the form

$$\ddot{s}(t) = \dot{v}(t) = F(v(t), \delta(t), v_p(t)) \tag{4.2}$$

or under the equivalent form

$$v(t + \Delta t) = F(v(t), \delta(t), v_p(t)) , \tag{4.3}$$

where Δt is a time step, v_p is the leading vehicle's speed, and δ is the *net distance* or *gap*, that is, the distance between the rear bumper of the preceding vehicle and the front bumper of the host vehicle. If the vehicle length is denoted by ℓ_p and s_0 is the desired residual gap when both vehicles are stopped, then

$$\delta(t) \triangleq s_p(t) - s(t) - \ell_p - s_0 , \tag{4.4}$$

see Fig. 4.5. Note that the role of the preceding vehicle can be easily generalized to other moving or stationary obstacles such as a stop sign etc.

The resulting acceleration has to be constrained to a "comfortable" range or, at least, to physically possible values, that is,

$$a_{min} \leq \ddot{s}(t) \leq a_{max} . \tag{4.5}$$

Most models actually make no distinction between theoretical limits and user-defined comfortable values, for which see Table 4.1.

Most car-following models distinguish between two driving behaviors: (i) a free-drive regime where the driving behavior is influenced only by the current speed and a desired speed v_d, that could be the maximum speed allowed in the road segment, and (ii) a gap-controlled regime, where the driving behaviour is dominated by the current gap and a desired gap δ_d. The so-called psycho-physical models have more than two regimes.

Fig. 4.5 Schematic representation of a car following scenario

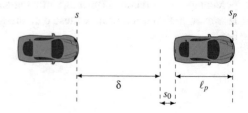

Table 4.1 Typical values of the car-following model parameters [11]

Parameter	Typical value
Time gap Δt	1 s
Minimum gap s_0	2 m
Acceleration exponent m	4
Acceleration a_{max}	1 m/s^2
Deceleration $-a_{min}$	1.5 m/s^2

4.2.1.1 Gipps' Model

Gipps' model [12] is the basis of car-following models implemented in commercial software such as AIMSUN and others [13]. It is of the form (4.3), where Δt is taken as the driver's reaction time (typically, 1 s), and the function F arbitrates between the free-flow and gap-controlled regimes,

$$v(t + \Delta t) = \min(v_{free}, v_{safe}) \tag{4.6}$$

with restrictions (4.5), where

$$v_{free} = v(t) + 2.5 a_{max} \Delta t \left(1 - \frac{v(t)}{v_d}\right) \sqrt{0.025 + \frac{v(t)}{v_d}} \tag{4.7}$$

is the free-drive speed, a heuristic function of the desired speed v_d, and the gap-controlled speed

$$v_{safe} = a_{min} \Delta t + \sqrt{a_{min}^2 \Delta t^2 - a_{min} \left[2\delta(t) - \frac{v_p^2(t)}{a_{min,p}} - v_p(t)\Delta t\right]} \tag{4.8}$$

is evaluated as the maximum speed to stop safely with respect to the preceding vehicle.

4.2.1.2 Krauss Model

The Krauss model [14], used in the software SUMO, is similar to the Gipps' model since it is of the form 4.3,

$$v(t + \Delta t) = \min(v_d, v_{safe}) - \eta(t) \tag{4.9}$$

with restrictions (4.5), where

$$v_{safe} = v_p(t) + \frac{\delta(t) - \delta_d(t)}{h_d + \dfrac{v(t) + v_p(t)}{2|a_{min}|}} \tag{4.10}$$

and η is a random perturbation. As proposed in [14], the desired gap is proportional to the leader's speed,

$$\delta_d(t) = h_d v_p(t) , \tag{4.11}$$

where h_d is the desired time headway.

4.2.1.3   Intelligent Driver Model

The time-continuous Intelligent Driver Model (IDM) is considered the simplest model producing realistic acceleration profiles. Unlike the Gipps' model, the IDM [15] combines the free-flow driving behavior and the gap-controlled behavior in one single equation,

$$\ddot{s}(t) = a_{max} \left[1 - \left(\frac{v(t)}{v_d} \right)^m - \left(\frac{\delta_d(t) + s_0}{\delta(t) + s_0} \right)^2 \right] , \tag{4.12}$$

where m is a tunable parameter (see Table 4.1) and the desired gap is

$$\delta_d(t) = h_d v(t) + \frac{v(t)(v(t) - v_p(t))}{2\sqrt{a_{max}|a_{min}|}} . \tag{4.13}$$

Improvements to this model aimed at yielding more realistic accelerations in some specific situations such as $v \geq v_d$ are discussed in [11].

4.2.1.4   Psycho-Physical Models

To this class of models belong the Wiedemann (74) model [16] used in VISSIM and the Fritzsche model [17] used in the software PARAMICS. In this framework, several regimes are defined as a function of the gap and the approaching rate. Wiedemann model defines four regimes, separated by variable thresholds: free driving, closing in (the driver perceives a slower leading vehicle), following, and emergency. The Fritzsche model has two different situation for the following and, consequently, five regimes.

4.2.2 Advanced Cruise Control Functions

Vehicles driven by an ACC can have a different behavior than human-driven vehicles [18] and thus are here modeled separately. Various automotive companies have already introduced such systems, while a large amount of research is still being conducted on some aspects such as the control law of the ACC system itself, the calibration of its parameters, the impact of the response time and its string stability properties. Additional issues concern safety implications, legal issues and technical restrictions, such as performance of ACC sensors in turning maneuvers, braking, hills, weather conditions etc. [19].

The ACC controller bypasses the usual driver-controlled torque structure and calculates the necessary acceleration, depending on the vehicle's net distance to the leading vehicle and the difference in the corresponding velocities (approaching rate).

Similarly to driving regimes discussed above, at least two control modes are present in common ACC systems [20]: (i) a speed-control mode, where the goal is to travel with a driver-specified speed v_d, if there are no leading vehicles within sensor range (or they exist but their velocities are higher than the set speed), and (ii) a gap-control mode, where the goal is to maintain the same speed of the leading vehicle at the desired gap δ_d. The transitions between the two aforementioned modes should be as smooth as possible, which can be difficult to obtain. For instance, the net distance may abruptly change during lane-changing or cut-in maneuvers.

In speed control mode, the vehicle behavior can be described as

$$\ddot{s}(t) = K_v \left(v_d - v(t) \right) , \qquad (4.14)$$

where K_v is a tunable parameter (see Table 4.2).

In gap-control mode, the acceleration is proportional to both the approaching rate and the gap error, i.e., the difference between desired and actual gap from the preceding vehicle. The desired gap is often defined as an affine function of speed, that is, $\delta_d(t) = h_d v(t)$, where h_d is the desired safe time headway. In summary,

$$\ddot{s}(t) = K_1 \left(v_p(t) - v(t) \right) + K_2 \left(\delta(t) - h_d v(t) \right) , \qquad (4.15)$$

where K_1 and K_2 are two tunable coefficient (see Table 4.2).

The ACC tuning parameters must guarantee both the individual vehicle stability, as well as the *string stability* [21, 22]. Individual vehicle stability is obtained as long

Table 4.2 Typical values of the ACC model parameters [20]

Parameter	Typical value
Speed coefficient K_v	0.4
Speed coefficient K_1	1.12
Gap coefficient K_2	1.7
Time headway h_d	1.5–2 s

as the gap error after a possible perturbation converges to zero, given that the leading vehicle travels at a constant speed. String stability concerns a platoon of vehicles and is obtained as long as all the spacing errors do not amplify as they travel upstream [23].

In the automatic control literature, proposals to implement ACC in terms of model predictive control (MPC) instead of the instantaneous feedback relations (4.14)–(4.15) have emerged [24–26]. This approach aims at minimizing a cost function that is composed of the cumulated approaching rate and gap error on a certain horizon. Adjoining energy-related terms to the cost function gives rise to the "eco-ACC" strategies that are discussed in Chap. 8.

4.2.3 Lane-Changing Models

Besides car-following models, a microscopic description of heterogeneous traffic flows must include multi-lane behavior, modeling lane changes and merges. In the literature these decisions are classified as either mandatory or discretionary. Mandatory changes are performed for strategic reasons, while the driver's motivation for discretionary lane changes is a perceived improvement of the driving conditions in the target lane compared with its actual situation.

Microscopic models often include only the operational stage of lane-changing decision, i. e., the choice if an immediate lane change is both safe and desirable. This choice is typically described by using *gap-acceptance models*, in which the available gaps are compared to the smallest acceptable gap (critical gap) and a lane change is executed if the available gaps are greater. These critical gaps vary not only among different individuals, but also for a given individual under different traffic conditions, for instance, with the relative speed to the lead and the following vehicles in the target lane and the type of lane change.

Several models have been proposed in the literature that are based on a probability distribution of critical gaps [27, 28]. Instead, we shortly present here a more recent lane-changing/lane-selection model based on the concept of utility.

4.2.3.1 MOBIL

In the "Minimizing Overall Braking Induced by Lane changes" model (MOBIL [29]) lane changes take place if the potential new target lane is more attractive (incentive criterion) and the change can be performed safely (safety criterion).

The model assumes that a driver makes a trade-off between the expected own advantage (utility) and the disadvantage imposed on other drivers. Intuitively, the own utility of a change increases with the gap to the new leader in the target lane, but also with its relative speed. If, for example, the velocity of the new leader is lower than that of the current leader, it may be favorable to stay in the present lane despite a smaller gap. Since in most car-following models (see Sect. 4.2.1 and Eq. (4.2))

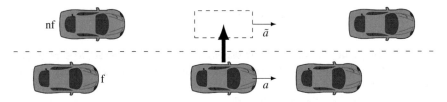

Fig. 4.6 Schematic representation of a lane change scenario

the acceleration increases with both the gap and the approaching rate, the utility is defined in MOBIL as the difference in the accelerations after and before the lane change, as calculated using a car-following model.

For symmetric lane usage rules,[2] the incentive criterion is

$$(\tilde{a} - a) + p \left(\tilde{a}_{nf} - a_{nf} + \tilde{a}_f - a_f \right) > \Delta a_{th} , \tag{4.16}$$

where accelerations after possible change are denoted with a tilde and the subscripts f and nf refer to the follower (lag) vehicle in the current or target lane, resp., see Fig. 4.6. The first term denotes the utility of a possible lane change for the driver. The *politeness factor* p weighs the total utility of the two immediately affected neighbors and varies between 0 (selfish behavior) to 1 (altruistic behavior). The threshold Δa_{th} models a certain inertia and prevents lane changes if the overall advantage is only marginal.

The safety criterion is

$$\tilde{a}_{nf} \geq a_{safe} , \tag{4.17}$$

where $a_{safe} < 0$ is a given safe limit. Table 4.3 summarizes some typical parameter values.

This utility-based method can be adapted to other discrete-choice decisions, such as deciding whether to cruise or stop at a yellow traffic light (setting a leader speed $v_p = 0 = v_d$), entering a priority road (similar to mandatory lane change) etc. [11].

Table 4.3 Typical values of the MOBIL parameters [11]

Parameter	Typical value
Changing threshold Δa_{th}	0.1 m/s^2
Safe deceleration limit a_{safe}	−2 m/s^2
Politeness factor p	0–1
Bias term for asymmetry	0.3 m/s^2

---

[2]Such as those valid in United States. In most European countries, the rightmost lane shall be preferred. This asymmetric situation is coped with in MOBIL by adding a bias term to the threshold Δa_{th}.

4.3   Macroscopic Modeling of Traffic Flows

Besides anticipating the microscopic driving behavior with respect to surrounding vehicles, predicting the average characteristics of traffic flows is also very relevant for energy efficiency optimization.

When modeling traffic, the main variables to consider are flow, density, and mean speed. The *traffic density* $\rho(s, t)$, typically measured in $[\text{km}^{-1}]$, is defined as the number of vehicles at one time instant per unit road length. It equals evidently the reciprocal of the mean headway (bumber-to-bumper distance) between two consecutive vehicles.

The *traffic flow* $Q(s, t)$, typically measured in $[\text{h}^{-1}]$, is defined as the number of vehicles crossing one road section per unit time. It equals the reciprocal of the mean time headway between two consecutive vehicles.

From these definitions, it becomes apparent that

$$Q(s, t) = \rho(s, t) V(s, t) , \qquad (4.18)$$

where $V(s, t)$ is the *mean traffic speed*. The latter quantity results from the speed of individual vehicles and can have various operational definitions. Usually, one considers N vehicles traveling different distances in the same time period, whence the time-mean speed[3] is evaluated as $V = \sum_j^N v_j/N$ (arithmetic average of individual speeds).

4.3.1   Fundamental Diagrams

Traffic is said to be stationary when mean speed does not change with time, and homogeneous when it does not change with the road section considered. Considering a stationary and homogeneous traffic flow, it is reasonable to assume that there exists some relationship between the density and speed or flow.

This fundamental relationship can be derived either by curve fitting to several empirical observations, or theoretically from microscopic models. Most car-following models introduced in Sect. 4.2.1 present a steady-state equilibrium when applied to a platoon of similar vehicles following each other along a one-dimensional pathway. Thus, they can be used to derive a theoretical flow-density relationship.

The microscopic equilibrium implies that $v(t + \Delta t) = v(t) = v_p(t) \equiv v_{eq}$ and $\delta(t) \equiv \delta_{eq}$. By setting these conditions in the Gipps' model, Eqs. (4.6)–(4.8), for instance, one obtains the *equilibrium speed*

---

[3]This definition should not be confused with the *space-mean speed* that considers N vehicles crossing the same segment with different travel times, and is evaluated as $N/\sum_j^N (1/v_j)$ (harmonic average of individual speeds).

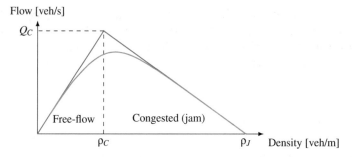

Fig. 4.7 Fundamental diagram of traffic flow from Gipps' microscopic model (blue) and IDM (orange). Critical-capacity point is labelled for the Gipps' model

$$v_{eq} = \min\left(v_d, \frac{2}{3} \cdot \frac{\delta_{eq}}{\Delta t}\right). \tag{4.19}$$

The equilibrium gap can be conveniently related to the traffic density. As observed above, the density is the reciprocal of the spacing between two consecutive vehicles, i.e.,

$$\rho_{eq} = \frac{1}{\delta_{eq} + s_0 + \ell}. \tag{4.20}$$

The equilibrium flow is evaluated from its definition as $Q_{eq} = \rho_{eq} v_{eq}$. From (4.19) and (4.20), it can be thus expressed as a function of the density as

$$Q_{eq} = \min\left(v_d \rho_{eq}, \frac{2}{3} \cdot \frac{1 - \rho_{eq}(\ell + s_0)}{\Delta t}\right). \tag{4.21}$$

The flow-density relationship is called *fundamental diagram* in the traffic engineering literature. Equation (4.21) reveals that the Gipps' model yields a triangular fundamental diagram, as shown in Fig. 4.7. The two car-following model regimes (free-drive, gap-controlled) describe the free-flow, resp., the congested traffic regime. Under the free-flow regime, the flow increases proportionally to density until the critical value $\rho_C \triangleq (\ell + s_0 + 3/2v_d \Delta t)^{-1}$ is reached. Then the flow decreases linearly with density under the congested regime, until the maximum congestion ("jam") density $\rho_J \triangleq (\ell + s_0)^{-1}$ is reached, at which speed becomes zero. The maximum flow (or capacity) is evaluated as $Q_C = v_d \rho_C$. A triangular fundamental diagram was first proposed by Daganzo based on empirical data [30].

Many other flow-density correlations have been proposed in the literature. For example, the fundamental diagram obtained with the IDM, equations (4.12)–(4.13), is smooth in the range between $\rho = 0$ and $\rho = \rho_J$, as shown in Fig. 4.7.

These theoretical fundamental diagrams generally do not fit real data particularly well. Refinements that have been introduced in the literature include "inverse-lambda"-shaped diagrams with a discontinuity around the capacity point, such as in Wu's model with capacity drop [31]. A substantially different approach is that of Kerner's three-phase theory, which considers three regimes: free flow, jam, and an intermediate synchronized flow, for which there is no flow-density line but a two-dimensional area [32].

4.3.2 Kinematic Models

In the *first-order* kinematic models, the relationship $Q_{eq}(\rho_{eq})$ obtained for equilibrium traffic conditions is generalized to all situations, including non-equilibrium ones, so that $Q(s, t) = Q_{eq}(\rho(s, t))$. This assumption was postulated in the '50s by Lighthill and Whitham and, independently, Richards, whence the name LWR also used for this class of models.

The fundamental diagram is coupled with the continuity equation that always holds,

$$\frac{\partial \rho(s, t)}{\partial t} + \frac{\partial Q(s, t)}{\partial s} = \phi(s, t) - \frac{Q(s, t)}{I(s)} \frac{dI(s)}{ds} , \qquad (4.22)$$

where ϕ is the net in- or out-flow density [$h^{-1}km^{-1}$], e.g., from on-, or to off-ramps, resp., and $I(s)$ is the number of lanes. The former quantity is defined as the vehicle flow (positive in the case of an on-ramp, negative in an off-ramp) divided by the length of the ramp and the number of lanes. The number of lanes is a non-integer,[4] location-dependent quantity that models situations such as lane merging and openings. The second right-hand term in (4.22) describes the net flow from ending lanes and to newly opening lanes.

Overall, the LWR models consist of a differential-algebraic system of two equations in the two unknowns Q and ρ. Such equations can be solved with numerical techniques once (i) the values of the unknowns at the boundaries of the geometrical region considered, (ii) their values at the start of the simulation, and (iii) the exogenous variables ϕ and I are specified. These first-order models allow a realistic description of simple phenomena such as shock wave propagation.

On the other hand, *second-order models* renounce the flow-density fundamental relation and introduce instead a momentum-conservation equation under the form of a differential equation in the variable $V(s, t)$. Consequently, these models take the form of a differential-algebraic system of three equations in the three variables Q, ρ, and V. Notable models of this type include Payne's [33]. Second-order models are the method of choice to macroscopically describe traffic waves and other complex phenomena.

---

[4]A value of $I = 2.2$, for example, means that in a fraction of the unit distance considered there are 2 lanes and in the remaining part there are 3 lanes.

4.4 Prediction of Energy Consumption on Road Networks

The goal of this section is to present methods to predicting energy consumption of various types of vehicles in road networks. This energy consumption prediction will be useful for implementing eco-routing (Chap. 5) and related strategies.

Energy cost of each link of the road network depends on the vehicle's velocity profile, road slope, vehicle parameters such as its coefficient of aerodynamic drag, rolling resistance, mass, and powertrain parameters. Models of specific or representative vehicles can be utilized to estimate the energy cost of each link [34]. Historically observed or real-time speed and traffic traces on each link can be used to estimate the expected velocity profile on a link [35]. Alternatively, sometimes link energy/emission costs are estimated based on regression fits to empirical road data as in the Comprehensive Modal Emission and Energy Model (CMEM) in [36], or relying on historically recorded consumption for each link as shown in [37]. In [38] a number of macroscopic and microscopic energy consumption and emission models are reviewed which can be used in eco routing.

Predicting the energy consumption of a vehicle on a given route can be achieved using the vehicle models of Chap. 2 and a prediction of the speed and altitude profiles followed by the vehicle. While altitude can be obtained from geographical information services (see Sect. 4.1), the estimation of speed profiles is particularly critical. The assumption that is usually made in routing is that the vehicle will follow the general characteristics of the traffic that it will find along its trajectory. In particular, its estimated speed and position are assumed to be

$$\hat{v}(t) = V(\hat{s}(t), t) , \quad \frac{d\hat{s}(t)}{dt} = \hat{v}(t) . \tag{4.23}$$

This approach would require that the local speed field $V(s, t)$ is known at any time and location. However, this information is generally not available, at least not completely.

On the one hand, real-time micro- or macroscopic simulations are not practical and have rarely been attempted [39]. On the other hand, the "operating speed" models, that are often used in the traffic engineering literature to design road infrastructure (Sect. 4.4.1), try to correlate the average speed on a segment V to geometrical and contextual characteristics of the segment and could be used for a first evaluation. It should be noted that these models are usually time-independent.

Similarly, commercial mapping web-services (see Sect. 3.2.3), generally provide aggregated traffic information in the form of an average speed V. Typical aggregation intervals are of the order of the road segment and, temporally, of several minutes.

In both latter cases, the provided average speed V is constant when seen by one vehicle on a road segment, and might vary only slowly with time as the general traffic conditions evolve. However, without any information on speed fluctuations around the average speed, contributions of speed moments to energy consumption, like those of speed variance and skewness in (2.11), cannot be evaluated. That could lead to underestimating energy consumptions especially in urban and/or suburban road networks. In fact, disruptions in the speed profiles and accelerations are caused not only by traffic, but also by the infrastructure. In particular, critical elements of the road infrastructure, such as traffic lights, intersections, and turning movements are very likely to induce stops or significant decelerations.

In order to take into account the effects of higher speed moments, in particular, infrastructure-induced accelerations and decelerations, and improve the energy consumption estimation, we describe next in Sect. 4.4.2 a simple method based on synthetic speed profiles.

4.4.1   Operating Speed Models

Operating speed models predict average speed on road segments as a function of fixed road attributes. The representative values are often the mean or, more frequently, the *85th percentile speed*. Usually free-flow speed only is considered, in order to suppress the influence of headway and reduce the number of exogenous variables.

These models can be expressed with correlations of the type

$$v_{FF,i} \quad \text{or} \quad V_i = f(R_i, \alpha_i, w_i, v_{lim}, n_l, \dots) , \tag{4.24}$$

where the subscript i refers to a road segment, R is its horizontal radius, α is road grade, w is road or lane width, v_{lim} is the posted speed limit, n_l is number of lanes. Additional dichotomic (0/1) parameters describe the presence of particular fetaures (school, parking, sidewalk, etc.). Such models have been derived, e.g., for rural road [40], for urban tangent roads with 30 km/h speed limits and other urban scenarios [41], and for suburban roads [42].

Two particular situations that have been studied in [43] are speed of heavy-duty vehicles (trucks) on turnings and on roundabouts. Statistical regression suggests that the lateral acceleration at turnings has a distribution centered around $0.15g$, such that the average cornering speed is

$$V_i \approx 0.39 v_{turn} = 0.39\sqrt{\mu g R_i} . \tag{4.25}$$

For roundabouts, speed at entrance, exit, and mean speed are crucial. The latter two have shown a strong correlation with the roundabout radius. For example, the average speed on the whole segment can be expressed as

$$V_i \approx c_1 R_i + c_0 , \tag{4.26}$$

where the values found are $c_1 = 0.43$ km/h/m and $c_0 = 16.3$ km/h. In contrast, the entrance speed is more severely affected by traffic conditions (for instance it can be zero) and thus less correlated to the road characteristics.

4.4.2 Synthetic Speed Profiles

For each road segment of the network, we assume that it is possible to know the link length ℓ_i, a prevailing average traffic speed V_i, and the road grade α_i which might vary within the considered link depending on the position.

Considering a road segment i, the speed profile on it is supposed to be composed of two phases: a transition phase to go from V_{i-1}, the cruising speed on the preceding segment, to V_i, and a cruising phase at constant speed V_i. Let us first introduce a transition speed at the interface between two segments defined as

$$v_{t,i} = \zeta_i \frac{V_i + V_{i-1}}{2} \, , \tag{4.27}$$

where $\zeta_i \in [0, 1]$ is a parameter depending on the type of interface (e.g. stop sign, traffic light, turning movement, etc.), which could be selected in a deterministic or stochastic fashion.

The speed change between two road segments is modeled as two distinct transients: a first transient from V_{i-1} to $v_{t,i}$, and a second transient from $v_{t,i}$ to V_i, both at constant acceleration/deceleration a_t (a model parameter), as shown on the right-hand side of Fig. 4.8. By taking $\tau = 0$ at the beginning of the transition, the predicted speed on the road segment i can be thus written as[5]

$$v_i(\tau) = \begin{cases} V_{i-1} + \text{sign}(v_{t,i} - V_{i-1}) \cdot a_t \tau, & \tau \in [0, \Delta \tau_{i-1,t}] \\ v_{t,i} + \text{sign}(V_i - v_{t,i}) \cdot a_t (\tau - \Delta \tau_{i-1,t}), & \tau \in (\Delta \tau_{i-1,t}, \Delta \tau_i] \, , \\ V_i, & \tau \in (\Delta \tau_i, \tau_i] \end{cases} \tag{4.28}$$

where the transient times are

$$\Delta \tau_{i-1,t} = \frac{|v_{t,i} - V_{i-1}|}{a_t}, \quad \Delta \tau_{t,i} = \frac{|V_i - v_{t,i}|}{a_t} \, . \tag{4.29}$$

Note that this synthetic speed model does not depict stop time at traffic lights or intersections, which anyway does not contribute to energy consumption.

By imposing that

$$\int_0^{\tau_i} v_i(\tau) d\tau = \ell_i \, , \tag{4.30}$$

the travel time of the road segment can be evaluated. In the common case where $v_t \leq V_{i-1}$, $v_t \leq V_i$, the result is

$$\tau_i = \frac{\ell_i}{V_i} + \frac{-V_{i-1}^2 + 2V_{i-1}V_i + V_i^2 - 4V_i v_t + 2v_t^2}{2a_t V_i} \tag{4.31}$$

---

[5]From here on we omit the hat symbol of the predicted speed for simplicity.

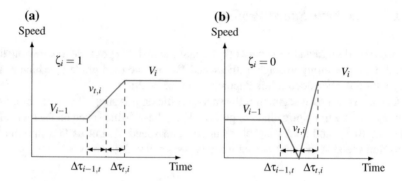

Fig. 4.8 Interface accelerations: on the left-hand side a standard link transition, on the right-hand side a link interface with a stop sign

and, consequently, the mean speed $\bar{v}_i \triangleq \ell_i/\tau_i$ is generally smaller than V_i.

The speed moments can be also evaluated from the synthetic speed profile for each road segment. For instance, we obtain

$$\int_0^{\tau_i} v_i^2(\tau)d\tau = \ell_i V_i + \frac{2V_{i-1}^3 - 3V_{i-1}^2 V_i - V_i^3 + 6V_i v_t^2 - 4v_t^3}{6a_t}, \qquad (4.32)$$

$$\int_0^{\tau_i} v_i^3(\tau)d\tau = \ell_i V_i^2 + \frac{V_{i-1}^4 - 2V_{i-1}^2 V_i^2 - V_i^4 + 4V_i^2 v_t^2 - 2v_t^4}{4a_t}, \qquad (4.33)$$

and consequently the moments $\sigma_{v,i}^2$, $b_{v,i}$ defined in Sect. 2.1.2.

4.4.3   Energy Consumption for Traction

After having predicted the synthetic speed profile $v_i(\tau)$ on a link, the energy consumption can be evaluated using the models presented in the previous chapters.[6] In particular, the energy consumption at the wheels can be evaluated with Eq. (2.11), that we rewrite here for the link i as

$$E_{W,i} = \frac{1}{2}m\left(V_i^2 - V_{i-1}^2\right) +$$

$$+ \left(mg\alpha_i + C_0 + C_1\bar{v}_i + C_2\bar{v}_i^2 + C_1\frac{\sigma_{v,i}^2}{\bar{v}_i} + 3C_2\sigma_{v,i}^2 + \frac{C_2 b_{v,i}\sigma_{v,i}^2}{\bar{v}_i}\right)\ell_i .$$

$$(4.34)$$

---

[6]A validation of this approach will be presented in the case study of Sect. 9.5.

Analogously, (2.29) and (2.43) hold when tank energy is considered, for ICEVs and EVs, respectively. For ICEVs and EVs, braking and traction phases over the link are identified from the sign of the wheel power $F_{w,i}(v_i(\tau))v_i(\tau)$. For ICEVs, gear engaged and thus gear ratio $\gamma_{e,i}(\tau)$ can be identified by coupling the synthetic speed profile with the gear shift law (2.17) of the particular transmission.

Although this procedure is straightforward and can be very accurate, the evaluation of instantaneous powers over each road link can be time consuming. Therefore, for the fast prediction of energy consumption over a large number of road links that is required by several functions (eco-routing, range estimation, etc.), a further simplified method is often necessary.

This method considers average speeds during the three periods of the synthetic speed profile (4.28), namely,

$$v_{i1} \triangleq (V_{i-1} + v_{t,i})/2 \,, \quad v_{i2} \triangleq (v_{t,i} + V_i)/2 \,, \quad v_{i3} \triangleq V_i \,, \tag{4.35}$$

and accordingly evaluates three constant levels of wheel power

$$P_{w,ij} = C_0 v_{in} + C_1 v_{ij}^2 + C_2 v_{ij}^3 + mg\alpha_i + ma_{ij}, \quad j = 1, \ldots, 3 \,. \tag{4.36}$$

We also define the three corresponding demand power levels $P_{d,ij} \triangleq P_{w,ij}\eta_t^{-\mathrm{sign}(P_{w,ij})}$.
For ICEVs, the corresponding fuel power levels are

$$P_{f,ij} = \begin{cases} a_{0,ij} + a_{1,ij}P_{d,ij}, & \text{if } P_{d,ij} > 0 \\ 0, & \text{otherwise} \end{cases} , \tag{4.37}$$

where $a_{0,ij} \triangleq k_{e,0}\omega_{e,ij} + k_{e,1}\omega_{e,ij}^2, a_{1,ij} \triangleq k_{e,2} + k_{e,3}\omega_{e,ij} + k_{e,4}\omega_{e,ij}^2$, and the resulting fuel energy consumption is

$$E_{f,i}^{(ICEV)} = \sum_{j=1}^{3} P_{f,ij}\tau_{ij} \,. \tag{4.38}$$

For EVs, the constant battery power levels are

$$P_{b,ij} = \begin{cases} \frac{1}{\eta_b}\left(b_{0,ij} + b_{1,ij}P_{d,ij} + b_{2,ij}P_{d,ij}^2\right), & \text{if } P_{d,ij} > 0 \\ \eta_b\left(b_{0,ij} + b_{1,ij}P_{d,ij} + b_{2,ij}P_{d,ij}^2\right), & \text{otherwise} \end{cases} , \tag{4.39}$$

where

$$b_{0,ij} \triangleq k_{m,0} + k_{m,1}\omega_{m,ij} + k_{m,2}\omega_{m,ij}^2, \quad b_{1,ij} \triangleq k_{m,3},$$

$$b_{2,ij} \triangleq \frac{k_{m,4}}{\omega_{m,ij}^2} + \frac{k_{m,5}}{\omega_{m,ij}} + k_{m,6} \tag{4.40}$$

and

$$E_{b,i}^{(EV)} = \sum_{j=1}^{3} P_{b,ij} \tau_{ij} , \tag{4.41}$$

respectively. Both the engine and the motor speed levels are evaluated from the corresponding v_{ij} and $\gamma_{e,ij}$. The times τ_{ij} are easily derived from (4.28)–(4.31).

For HEVs, Eq. (2.43) could be similarly used. However, instead of an overall energy consumption, one is more often interested in evaluating the *minimal* fuel consumption $E_{f,i}$ that corresponds to a *given* electric consumption $E_{b,i}$ or SoC variation. The optimal energy management strategy (see Sect. 2.4.2) can be defined by assuming that each period of the synthetic speed profile is characterized by one value of the torque or power split ratio u_{ij}. Consequently, for parallel HEVs,

$$P_{f,ij}(u_{ij}) = \begin{cases} a_{0,ij} + a_{1,ij} u_{ij} P_{d,ij}, & \text{if } u_{ij} P_{d,ij} > 0, \\ 0 & \text{otherwise} \end{cases} , \tag{4.42}$$

$$P_{b,ij}(u_{ij}) = \begin{cases} \frac{1}{\eta_b} \left(b_{0,ij} + b_{1,ij} \bar{u}_{ij} P_{d,ij} + b_{2,ij} \bar{u}_{ij}^2 P_{d,ij}^2 \right), & \text{if } \bar{u}_{ij} P_{d,ij} > 0 \\ \eta_b \left(b_{0,ij} + b_{1,ij} \bar{u}_{ij} P_{d,ij} + b_{2,ij} \bar{u}_{ij}^2 P_{d,ij}^2 \right), & \text{otherwise} \end{cases} , \tag{4.43}$$

where $\bar{u}_{ij} \triangleq (1 - u_{ij})$.

The values u_{ij} can be found with numerical methods as the solution of the quadratically constrained linear program

$$\min_{u_i = \{u_{i1}, u_{i2}, u_{i3}\}} \sum_{j=1}^{3} P_{f,ij}(u_i) \tau_{ij} , \tag{4.44}$$

$$\text{s.t. } \sum_{j=1}^{3} P_{b,ij}(u_i) \tau_{ij} = E_{b,i} . \tag{4.45}$$

Finally, the optimal fuel consumption associated with the considered synthetic speed profile is

$$E_{f,i}^{(HEV)}(E_{b,i}) = \sum_{j=1}^{3} P_{f,ij}(u_{ij}) \tau_{ij} . \tag{4.46}$$

A more detailed procedure, including variable power levels and a careful enforcing of power limits, is described in [44].

The battery energy consumptions $E_{b,i}$ or SoC variations per road segments are free parameters in HEV and are usually determined by a further optimization of the energy management strategy over the whole trip, see Sect. 6.6. A rough prediction, that is only dependent on the road characteristics and the speed profile assumed, is possible using the method of [44]. This model states that the optimal battery energy

Fig. 4.9 Typical map of the auxiliary power absorption of an electric car as a function of ambient temperature

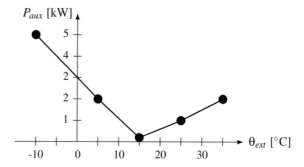

consumption is a function of the predicted kinetic and potential energy to travel on the road segment, and of the vehicle parameters, and is defined as

$$E_{b,i} \approx \rho \left(\frac{1}{2} m (V_i^2 - V_{i-1}^2) + mg(z_i - z_{i-1}) \right) , \qquad (4.47)$$

where z is the altitude and ρ is a tuning parameter. This method also provides an a-priori estimation of the bounds of achievable ΔSoC for each road segment, which can be useful for a more precise optimization (Sect. 5).

4.4.4 Energy Consumption for Thermal Comfort

To improve the estimation of energy consumption in real-life driving, it is important to take into account also the auxiliary power requirements. In particular for electric vehicles, power requirement for cabin heating, ventilating, and air conditioning has a major impact on the driving range. Such an impact increases with the trip duration and the ambient temperature difference with respect to the desired cabin temperature.

To obtain a desired thermal comfort level, a power P_{aux} must be ultimately drained from the onboard source (fuel for ICEVs, battery for EVs and HEVs), accounting for the various efficiencies in the conversion process from electric or fuel energy into useful thermal energy. For our purposes, this auxiliary power can often be simply modeled as a function of the ambient temperature θ_{ext} and considered constant over the trip. Consequently, the auxiliary energy consumption over the link i is evaluated as

$$\Delta E_{T,i} = \int_0^{\tau_i} P_{aux} dt = \int_0^{\tau_i} f(\theta_{ext}) dt = f(\theta_{ext})\tau_i . \qquad (4.48)$$

Typical values of the function $f(\cdot)$ for an electric city car are shown in Fig. 4.9 [45]. These data suggest a parametrization of the type

$$\Delta E_{T,i} = K_{th} |\theta_{ext} - \theta_0| \tau_i , \qquad (4.49)$$

where θ_0 is a reference temperature around 15 °C and the coefficient K_{th} takes different constant values for heating ($\theta_{ext} < \theta_0$) and air conditioning ($\theta_{ext} > \theta_0$).

The introduction of thermal storage systems, based on various technologies such as Sensible Heat Storage (SHS) and Phase-Change Materials (PCM) offers an additional degree of freedom. In the context of CAVs, anticipating the thermal flows released by the powertrain components as a function of the vehicle usage (*predictive thermal management*), will enable the optimization of storage system charging/discharging and allow substantial reductions of P_{aux} [46], see also Sect. 3.3.5.

References

1. Google. Google maps platform (2018). https://cloud.google.com/maps-platform/
2. Bing Maps (2019). https://www.bing.com/maps
3. Here We Go (2019). https://wego.here.com
4. MapQuest (2019). https://www.mapquest.com/
5. OpenStreetMap (2019) https://www.openstreetmap.org
6. Koonce P, Rodegerdts L (2008) Traffic signal timing manual. Technical report, United States. Federal Highway Administration
7. Musavi F, Eberle W (2014) Overview of wireless power transfer technologies for electric vehicle battery charging. IET Power Electron 7(1):60–66
8. Parvini Y, Vahidi A (2015) Maximizing charging efficiency of lithium-ion and lead-acid batteries using optimal control theory. In: American control conference (ACC), 2015, pp 317–322. IEEE
9. Montoya A, Guéret C, Mendoza JE, Villegas JG (2017) The electric vehicle routing problem with nonlinear charging function. Transp Res Part B: Methodol 103:87–110
10. Varaiya P (2013) The max-pressure controller for arbitrary networks of signalized intersections. In: Advances in dynamic network modeling in complex transportation systems, pp 27–66. Springer
11. Treiber M, Kesting A (2014) Traffic flow dynamics: data, models and simulation. Phys Today 67(3):54
12. Gipps PG (1981) A behavioural car-following model for computer simulation. Transp Res Part B: Methodol 15(2):105–111
13. Ciuffo B, Punzo V, Montanino M (2012) Thirty years of gipps' car-following model: applications, developments, and new features. Transp Res Rec: J Transp Res Board 2315(1):89–99
14. Krauß S (1998) Microscopic modeling of traffic flow: Investigation of collision free vehicle dynamics. PhD thesis, Universitat zu Koln
15. Kesting A, Treiber M, Helbing D (2010) Enhanced intelligent driver model to access the impact of driving strategies on traffic capacity. Philos Trans R Soc Lond A: Math Phys Eng Sci 368(1928):4585–4605
16. Wiedemann R (1974) Simulation des Strassenverkehrsflusses. PhD thesis, University of Karlsruhe
17. Fritzsche H-T (1994) A model for traffic simulation. Traffic Eng+Control 35(5):317–321
18. Kesting A, Treiber M, Schönhof M, Helbing D (2008) Adaptive cruise control design for active congestion avoidance. Transp Res Part C: Emerg Technol 16(6):668–683
19. Gurulingesh R (2004) Adaptive cruise control. Indian Institute of Technology Bombay
20. Shladover S, Dongyan S, Xiao-Yun L (2012) Impacts of cooperative adaptive cruise control on freeway traffic flow. Transp Res Rec: J Transp Res Board 2324(1):63–70
21. Naus GJL, Vugts RPA, Ploeg J, van de Molengraft MJG, Steinbuch M (2010) String-stable CACC design and experimental validation: a frequency-domain approach. IEEE Trans Veh Technol 59(9):4268–4279

22. Oncu S, Van de Wouw N, Heemels WMH, Nijmeijer H (2012) String stability of interconnected vehicles under communication constraints. In: Proceedings of conference on decision and control (CDC), pp 2459–2464. IEEE
23. Swaroop D, Hedrick JK (1996) String stability of interconnected systems. IEEE Trans Autom Control 41(3):349–357
24. Luo L, Liu H, Li P, Wang H (2010) Model predictive control for adaptive cruise control with multi-objectives: comfort, fuel-economy, safety and car-following. J Zhejiang Univ Sci A 11(3):191–201
25. Stanger T, del Re L (2013) A model predictive cooperative adaptive cruise control approach. In: Proceedings of American control conference (ACC), pp 1374–1379. IEEE
26. Vajedi M, Azad NL (2016) Ecological adaptive cruise controller for plug-in hybrid electric vehicles using nonlinear model predictive control. IEEE Trans Intell Transp Syst 17(1):113–122
27. Gipps PG (1986) A model for the structure of lane-changing decisions. Transp Res Part B: Methodol 20(5):403–414
28. Ahmed KI (1999) Modeling drivers' acceleration and lane changing behavior. PhD thesis, Massachusetts Institute of Technology
29. Kesting A, Treiber M, Helbing D (2007) General lane-changing model mobil for car-following models. Transp Res Rec: J Transp Res Board 1999(1):86–94
30. Daganzo CF (1994) The cell transmission model: a dynamic representation of highway traffic consistent with the hydrodynamic theory. Transp Res Part B: Methodol 28(4):269–287
31. Ning W (2002) A new approach for modeling of fundamental diagrams. Transp Res Part A: Policy Pract 36(10):867–884
32. Kerner BS, Konhäuser P (1994) Structure and parameters of clusters in traffic flow. Phys Rev E 50(1):54
33. Payne HJ (1971) Models of freeway traffic and control. Mathematical models of public systems. Simulation councils
34. Jurik T, Cela A, Hamouche R, Natowicz R, Reama A, Niculescu S-I, Julien J (2014) Energy optimal real-time navigation system. IEEE Intell Transp Syst Mag 6(3):66–79
35. Boriboonsomsin K, Barth MJ, Zhu W, Vu A (2012) Eco-routing navigation system based on multisource historical and real-time traffic information. IEEE Trans Intell Transp Syst 13(4):1694–1704. ISSN 1524-9050. https://doi.org/10.1109/TITS.2012.2204051
36. Barth M, Boriboonsomsin K, Vu A (2007) Environmentally-friendly navigation. In: Proceedings of intelligent transportation systems conference (ITSC), pp 684–689. IEEE
37. Andersen O, Jensen CS, Torp K, Yang B (2013) Ecotour: reducing the environmental footprint of vehicles using eco-routes. In: Proceedings of international conference on mobile data management (MDM), vol 1, pp 338–340. IEEE
38. Demir E, Bektaş T, Laporte G (2014) A review of recent research on green road freight transportation. Eur J Oper Res 237(3):775–793
39. Kamal MAS, Imura J, Hayakawa T, Ohata A, Aihara K (2014) Smart driving of a vehicle using model predictive control for improving traffic flow. IEEE Trans Intell Transp Syst 15(2):878–888
40. Bysveen M (2017) Vehicle speed prediction models for consideration of energy demand within road design. Master's thesis, NTNU
41. Dinh DD, Kubota H (2013) Profile-speed data-based models to estimate operating speeds for urban residential streets with a 30 km/h speed limit. IATSS Res 36(2):115–122
42. Fitzpatrick K, Shamburger CB, Krammes RA, Fambro DB (1997) Operating speed on suburban arterial curves. Transp Res Rec 1579(1):89–96
43. Ojeda LL, Chasse A, Goussault R (2017) Fuel consumption prediction for heavy-duty vehicles using digital maps. In: Proceedings of international conference on intelligent transportation systems (ITSC), pp 1–7. IEEE
44. De Nunzio G, Sciarretta A, Gharbia IB, Ojeda LL (2018) A constrained eco-routing strategy for hybrid electric vehicles based on semi-analytical energy management. In: Proceedings of international conference on intelligent transportation systems (ITSC), pp 355–361. IEEE

45. Sciarretta A, di Domenico D, Pognant-Gros P, Zito G (2014) Optimal energy management of automotive battery systems including thermal dynamics and aging. In: Optimization and optimal control in automotive systems, pp 219–236. Springer
46. De Nunzio G, Sciarretta A, Steiner A, Mladek A (2018) Thermal management optimization of a heat-pump-based hvac system for cabin conditioning in electric vehicles. In: Proceedings of international conference on ecological vehicles and renewable energies (EVER), pp 1–7. IEEE

Chapter 5
Energy-Efficient Route Navigation (Eco-Routing)

Eco-routing methods are the strategies and tools aimed at minimizing a vehicle's energy consumption by route selection. Given some origin and destination, which are typically chosen by the driver or user, eco-routing plans an energy-minimal route.

The set of possible routes from origin to destination constitutes a graph, where nodes and links represent junctions and roads. In the next Sect. 5.1 we will define a weighting function, which associates each link of the graph with a weight. In conventional routing graphs, the weight associated with each arc is either the length of the arc or its travel time. In the eco-routing framework, each link of the graph is assigned a weight that represents the travel energy expenditure. Differently from length or travel time weights, energy weights can be negative when EVs or HEVs with regenerative braking are concerned. Then in Sect. 5.2 we present how the eco-routing algorithms can be used to predict the maximal driving range of a vehicle. Finally, Sect. 5.3 will discuss some practical implementation issues.

5.1 Eco-Routing as a Shortest-Path Problem

The energy-minimal navigation problem introduced above is treated by formulating an *eco-routing shortest-path problem* (ER-SPP) in Sect. 5.1.1, while Sect. 5.1.2 presents various techniques to solve such a problem.

5.1.1 Problem Formulation

The eco-routing as a shortest path problem can be stated in the following way. For a given directed graph $G = (V, A)$, where V is the set of nodes n_i, and A is the set of

© Springer Nature Switzerland AG 2020
A. Sciarretta and A. Vahidi, *Energy-Efficient Driving of Road Vehicles*,
Lecture Notes in Intelligent Transportation and Infrastructure,
https://doi.org/10.1007/978-3-030-24127-8_5

links (or edges) e_k connecting these nodes, find the path $\mathbf{p} = (e_1, e_2, \ldots) \subset \mathcal{P}$, where $\mathcal{P}$ is the set of all simple[1] paths in $\mathcal{G}$, such as to minimize the objective function

$$J(\mathbf{p}) \triangleq \sum_{e_k \in \mathbf{p}} w_k(t, b_k(t)) , \tag{5.1}$$

where w_k is the weight attributed to the link k, possibly as a function of time and of additional *decision variables* b_k. Minimization of (5.1) is subject to (i) initial and terminal conditions (*single-source* shortest path)

$$n_1 = n_O, \quad n_{(|\mathbf{p}|)} = n_D , \tag{5.2}$$

where n_O and n_D are the origin and the destination sought, respectively, and $| \cdot |$ denotes cardinality, (ii) first-order dynamic constraints on the *state vector* x_k for each node n_k belonging to $\mathbf{p}$,

$$x_{k+1} = x_k + f_k(t, w_k(t, b_k(t)), b_k(t)) \quad x_1 = x(n_O) , \tag{5.3}$$

(iii) algebraic constraints on the state,

$$g_i(x_k) \leq 0, \quad i = 1, \ldots, \ell , \tag{5.4}$$

terminal inequality constraints over (iv) the state,

$$h(x(n_D)) \leq 0 , \tag{5.5}$$

and (v) possibly, over a certain *resource*,

$$R(\mathbf{p}) \triangleq \sum_{e_k \in \mathbf{p}} r_k(t, b_k(t)) \leq R_f , \tag{5.6}$$

where r_k is the resource consumption over link k.

How the quantities V, A, w_k, b_k, x_k, f, g, h, and r_k, are particularized for our ER-SPP will be discussed in the following sections. Although costs, state variations, and resource consumptions might change with time, we will consider only the *time-independent* SPP. Similarly, although the situation in the road network is continually evolving and predicting its state in future is arguably difficult, we shall not consider stochastic SPP.

5.1.1.1   Graph

The most intuitive choice to set $\mathcal{G}$ for a given road network is such that nodes represent road intersections and links represent roads. However, the use of this *primal*

---

[1] A simple path is a path without cycles: its edges are distinct.

graph presents some major difficulties. In particular, consider the case of an intersection with two or more incoming roads and one outgoing road i. The weight of the latter clearly depends on which upcoming link the vehicle comes from because each movement to enter i (e.g., left-turn, right-turn, through) might be accompanied by a different speed transition. In fact, the synthetic profile $v_i(\tau)$ introduced in Sect. 4.4 is not unique since it depends on the upstream average speed V_{i-1} that is not unique in this case.

This difficulty is resolved by modeling the road network as an *adjoint graph*, which is defined as follows. The adjoint graph $G = (A, A^*)$ of a directed graph $G' = (V, A)$ has a node for each link in G', and a link connecting two nodes if the corresponding two links in G' share a common node.

In other words, each link $i \in A$ of G' becomes a node of G, and each link $k \in A^*$ of G encompasses two generic links $i - 1, i \in A$ of G'. Thus, the adjoint graph allows to correctly assign unique weights to all the possible maneuvers in the original graph G. Figure 5.1 illustrates this concept. Useful properties of adjoint graphs allow to compute the number of links of G, and therefore its size, based on the connectivity

(a)

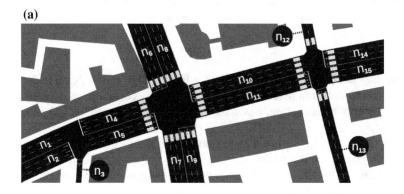

(b)

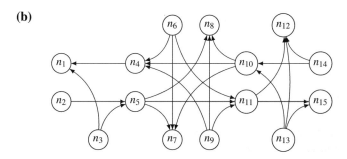

Fig. 5.1 Relation between a road network (**a**) and the corresponding road network graph G (**b**) [2]. The nodes are shown as circles, the edges as arrows. There are fifteen nodes n_1 to n_{15} representing the roads in the road network. The edges (arrows) spanning from nodes indicate to which nodes (roads) can the vehicle turn to at the downstream intersection

properties of $\mathcal{G}'$ [1]. Note that the origin and destination nodes of (5.2) represent in $\mathcal{G}$ the initial and target links (i_O, i_D) of the primal graph.

In this context, battery recharge stations (of either type, swapping, full/partial refill, continuous wireless charging, etc.) can be attributed to some nodes of the primal graph, thus to a subset $A_c^* \subset A^*$ in the adjoint graph.

5.1.1.2   Objective Function

In standard navigation systems, the objective function $J(\mathbf{p})$ is chosen such as to represent the trip length or the travel time. In eco-routing, the focus is on energy saving and therefore the weight assigned to each link of the graph represents the associated (tank) energy consumption E_T.

Using the synthetic energy consumption defined on the links of the primal graph $\mathcal{G}'$, we can now attribute energy weights to each link $e_k \in A^*$. Reminding that the link k of the adjoint graph encompasses the road segments $i - 1$ and i of the primal road network, the most natural choice is to attribute to e_k the energy consumption of the downstream road segment (i) that already accounts for the transition from road segment $i - 1$, see Sect. 4.4.3. In other words, we set

$$w_k = E_{T,i} \, , \tag{5.7}$$

where $E_{T,i}$ can be evaluated with (4.38) for ICEVs, (4.41) for EVs, and with (4.46) for HEVs. Note that in the latter case, the fuel energy consumption depends on the electric energy consumption, which thus plays the role of a decision variable b_k in (5.1).

The shortest path (SP) and the fastest path (FP) routing can be defined analogously such that the SP routing minimizes the distance and the FP routing minimizes the travel time. In these cases, the link weights w_k are represented by the length of the outgoing road segment or its travel time, respectively,

$$w_k^{(SP)} = \ell_i \, , \quad w_k^{(FP)} = \tau_i \, . \tag{5.8}$$

This travel time can be evaluated with (4.31).

5.1.1.3   Decision Variables

Decision variables b add degrees of freedom to the eco-routing problem besides the choice of the link sequence in the path.

In HEVs, the additional decision variables are the electricity consumption on links k,

$$b_k^{(HEV)} = E_{b,i} \, , \tag{5.9}$$

whereas the cost is represented by the minimal fuel consumption. Note that the b_k's can be negative in this case.

When recharge facilities for electrified vehicles (a scenario denoted here as xEVR) are taken into account, that is, the path is not supposed to be covered with one single battery recharge, the additional decision variable is the quantity of electricity refilled at links where a recharge facility is available,

$$b_k^{(xEVR)} = \Delta E_{c,i}, \quad e_k \in A_c^* . \tag{5.10}$$

In both aforementioned cases, the kth graph weight w_k can take on infinitely-many values depending on the decision variable over the graph link. The problem of finding the optimal b_k for each link of the graph while searching the optimal route from an origin to a destination has been addressed in the literature [3]. However, the proposed solution strategies are often impractical due to large computational cost. A practical approach therefore consists of augmenting the adjoint graph $\mathcal{G} = (A, A^*)$ by creating as many *copies* of each link of the graph as the number of pre-defined possible values of b_k. Using the discretized values $b_k^{(j)} \in [b_{k,min}, b_{k,max}], j = 1, \ldots, N_b$, let us define a b-augmented graph $\mathcal{G}_B = (A, A_B^*)$ with link copies $k^{(j)} \in A_B^*$, and a new cost function

$$w_k^{(j)} \triangleq w_k : b_k = b_k^{(j)}, \quad j = 1, \ldots, N_b , \tag{5.11}$$

that represents the cost to perform the maneuver $k^{(j)} \in A_B^*$ for a given decision variable $b_k^{(j)}$. The number of copies, or decision variable levels, N_b is a design parameter, in a trade-off between discretization accuracy and computational burden. Furthermore, the variation range of $b_k^{(j)}$ depends on the physical properties of the maneuver $k \in A^*$.

Another possibility to treat the problem of eco-routing with additional decision variables is to make copies of the links w.r.t. discretized values of the state (see below) at the end of the link [3, 4]. However, augmenting the graph in terms of decision variables offers a much higher precision as compared to these alternative graph expansions, while the best choice in terms of algorithmic complexity is a matter of debate.

5.1.1.4 State Dynamics and Constraints

In both scenarios where states are relevant, that is, HEVs and xEVs with battery recharge, the role of state variable is played by the electrical energy stored in the battery, $x = \{\varepsilon_b\}$. Given (5.7) and (5.9)–(5.10), the state dynamics (5.3) reads

$$f_k^{(HEV)} = -E_{b,i} = -b_k \tag{5.12}$$

for HEVs and

$$f_k^{(xEVR)} = -E_{b,i} = -w_k + b_k \tag{5.13}$$

for xEVs with battery recharge, with $x(n_O) = \varepsilon_{b,0}$ in both cases. These dynamics are further subject to the physical limits of the battery, that is, the local constraints (5.4) that are to be applied at each link and for partial paths,

$$0 \leq \varepsilon_b \leq \varepsilon_{b,max} , \tag{5.14}$$

with $\varepsilon_{b,max} \triangleq Q_b V_{b0}$ being the battery maximum capacity, while it has been assumed for simplicity that the battery can be completely depleted.[2]

Depending on the type of HEV, the battery energy at the end of the trip should match a prescribed value. For charge-sustaining HEVs, the final SoC should match the initial value, thus $\varepsilon_b(n_D) - \varepsilon_{b,0} = 0$ in (5.5). For plug-in HEVs, if a recharge is possible at the destination, the battery should be depleted. Thus the desired value for $\varepsilon_b(n_D)$ is usually close to zero, that is, a battery fully discharged.

5.1.1.5   Resource Constraints

Although energy efficiency is the main objective in ER-SPP, neglecting travel time might reduce the appeal of the selected route and the drivers' compliance to the routing suggestion. Hence, it is important to optimize energy consumption while providing the opportunity to the drivers to define their preferred trade-off with travel time. This trade-off can be achieved by selecting the resource consumption r_k in (5.6) as to represent the travel time of the link k,

$$r_k = \tau_i , \tag{5.15}$$

where τ_i is the travel time of the downstream link of the primal graph. The term R_f in (5.6) thus plays the role of the maximum time allowed for the route.

When battery recharge is taken into account, the travel time must include the waiting time at stations and the recharging time. Therefore, the resource consumptions become a function of the electricity recharged, thus of the decision variable (5.10),

$$r_k^{(xEVR)} = \tau_i + \Delta\tau_{c,i}(\Delta E_{c,i}, \varepsilon_{b,k}) . \tag{5.16}$$

For the charging function $\Delta\tau_{c,i}(\Delta E_{c,i}, \varepsilon_{b,k})$, the models introduced in Sect. 4.1.4 can be used.

*
* *

The presence of terminal state and resource constraints makes the eco-routing problem a resource-constrained SPP (RCSPP), which is known to be NP-hard. In order to overcome this complexity, which makes the problem impractical for end-

---

[2]In practice, that is not true, and a practical SoC window must be considered instead.

user applications and driving assistance, a more tractable problem formulation can be introduced under the form of a *Multi-Objective Optimization* Problem (MOOP), where resources and state constraints are transformed into additional objective functions to be minimized alongside with $J(\mathbf{p})$.

Solving a MOOP requires an external decision making process to give relative preference to the various objectives. In a-posteriori methods, preference is given, and a decision is made, once a representative set of optimal solutions is found. In contrast, a-priori methods require that preference information is available before searching for optimal solutions. A standard a-priori method is scalarization, that is, combining the various objectives into a single objective function. Several approaches to scalarization exist, depending on the form of the single objective function. In particular, the standard approach of the weighted-sum scalarization [5] casts the MOOP as a single-objective problem by appending the terminal constraints as additional terms of the objective function,

$$(1 - \lambda_1 - \lambda_2) \sum_{e_k \in \mathbf{p}} w_k + \lambda_1 \sum_{e_k \in \mathbf{p}} r_k + \lambda_2 \sum_{e_k \in \mathbf{p}} f_k , \qquad (5.17)$$

where the first term of (5.17) is the original objective function, the second is the terminal value of the resource that is constrained by (5.6), and the third is the terminal value of the state constrained by (5.5), where for simplicity a hard constraint has been assumed. The optimization weights $\lambda \in [0, 1]$ define the trade-off between the objectives, and should be sought such that the terminal constraints for the optimal path $\mathbf{p}$ are met, see below Sect. 5.1.2.3.

The presence of local state constraints (5.4) could be treated by making several copies of each link for several state levels in a discretized set and thus directly enforce the constraints. However, this problem makes the graph complexity grow exponentially. Therefore, a common approach to deal with these constraints is to relax them in the problem formulation, and verify them a posteriori on the optimal path [6]. Such a path is feasible if $\forall e_k \in \mathbf{p}$, x_k fulfills (5.4). If it is not, it is necessary to select the next-best path, which in turn requires a routing algorithm that is capable to sort paths according to the objective function chosen (see Sect. 5.3).

5.1.2 Routing Algorithms

Graphs modeling road networks are *directed* and *cyclic* by nature, and, due to limited connectivity and the presence of one-way roads, the adjacency matrix representing road networks graphs tends to be highly sparse. Furthermore, the consideration of electric vehicles and energy recuperation phenomena implies that the graph links may be weighted with *negative costs*. In theory, negative weights on a cyclic graph may lead to the criticality of negative cycles: there might be cycles in the road network graph whose sum of costs is negative. Routing in such graphs is not possible, as the vehicle can gain any amount of energy simply by running along these cycles sufficiently many times. However, this issue does not make physical sense in our framework, provided that the modeling approach is correct.

In the next sections two common shortest-path algorithms are presented, namely Dijkstra's and Bellman-Ford (BF). Then, algorithms are presented that can be used to effectively find the Pareto front for the bi-objective optimization (5.17).

5.1.2.1   Dijkstra's Algorithm

Dijkstra's algorithm, conceived by Edsger Dijkstra in the 1950s [7], is perhaps the most commonly used algorithm in routing context. A basic implementation is described in Algorithm 1. The procedure starts marking all nodes of the graph unvisited and assigning a cost, denoted with J, equal to zero to the origin node, infinity to the others. Then an iteration over the unvisited nodes is launched, starting from the origin node. For the current node, all of its unvisited neighbors are considered and their link costs through the current node are evaluated. This value is compared to the current assigned cost and the smaller one is assigned. When all of the unvisited neighbors of the current node are considered, the current node is marked as visited (it will never be checked again). If the destination node has been marked visited, then the procedure has finished. Otherwise, the unvisited node with the smallest cost is selected as the new current node and a new iteration begins. Once this procedure is completed, a reverse iteration allows then reconstructing the optimal path.

The computational complexity of Dijkstra's algorithm is $O(||A||^2)$, where $||\cdot||$ denotes the set cardinality, which makes its use particularly attractive. However, Dijkstra's algorithm requires that the costs are non-negative to provide an optimal solution. Therefore, it cannot be generally used for the eco-routing problem with EVs or HEVs.

A widely used extension of Dijkstra's algorithm is the A^* algorithm. This search method differs from Dijkstra's essentially because it uses a heuristic function to evaluate the current node's neighbors in addition to the "real" cost w. This heuristic is an estimate of the remaining cost from the neighbor to the target node.

5.1.2.2   Bellman-Ford Algorithm

The well-known Bellman-Ford (BF) algorithm [8] can be used to route on graphs with negative costs. Similarly to Dijkstra's algorithm, BF can be considered fast since their runtime is polynomially bounded. However, while Dijkstra's method visits only those nodes that can potentially be on the shortest path, Bellman-Ford operates on every node in the graph. Hence the computational effort associated with Dijkstra's method is dominated by trip properties while the road network size dominates the computational effort associated with the Bellman-Ford algorithm. In particular, its computational complexity is $O(||A|| \cdot ||A^*||)$.

For large road networks, such a computational complexity leads to significantly high computation time, which is not suitable for user-oriented applications and real-time use. However, early termination conditions [9] can be introduced in order to stop the search when an iteration of the algorithm main loop ends without making any link relaxation. When this happens, it means that the algorithm has already found all the shortest paths from the origin and the following iterations would not modify them. This does not improve the worst-case performance of the algorithm, but it performs extremely well on real road networks [10]. In practice, the Algorithm 2 drastically reduces the computation time.

Algorithm 1 Dijkstra algorithm

Require: $\mathcal{G}, w_k, n_O, n_D$
Ensure: $\mathbf{p}, J$

▷ Initialization

$\mathbf{p} \leftarrow \emptyset, \quad J \leftarrow \infty, \quad \text{pred}(n_O) \leftarrow 0, \quad J(n_O) \leftarrow 0$
$Q \leftarrow A$
while $Q \neq \emptyset$ **do**

▷ Find current node

    $J_{opt} \leftarrow \infty$
    for $u \in Q$ **do**
      if $J(u) < J_{opt}$ **then**
        $J_{opt} \leftarrow J(u)$
        $u_{opt} \leftarrow u$
      end if
    end for
    $u \leftarrow u_{opt}$

▷ Remove u from Q

    $Q \leftarrow Q \setminus u_{opt}$

▷ Termination condition

    if $u_{opt} = n_D$ **then**
      return
    end if

▷ Evaluate neighbors

    for $v \in \text{neighbors}(u)$ **do**
      if $J(u) + w(u, v) < J(v)$ **then**
        $J(v) \leftarrow J(u) + w(u, v)$
        $\text{pred}(v) \leftarrow u$
      end if
    end for
end while

▷ Reverse iteration

$u \leftarrow n_D, \quad \mathbf{p} \leftarrow \emptyset$
repeat
    $\mathbf{p} \leftarrow u \cup \mathbf{p}$
    $u \leftarrow \text{pred}(u)$
until $u = n_O$

Algorithm 2 Bellman-Ford algorithm

Require: $\mathcal{G}, w_k, n_O, n_D$
Ensure: $\mathbf{p}, J$

$\qquad\qquad\qquad\qquad\qquad\qquad\qquad\qquad\qquad\qquad\qquad\qquad\quad$ ▷ Initialization

$\quad \mathbf{p} \leftarrow \emptyset, \quad J \leftarrow \infty, \quad \text{pred}(n_O) \leftarrow 0, \quad J(n_O) \leftarrow 0$

$\qquad\qquad\qquad\qquad\qquad\qquad\qquad\qquad\qquad\qquad\qquad\qquad\qquad\quad$ ▷ Cycle

$\quad$ **for** $i \in \{1, \ldots, \|A\|\}$ **do**
$\qquad$ optimal $\leftarrow$ **True**
$\qquad$ **for** $arc \in \{1, \ldots, \|A^*\|\}$ **do**
$\qquad\quad u \leftarrow \text{tail}(arc)$
$\qquad\quad v \leftarrow \text{head}(arc)$
$\qquad\quad$ **if** $J(v) > J(u) + w(arc)$ **then**
$\qquad\qquad J(v) \leftarrow J(u) + w(arc)$
$\qquad\qquad \text{pred}(v) \leftarrow u$
$\qquad\qquad$ optimal $\leftarrow$ **False**
$\qquad\quad$ **end if**
$\qquad$ **end for**
$\qquad$ **if** optimal **then**
$\qquad\quad$ **return**
$\qquad$ **end if**
$\quad$ **end for**

$\qquad\qquad\qquad\qquad\qquad\qquad\qquad\qquad\qquad\qquad\qquad\qquad\quad$ ▷ Reverse iteration

$\quad u \leftarrow n_D, \quad \mathbf{p} \leftarrow \emptyset$
$\quad$ **repeat**
$\qquad \mathbf{p} \leftarrow u \cup \mathbf{p}$
$\qquad u \leftarrow \text{pred}(u)$
$\quad$ **until** $u = n_O$

5.1.2.3 Bi-objective Optimization

We consider a particular case of the MOOP defined in Sect. 5.1.1 where only a resource constraint is present (bi-objective optimization), weighted with $\lambda_1 = 1 - \lambda$, $\lambda_2 = 0$.[3] The goal is to find the paths that do not allow to improve one component of the objective function without deteriorating the other one (*non-dominated* solutions). The corresponding set in the objective space $\{J(\mathbf{p}), R(\mathbf{p})\}$ is generally called *Pareto front*, see Fig. 5.2. Given the nature of the eco-routing problem, the Pareto front is intrinsically made of a discrete set of solutions. Spanning the entire front thus provides the exhaustive list of possible solutions compromising between the selected cost (energy) and resource consumption (often, travel time). On the one hand, that allows a decision maker to arbitrarily select a desired trade-off. On the other hand, the particular non-dominated solution that minimizes the original objective function J without violating the resource constraint (5.6) can be easily identified.

The combinatorial nature of such problems makes the search of all Pareto solutions a very time-consuming task. Thus the practical goal is often to find the most diverse set of solutions, i.e., solutions that are sufficiently varied in the true Pareto set. An

---

[3]The same approach could be of course applied to situations where the state constraints but not the resource constraints are relevant, just by replacing R with h.

Algorithm 3 Scalarization bi-objective optimization

Require: w_k, r_k, n_O, n_D
Ensure: S
  $S \leftarrow \emptyset$
                                                                    ▷ Initialize binary search
  $\lambda \leftarrow 0$
  $\mathbf{p}^1 \leftarrow$ BFalgorithm $(\mathcal{G}, \lambda w_k + (1 - \lambda)r_k, n_O, n_D)$
  $S \leftarrow S \cup \left\{ J(\mathbf{p}^1), R(\mathbf{p}^1) \right\}$
  $\lambda \leftarrow 1$
  $\mathbf{p}^2 \leftarrow$ BFalgorithm $(\mathcal{G}, \lambda w_k + (1 - \lambda)r_k, n_O, n_D)$
  $S \leftarrow S \cup \left\{ J(\mathbf{p}^2), R(\mathbf{p}^2) \right\}$
                                                                              ▷ Compute S
  $S \leftarrow$ solveRecursion $\left(J(\mathbf{p}^2), R(\mathbf{p}^2), J(\mathbf{p}^1), R(\mathbf{p}^1), S, 1, 0 \right)$
                                                                          ▷ Recursive function
  function SOLVERECURSION($z_1^l, z_2^l, z_1^r, z_2^r, S, \lambda_l, \lambda_r$)
    $\lambda \leftarrow (\lambda_l + \lambda_r)/2$
    $\mathbf{p} \leftarrow$ BFalgorithm $(\mathcal{G}, \lambda w_k + (1 - \lambda)r_k, n_O, n_D)$
    $z_1 \leftarrow J(\mathbf{p})$
    $z_2 \leftarrow R(\mathbf{p})$
    if $\{z_1, z_2\} \notin S$ **then**
      $S \leftarrow S \cup \{z_1, z_2\}$
      if $|\lambda - \lambda_r| \geq \gamma_\lambda$ AND $| (\lambda z_1 + (1 - \lambda)z_2) - (\lambda z_1^r + (1 - \lambda)z_2^r) | \geq \gamma_d$ **then**
        $S \leftarrow$ solveRecursion $\left(z_1, z_2, z_1^r, z_2^r, S, \lambda, \lambda_r \right)$
      end if
      if $|\lambda - \lambda_l| \geq \gamma_\lambda$ AND $| (\lambda z_1 + (1 - \lambda)z_2) - (\lambda z_1^l + (1 - \lambda)z_2^l) | \geq \gamma_d$ **then**
        $S \leftarrow$ solveRecursion $\left(z_1^l, z_2^l, z_1, z_2, S, \lambda_l, \lambda \right)$
      end if
    else
      if $\{z_1, z_2\} = \left\{ z_1^l, z_2^l \right\}$ **then**
        if $|\lambda - \lambda_r| \geq \gamma_\lambda$ **then**
          $S \leftarrow$ solveRecursion $\left(z_1, z_2, z_1^r, z_2^r, S, \lambda, \lambda_r \right)$
        end if
      else
        if $|\lambda - \lambda_l| \geq \gamma_\lambda$ **then**
          $S \leftarrow$ solveRecursion $\left(z_1^l, z_2^l, z_1, z_2, S, \lambda_l, \lambda \right)$
        end if
      end if
    end if
    return S
  end function

---

example of efficient dichotomic algorithm for the search of the non-dominated solutions is Aneja's [11], which was proved to find all the non-dominated solutions within a finite number of iterations. In [10], a new binary search algorithm (see Algorithm 3) has been proposed to significantly reduce the computation time, while computing a representative sub-set of non-dominated solutions.[4] The main difference between

---

[4]In theory, the algorithm only finds a subset of extreme supported non-dominated solutions, that is, a subset of those non-dominated solutions that lie on vertices of the convex hull of the Pareto front, see Fig. 5.2.

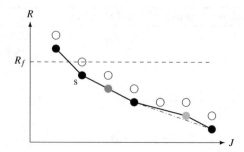

Fig. 5.2 Objective space of a bi-objective optimization problem. Solid line: Pareto front. Dash-dot line: convex hull. Black circles: extreme supported non-dominated solutions. Gray circle: non-extreme supported non-dominated solution. Light gray circle: non-supported non-dominated solution. White circles: dominated solutions. Solution s is the one that minimizes energy consumption without violating the constraint on resource consumption

the two algorithms consists of the space where the solutions are searched: Aneja's algorithm recursively explores the objective space selecting the decision weight λ based on the known non-dominated solutions, the modified algorithm explores the decision space performing a recursive binary search of the decision weight. The latter algorithm presents a standard initialization phase in which the two single-objective solutions for $\lambda = 0$ (minimizing the resource consumption, e.g., the fastest route) and $\lambda = 1$ (the energy-optimal route) are computed. The parameters γ_λ and γ_d in Algorithm 3 are set in such a way to find the right trade-off between number of algorithm iterations and number of solutions found.

5.1.3   Numerical Solutions

This section presents some results obtained using the modified BF algorithm (Algorithm 2) for a simple route graph. The primal graph is shown in Fig. 5.3a, which has been obtained from the road network of Fig. 4.2 by further considering separate directions of the original database links as different segments. Also indicated in the figure are average speeds for each route segment and altitude at each node. The adjoint graph is shown in Fig. 5.3b, with energy cost and travel time (resource consumption) indicated at each link. In this illustration, wheel energy is considered as the energy cost, which is independent from the particular powertrain used. No recharge facilities are considered. Extra energy and travel time are assigned to links, according to the method in Sect. 4.4, depending on the value of the parameter ζ defined in (4.27), that is randomly assigned to the links to represent traffic signals ($\zeta = 1$ is for free flow, $\zeta = 0$ is for a stop).

Figure 5.4 shows the eco-route and the space of energy-time solutions for a given sequence of stop signals. Note that, for such a small graph, the distinct routes with no loops are easily enumerated and the Pareto front explicitly calculated.

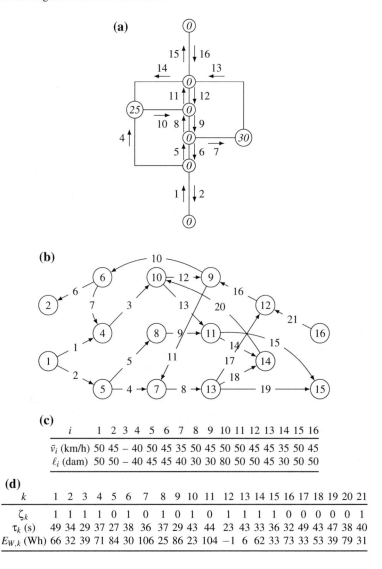

Fig. 5.3 Simple route graph, with directions on route segments numbered from 1 to 16 and altitude values on nodes (**a**). Corresponding adjoint graph with nodes numbered from 1 to 16 and links numbered from 1 to 21 (**b**). Values of average speed and length for each road segment/graph node (**c**). Values of parameter γ, travel time, and wheel energy for each graph link, for a vehicle with $C_0 = 161.7$, $C_2 = 0.409$, $m_v = 1100$ kg (**d**)

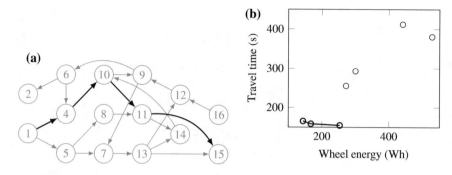

Fig. 5.4 Minimal-energy path (eco-route) in red for the graph of Fig. 5.3 and $n_O = 1$, $n_D = 15$ (**a**). Results for the the seven possible solutions with no loops (1-5-7-13-15, 1-5-8-11-15, 1-4-10-11-15, 1-4-10-9-7-13-15, 1-5-7-13-14-10-11-15, 1-5-8-11-14-10-9-7-13-15, 1-5-7-13-12-9-6-4-10-11-15), in terms of wheel energy and travel time; Pareto set (non-dominated solutions) in blue (**b**)

5.2   Energy-Optimal Driving Range Estimation

Vehicles' owners often wonder how far they can drive with the on-board stored energy. Such a question is particularly relevant for electric vehicles, where the limited amount of energy stored in the battery is often perceived as a strong limitation to the penetration of this powertrain technology ("range anxiety").

Driving range is often estimated in terms of distance making assumptions on the average energy consumption per kilometer. Such estimations are made either before the trip based on worst-case average energy consumption assumptions, or during the trip based on measured consumption [12]. Because the range output to the user is typically conservative to avoid running out of charge, imprecise range estimates may further increase range anxiety.

In order to be insightful and effective, a driving range estimation strategy should be accurate by specifically considering the characteristics of the vehicle and the road transportation network. Also, when predicting driving range, it is of paramount importance to clearly state the route types allowing the driver to reach the locations within the driving range. For instance, routes favoring lower travel time could be more energy-expensive than other types of routes, therefore the route choice directly affects the size of the driving range. This aspect, although very important, is often neglected [13]. In only few works [14, 15], the authors realize the importance of choosing eco-routes to determine the driving range, however the simplified energy consumption and road network model do not allow for a satisfactory precision in urban and suburban environments.

In the next sections, it will be shown how the eco-routing methods of Sect. 5.1 can be used to provide an energy-optimal driving range, by predicting the optimal energy consumption to reach all the possible destinations within the driving range. This allows to relax the typical unrealistic assumptions of a worst-case average energy consumption and to give a clearer insight into the energy characteristics of the road network [16].

5.2.1 Problem Formulation

The energy-optimal driving range to find may be defined as

$$\text{range}(n_O) \triangleq \left\{ \Delta \subseteq A : \forall n \in \Delta,\ E_T(n|n_O) \leq E_T^* \right\}, \tag{5.18}$$

where n_O is the origin node, A is the set of nodes in the graph, Δ is the subset of nodes within the energy-optimal driving range, $E_T(n|n_O)$ is the predicted energy consumption from the origin node to node n, and E_T^* is a desired (tank) energy consumption. Typically, for EVs, this quantity is the electric energy stored in the battery at the origin node.

5.2.2 Solution Method

The single-source shortest-path algorithms used for eco-routing, typically Dijkstra's or Bellman-Ford algorithms, have the property of returning within the same execution not only the shortest path from origin to destination, denoted in Sect. 5.1 with **p**, but also the optimal costs to reach all the nodes in the graph, denoted, e.g., in Algorithms 1 and 2, with J. Such a characteristic may be exploited to identify all the destinations reachable with a certain desired energy consumption.

In practice, it suffices to replace $E_T(n|n_O)$ in (5.18) with $J(n)$, the minimal cost function to reach node n. Note that the driving range obtained by means of a single-source SP algorithm is optimal in the sense that each node within the range is reachable by following the shortest path from the origin to that node. In other words, since the routing graph is weighted with energy costs (for a choice of $\lambda = 1$), the nodes in the energy-optimal driving range are reachable via eco-routes.

This approach enables a high level of precision in the driving range calculation. As opposed to the techniques providing only the polygonal curve delimiting the driving range, this strategy allows to analyze the energy consumption characteristics of the region within the driving range and, more importantly, to determine whether such a region is simply connected. If the properties of simple connectedness do not hold, then some nodes within the driving range are unreachable even by means of an eco-route.

The strategy may be easily extended to compute a round-trip driving range, which may be of interest especially in the case of electric vehicles. The round-trip driving range can provide an insight into the reachable destinations that also allow to return to the departure point within a desired energy consumption threshold. In order to do so, the single-destination shortest path problem is simply solved on the same routing

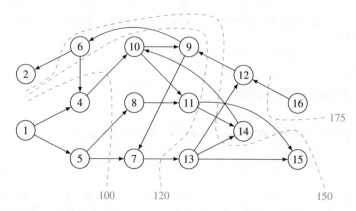

Fig. 5.5 Energy range evaluated for the problem of Fig. 5.3 with $n_O = 1$. Gray dashed lines are the contour lines at the wheel energy consumption indicated (Wh)

graph with reversed links orientation. The Bellman-Ford algorithm is run on such a graph using again the origin (i.e. the destination of the round-trip itinerary) as the starting node.

5.2.3 Numerical Solutions

An example result of energy-optimal driving range estimation is presented in Fig. 5.5. The graph considered is the simple road network of Fig. 5.3. Taking $n_O = 1$ as the origin node, the figure shows some contour lines of equal energy consumption.

More extensive experiments [16] have demonstrated that the methods of this section allow to correctly capture important characteristics of the energy driving range for EVs and HEVs, such that:

- the driving range may be asymmetric about the origin;
- the driving range region may be not simply connected;
- auxiliary power demand may have a significant impact on the driving range;
- the round-trip driving range may be able to restore symmetry about the origin.

A detailed case study is presented in Sect. 9.5.

5.3 Practical Implementation

In principle, the eco-routing algorithm is run at the demand of the user and should start with an update of the energy weights and resource consumptions (w_k and r_k in this chapter) for each link of the road network as a function of the current traffic conditions. In other terms, w_k and r_k should be evaluated from the speed data v_k in order to implement a procedure like the one introduced in Sect. 4.4. This calculation could be even projected for the future time when the host vehicle will cross that particular link. However, this complete update is seldom practical, due to the large computing time required.

A more practical implementation of an eco-routing system aimed at reducing the computing time, and thus the service time for the user, distributes the calculations between an offline and an online layer, as illustrated by the flowchart in Fig. 5.6.

A remote server (cloud-computing) stores the road network of the preferred region, with N sets of different energy weights and resource consumptions for each link. These sets have been evaluated offline, based on historical traffic information. Typically, they correspond to different daytimes and weekdays to represent the most diverse possible set of traffic conditions.

Through an HMI (see Sect. 8.31), the user enters either an address or coordinates for its destination. In the former case, street addresses are converted by most GIS into geographic coordinates, in a process called *geocoding*. As for the origin of the trip, it is provided by a GPS signal.

These pieces of information are sent to the remote server where calculations are performed. At first, origin and destination locations must be converted into useful nodes or links in the road graph in a process called *map-matching*. For example, the point-to-point method matches a given location to the nearest node in A. Conversely,

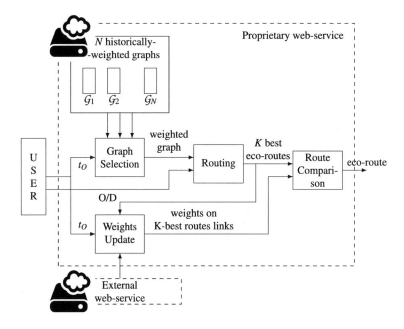

Fig. 5.6 Conceptual sketch of an eco-routing system

the point-to-curve method matches a given location to the nearest point on the nearest road in A^*. When entire trajectories, i.e., sequences of locations, are to be matched, curve-to-curve methods try to match them to the most similar route in the road network.

Based on current time of request (t_O in Fig. 5.6), the most suitable among the N weighted graphs stored is selected. A routing algorithm is then executed on this graph to output the K best routes according to the objective function implemented. For single-objective routing problems, the K best routes can be evaluated using Yen's algorithm, a variant of the Belmann-Ford algorithm. For multi-objective cases, the spanning of the Pareto front described in Sect. 5.1.2.3 already provides several alternative routes.

The actual traffic conditions are provided by an external web-service and transformed in weights and resource consumptions for the selected routes only using, for example, the model of Sect. 4.4.

The K best routes are then evaluated using these updated weights. In other terms, the objective function is evaluated for each of these routes and the constraints on the state are verified to discard possible unfeasible routes. Finally, the best ("eco") route is found as the optimal sequence of links in A^*. To transform this sequence into a sequence of geographical coordinates suitable to be displayed by the HMI, a new map-matching stage is required.

References

1. De Nunzio G, Thibault L, Sciarretta A (2016) A model-based eco-routing strategy for electric vehicles in large urban networks. In: Proceedings of international conference on intelligent transportation systems (ITSC), pp 2301–2306. IEEE
2. Kubička M (2017) Constrained time-dependent adaptive eco-routing navigation system. PhD thesis, Université Paris-Saclay
3. Strehler M, Merting S, Schwan C (2017) Energy-efficient shortest routes for electric and hybrid vehicles. Transp Res Part B: Methodol 103:111–135
4. Fontana MW (2013) Optimal routes for electric vehicles facing uncertainty, congestion, and energy constraints. PhD thesis, Massachusetts Institute of Technology
5. Caramia M, Dell'Olmo P (2008) Multi-objective optimization. Springer, Berlin
6. De Nunzio G, Sciarretta A, Gharbia IB, Ojeda LL (2018) A constrained eco-routing strategy for hybrid electric vehicles based on semi-analytical energy management. In: Proceedings of international conference on intelligent transportation systems (ITSC), pp 355–361. IEEE
7. Dijkstra EW (1959) A note on two problems in connexion with graphs. Numer Math 1(1):269–271
8. Bellman R (1958) On a routing problem. Q Appl Math 16(1):87–90
9. O'Connor D (2012) Notes on the Bellman-Ford-Moore shortest path algorithm and its implementation in MATLAB. Dublin University College, technical report
10. De Nunzio G, Thibault L, Sciarretta A (2017). Bi-objective eco-routing in large urban road networks. In: Proceedings of international conference on intelligent transportation systems (ITSC), pp 1–7. IEEE
11. Aneja YP, Nair KPK (1979) Bicriteria transportation problem. Manag Sci 25(1):73–78
12. De Cauwer C, Van Mierlo J, Coosemans T (2015) Energy consumption prediction for electric vehicles based on real-world data. Energies 8(8):8573–8593

13. Ferreira JC, Monteiro V, Afonso JL (2013) Dynamic range prediction for an electric vehicle. In: Proceedings of world electric vehicle symposium and exhibition (EVS27), pp 1–11. IEEE
14. Neaimeh M, Hill GA, Hübner Y, Blythe PT (2013) Routing systems to extend the driving range of electric vehicles. IET Intell Transp Syst 7(3):327–336
15. Stankoulov P (2015) Vehicle range projection, September 1 2015. US Patent 9,121,719
16. De Nunzio G, Thibault L (2017) Energy-optimal driving range prediction for electric vehicles. In: Proceedings of intelligent vehicles symposium (IV), pp 1608–1613. IEEE

Chapter 6
Energy-Efficient Speed Profiles (Eco-Driving)

The adoption of an energy-efficient driving style is the goal of "eco-driving" techniques. Drivers that follow these techniques (often called "hypermilers") are obviously motivated by the fuel or energy savings that can be achieved. Drivers can be assisted by specific ADAS tools that calculate energy-efficient speed profiles. With the advent of CAVs, eco-driving can be enforced more effortlessly than with human drivers.

General techniques, often adopted without any specific tool to support the driver, are the subject of Sect. 6.1. In Sect. 6.2 it will be shown that eco-driving can be formulated as an optimal control problem, with an energy-based objective function that is minimized. Depending on which energy amount is minimized, several distinct eco-driving strategies can be obtained, some of which are described in Sect. 6.3 (powertrain energy), Sect. 6.4 (fuel energy in ICEVs), Sect. 6.5 (battery energy in EVs), and Sect. 6.6 (equivalent fuel energy in HEVs).

6.1 Eco-Driving Techniques

Often listed among "eco-driving" techniques are general common-sense practices such as mechanically maintaining the vehicle (tire inflation pressure, wheel alignment, engine lubrication, etc.), reducing transported mass, removing unnecessary equipment that would increase aerodynamic drag, and reducing ancillary loads (air conditioning, heating). However, only proper eco-driving techniques that act on the driving style are considered here, that is, those acting on vehicle's speed and acceleration.

Eco-driving can be applied in several driving scenarios, that are introduced in Sect. 6.1.1. Most current approaches to eco-driving are based on heuristic rules of thumb or good practices that are associated with an energy-efficient drive. These practices are reviewed in Sect. 6.1.2.

© Springer Nature Switzerland AG 2020
A. Sciarretta and A. Vahidi, *Energy-Efficient Driving of Road Vehicles*,
Lecture Notes in Intelligent Transportation and Infrastructure,
https://doi.org/10.1007/978-3-030-24127-8_6

6.1.1   Eco-Driving Scenarios

Several driving scenarios can be the object of eco-driving. A non-exhaustive list includes:

1. Accelerating to a cruise speed: from a given speed v_i, reach a target speed $v_f > v_i$, in a free time t_f and covering a free distance s_f, minimizing the energy spent per unit distance.
2. Decelerating to a stop: from a given speed v_i, decelerate to $v_f = 0$ while covering a distance s_f in a prescribed or free time t_f.
3. Driving between stops: cover a distance s_f in a time t_f starting from $v_i = 0$ and terminating at $v_f = 0$.
4. "Eco-approaching" a signalized intersection: cover a distance s_f in a time t_f bounded within the "green light" window, starting from a given speed v_i and terminating at a free passage speed v_f.
5. "Green waving": the same as scenario 4 but repeated for multiple traffic lights in a sequence.
6. Urban trip: a combination of the scenarios 1–5 above.
7. Highway trip: cover a (large) distance s_f without stops, by minimizing a given compromise between energy and trip time, in the presence of speed limits, altitude variations, and lane changes (or merging).
8. Cruising: minimize the energy per distance within an admissible speed band, and with $v_f = v_i$.
9. Car following: same as scenarios 6–8 with in addition a safe gap to be maintained with respect to a leading vehicle.

We will make reference to these scenarios later in the book and in particular in Chap. 7.

6.1.2   Eco-Driving Rules

Perhaps the simplest "eco-driving rule" concerns keeping a low and constant speed. Intuitively, lower speeds reduce the aerodynamic and rolling resistance losses described in Chap. 2, while substantially constant speeds are aimed at suppressing the energy spent for accelerating that is usually impossible to completely recover during decelerations. The wheel energy per distance at a constant cruising speed is easily obtained from (2.11), from where it is apparent that, the lower the speed, the lower this expenditure. Of course, lowering the speed also generally increases the time that is necessary to cover a given distance, therefore a compromise must be made between energy consumption and trip time. If, for example, the trip time is converted into an equivalent energy expenditure through a tunable coefficient β,[1] the

---

[1]This approach leads to a bi-objective optimization by scalarization similar to the one described in Chap. 5.

optimal cruising speed could be defined as the value that minimizes the equivalent wheel energy per distance,

$$v_{cr,opt} = \arg \min_v \left(C_0 + C_1 v + C_2 v^2 + \frac{\beta}{v} \right), \tag{6.1}$$

that is, a generally positive speed level. For instance, using the numerical values $C_0 = 162$, $C_1 = 0$, $C_2 = 0.410$, the optimal cruising speed would be $(\beta/2C_2)^{1/3}$, that is, 38 km/h for $\beta = 10^3$, 66 km/h for $\beta = 5 \cdot 10^3$, and 83 km/h for $\beta = 10^4$.

Seen from another perspective, the optimal cruising speed can be defined as the value that minimizes the "tank" energy expenditure, rather than the wheel energy per distance.[2] This concept is particularly used for ICE vehicles, where the energy consumption further depends on the transmission ratio, that is, the gear engaged. Using the models of Sect. 2.2.2, (6.1) is modified as

$$
\begin{aligned}
(v_{cr,opt}, \gamma_{e,cr,opt}) = \arg \min_{v,\gamma_e} &\left\{ \left(\frac{k_{e,0}\gamma_e}{r_w} + \frac{k_{e,2}C_0}{\eta_t} \right) + \right. \\
&+ \left(\frac{k_{e,1}\gamma_e^2}{r_w^2} + \frac{k_{e,2}C_1}{\eta_t} + \frac{k_{e,3}C_0\gamma_e}{r_w\eta_t} \right) v + \\
&+ \left(\frac{k_{e,2}C_2}{\eta_t} + \frac{k_{e,3}\gamma_e C_1}{r_w\eta_t} + \frac{k_{e,4}\gamma_e^2 C_0}{r_w^2\eta_t} \right) v^2 + \\
&+ \left(\frac{k_{e,3}\gamma_e C_2}{r_w\eta_t} + \frac{k_{e,4}\gamma_e^2 C_1}{r_w^2\eta_t} \right) v^3 + \left(\frac{k_{e,4}\gamma_e^2 C_2}{r_w^2\eta_t} \right) v^4 + \frac{\beta}{v} \right\}.
\end{aligned}
\tag{6.2}
$$

For instance, using the numerical values above and $k_{e,1} = 0.0396$, $k_{e,2} = 2.55$, $k_{e,3} = -0.0016$, $k_{e,4} = 2.93 \cdot 10^{-8}$, $r_w = 0.32$, $\eta_t = 1$, $\gamma_e = 3.5$, the optimal cruising speed is generally lower than the previous case: 27 km/h for $\beta = 10^3$, 47 km/h for $\beta = 5 \cdot 10^3$, 61 km/h for $\beta = 10^4$, and 105 km/h for $\beta = 5 \cdot 10^4$, see Fig. 6.1. When $\beta = 0$, also in this case the optimal cruising speed would be the lowest obtained with the given transmission ratio.

Equation (6.2) also shows that the steady-state engine consumption is minimized by lowering the transmission ratio as much as possible, that is, using the highest gear available. Overall, since in ICEVs each gear can be used only within a certain vehicle speed range, such that the resulting engine speed lays in its admissible window, practically the best cruising strategy is the lower speed that can be reached using the highest gear. Note that this gear shifting strategy would certainly be optimal from a purely energetic viewpoint, but it is generally not desirable for drivability, because it leaves little torque reserve for, e.g., overtaking or emergency maneuvers. In the rest of this book, we will generally not treat the optimization of gear shifting as a part

---

[2]Some sources tend to define the best cruising speed as the vehicle speed that maximizes the powertrain efficiency at a given transmission ratio. For an ICE powertrain, this value corresponds to the engine speed at which the engine efficiency has its maximum. It is important to note, however, that a higher powertrain efficiency does not imply an overall lower energy consumption.

Fig. 6.1 Energy per unit
distance (6.2) as a function
of speed, for a given
transmission ratio ($\gamma_e = 3.5$)
and four values of the
parameter β

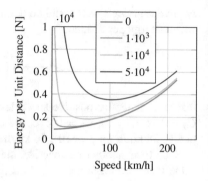

of eco-driving, unless otherwise specified, and shall consider the gear shift law as imposed by drivablity considerations.

The analysis above is based on a constant cruising speed. Actually, it will be shown in the next sections that, with ICE powertrains, periodically operating the engine at high load and then shutting it down, for a constant *average* vehicle speed, yields a lower energy consumption than keeping a constant cruising speed. These considerations form the theoretical basis of the *"pulse-and-glide"* (P&G) strategy that is often advised as an eco-driving mechanism.

The constant-speed strategy (or the P&G) can be only applied to a limited number of situations, like the "cruising" or the "highway trip" scenarios listed in Sect. 6.1.1. When the initial and final (target) velocity are different, the velocity cannot be constant, and at least one acceleration and/or deceleration maneuver has to be performed. Generic eco-driving rules often advise the driver to accelerate or decelerate as "smoothly" as possible. However, the methods presented in the next sections will show that the best acceleration profiles often use the powertrain capabilities at their maximum, in order to reach the cruising speed as quickly as possible.

In contrast, decelerating requires either application of the friction brakes, i.e., energy dissipation as heat, or reliance on vehicle/powertrain braking. Even in hybrid and electric vehicles with regenerative (powertrain) braking, a part of the available energy is lost as heat. Therefore, it is best to avoid these situations when possible. This means that when deciding to slow down or stop on a level road, it is often more energy efficient to glide with the powertrain providing no force, and using rolling resistance and aerodynamic drag to slow the vehicle down. However, the resulting operation, called *coasting*, may not be always safe or desirable. Therefore, coasting is often replaced by powertrain braking or *overrun*. In ICE vehicles, that corresponds to using the "engine brake", with the engine connected to the wheels and the fuel injection cut off.

Also, when descending down a steep road and in a hypothetical scenario when there are no upper bounds on velocity, it is more energy efficient to coast down the hill allowing the velocity to increase toward the equilibrium imposed by rolling resistance and aerodynamic drag. Unfortunately this is very unsafe and often impractical due to road speed limits and the bounds imposed by preceding vehicles.

Overall, the implementation of coasting strategies needs the anticipation of imminent slow downs or descents, so that the vehicle can start gliding at the right time, together with a precise knowledge of the speed limitations that could narrow the admissible speed band.

More generally, effective eco-driving strategies cannot be solely based on current driving parameters, but have to be *predictive*, that is, based on estimations of future external conditions, like anticipation of traffic and route characteristics. These pieces of information can be obtained through the knowledge of the road profile and the monitoring of surrounding vehicles and other road occupants. In this respect, vehicle connectivity is a major lever to enable predictive eco-driving.

While the heuristic rules illustrated above are generally intuitive and relatively easy to implement, the ultimate potential of eco-driving requires a more rigorous framework. That can be achieved if eco-driving is regarded as an optimal control problem where the drive commands are sought that minimize the energy consumption for a given trip. This approach is the subject of Sect. 6.2.

6.1.3 Eco-Driving Systems

Both the rules introduced in Sect. 6.1.2 and the optimal driving profiles that are to be discussed starting from Sect. 6.2 could be in principle implemented in an automatic way. While this possibility is actually envisaged for autonomous vehicles (see Chap. 3), currently eco-driving must be realized by vehicles with human drivers.

A first level of support to the driver is provided by eco-driving training courses. However, it is common opinion, supported by assessment campaigns, that the good practices acquired with training courses are quickly forgotten if drivers have no online tool that reminds them of those practices [1].

Therefore, software tools and systems that help the driver in performing eco-driving have emerged in the last decade. Most of these systems provide some kind of advice to the driver, solely based on current driving information that is typically extracted from vehicle's network data. The provided suggestions may concern, for instance, gear shifting as a function of the current speed, or a judgment on acceleration/deceleration intensity [2]. Even more rudimentary concepts, essentially consisting of alerts based on the acceleration sensor of smartphones, are typically found among mobile application software labeled "eco-drive".

A different class of tools advises the driver about the energy-optimal speed profiles to follow, using the formulation and the methods described in next sections. These tools can either provide a comparison between the actually performed drive and the optimal one (eco-coaching), or predictively compute the optimal drive to follow, as will be described in a more detail in Chap. 8.

6.2 Eco-Driving as an Optimal Control Problem

Although energy-optimal driving can be applied to several distinct scenarios, these situations can be treated in a similar way by defining an *eco-drive optimal control problem* (ED-OCP). This problem is formulated in Sect. 6.2.1, while Sect. 6.2.2 presents two of the main techniques to solve such an OCP.

6.2.1 Problem Formulation

A generic optimal control problem where (continuous) time is the independent variable can be stated in the following way: for each time $t \in [0, t_f]$ in the *optimization horizon* t_f, find the *control* vector $\mathbf{u}(t) \subset \mathbb{R}^m$ such as to minimize the performance index or *objective function*

$$J = \Phi(\mathbf{x}(t_f), t_f) + \int_0^{t_f} L(\mathbf{x}(t), \mathbf{u}(t), t)dt , \qquad (6.3)$$

where L is the running cost and Φ is the terminal cost, subject to (i) first-order dynamic constraints on the *state vector* $\mathbf{x}(t) \subset \mathbb{R}^n$,

$$\dot{\mathbf{x}}(t) = f(\mathbf{x}(t), \mathbf{u}(t), t), \quad \mathbf{x}(0) = \mathbf{x}_i , \qquad (6.4)$$

(ii) algebraic constraints on the control and state vectors (pure) and on combinations of them (mixed),[3]

$$g_i(\mathbf{x}(t), \mathbf{u}(t), t) \leq 0, \quad i = 1, \ldots, \ell , \qquad (6.5)$$

(iii) terminal (equality) constraints,

$$h(\mathbf{x}(t_f)) = 0 , \qquad (6.6)$$

and (iv) interior-point constraints

$$\mathbf{x}(t_{B,j}) = \mathbf{x}_{B,j}, \quad j = 1, \ldots, n_B . \qquad (6.7)$$

Note that the function Φ can penalize both final states and time, if the latter is not specified but free. When applied to final states, it can be regarded as a "soft" constraint, as opposed to the function h that strictly forces the final state to a given point of the state space. In some cases, only q of the state variables are fixed at the terminal time, while the other $n - q$ are generally associated to a terminal cost.

---

[3] Sometimes in the rest of this book we shall use a different notation for mixed control constraints, $\mathbf{u}(t) \in U(\mathbf{x}(t), t) \subset \mathbb{R}^m$, and reserve formulation (6.5) for pure state constraints.

How the quantities t_f, L, Φ, $\mathbf{x}$, $\mathbf{u}$, f, g, and h are particularized for our ED-OCP will be discussed in the following sections.

6.2.1.1 Optimization Horizon

The main objective of the ED-OCP is to minimize fuel or energy consumption over a certain time and distance horizon. Using the same nomenclature of Chap. 3 and 5, we shall refer to this horizon as *trip*. A trip is thus defined by its duration or *time horizon*, t_f, and distance or *spatial horizon* s_f.

Often, both horizons are prescribed, that is, it is expected that the distance s_f will be covered exactly in a time t_f. In such cases, while the enforcement of t_f is implicit in formulation (6.3) since time is the independent variable, the enforcement of s_f plays the role of a terminal state constraint, see Sect. 6.2.1.6.

However, in some scenarios the trip duration is not to be specified but considered as a free parameter. That final time, to which the minimum objective function corresponds, is the solution of the problem.[4]

6.2.1.2 Objective Function

Given the energy-oriented nature of the ED-OCP, a natural choice for the objective function is the energy consumption over the horizon chosen. That leads to choosing $\Phi = 0$ and the integral in (6.3) representing either the powertrain energy E_p (2.13) or the tank energy E_T. In the former case, the running cost L is represented by the power $F_p v$ introduced in Chap. 2. In the latter case, the running cost coincides with P_f (the fuel power) for ICEVs and HEVs, or P_b (the battery power) for EVs.

In addition to energy, trip time may be the subject of minimization for a given trip distance, or constrained in an admissible window, as stated above. These requirements lead to a multi-objective optimization problem, which can be treated with the scalarization approach (see Chap. 5) by adjoining the trip duration to the energy objective function as a terminal cost,

$$\Phi(\mathbf{x}, t_f) = \cdots + \beta t_f \, , \qquad (6.8)$$

where β is a tuning parameter. Changing β clearly changes the compromise between energy expenditure and trip time.

In a similar way, additional terms might be appended either to the integral part of the objective function or to the terminal cost, to penalize, e.g., deviations from a reference average speed, vehicle acceleration, number of gear shifts or other drivability measures, as well as battery aging.

---

[4]That would be determined using the "transversality condition" or through an iterative procedure. To avoid ambiguity, we shall consider in the rest of the book, while not otherwise specified, ED-OCP with prescribed final time.

6.2.1.3 Controlled Variables

From the discussion of the previous sections, it should be clear that, if the wheel energy is chosen as the objective function, the control vector $\mathbf{u}(t)$ must comprise of the forces F_p and F_b.

In the case that the objective function is the tank energy, the control vector may have additional components. In ICEVs the fuel consumption is unambiguously defined by the net wheel force and the transmission ratio γ_e. However, the latter quantity is usually not the object of energy-based optimization, but rather determined by drivability considerations (e.g., torque reserve) implemented in a gearshift map of the type (2.17). Clutch control is another possible discrete control but it can be conveniently lumped either with the gear control input or with the engine control input. As a consequence, we shall choose

$$\mathbf{u}^{(ICEV)} = \{F_p, F_b\} , \tag{6.9}$$

in the rest of this book. The same considerations apply to EVs and HPVs, leading to the choices

$$\mathbf{u}^{(EV)} = \{F_p, F_b\} . \tag{6.10}$$

and

$$\mathbf{u}^{(HPV)} = \{F_c, F_b\} , \tag{6.11}$$

respectively. Note that other variables possibly chosen as control inputs, such as acceleration, fuel quantity for ICEVs, or motor voltage for EVs, all map to powertrain force in single-source vehicles.

Conversely, in HEVs the internal degrees of freedoms that have been described in Chap. 2 must be taken into account. In parallel HEVs, the powertrain force is the sum of two contributions provided by the engine, $F_{p,e} = uF_p$, and the motor, $F_{p,m} = (1 - u)F_p$, where the torque split factor u is the internal degree of freedom introduced in (2.47). Consequently, the control vector for the ED-OCP is

$$\mathbf{u}^{(PHEV)} = \{F_{p,e}, F_{p,m}, F_b\} . \tag{6.12}$$

In series HEVs, u is the power split ratio between the battery and the APU, see (2.49), while the second internal degree of freedom, e.g., the APU speed, is resolved with the optimal operating line (OOL) approach illustrated in Fig. 2.16. Consequently, the pair u and F_p as control variables for the ED-OCP can be equivalently replaced by the two power levels, such that

$$\mathbf{u}^{(SHEV)} = \{P_g, P_b, F_b\} . \tag{6.13}$$

6.2.1.4 State Dynamics

In principle there are many dynamics to consider when representing a vehicle system, and thus many state variables. However, following the quasistatic modeling approach of Chap. 2, the cardinality of the state vector shall be generally limited to two or three.

In single-source vehicles, such as ICEVs and EVs, the state vector is defined as

$$\mathbf{x}^{(ICE,EV)} = \{s, v\} . \tag{6.14}$$

The speed dynamics is described by (2.1), that is rewritten here as

$$\dot{v}(t) = \frac{F_p(t)}{m} - \frac{C_2}{2m}v(t)^2 - \frac{C_1}{m}v(t) - \frac{C_0}{m} - g\sin(\alpha(s(t))) - \frac{F_b(t)}{m}, \quad v(0) = v_i . \tag{6.15}$$

The position dynamics is simply given by (6.16),

$$\dot{s}(t) = v(t), \quad s(0) = 0 . \tag{6.16}$$

In HEVs, an additional state variable is the battery state of charge,

$$\mathbf{x}^{(HEV)} = \{s, v, \xi_b\} , \tag{6.17}$$

whose dynamics is given by (2.41),

$$\dot{\xi}_b(t) = -\frac{P_b(t)}{V_{b0}Q_b}, \quad \xi_b(0) = \xi_0 . \tag{6.18}$$

In human-powered vehicles, the maximum cyclist force varies dynamically, as described in Sect. 2.5, and thus must be represented by a state variable,

$$\mathbf{x}^{(HPV)} = \{s, v, F_{c,max}\} , \tag{6.19}$$

whose dynamics is given by (2.65) with $F_{c,max}(0) = \bar{F}_{c,max,0}$.

6.2.1.5 Control and State Constraints

Both control and state variables are subject to inequality constraints of the type (6.5). Control inputs $F_{p,e}$, $F_{p,m}$, F_c, P_g, P_b defined in Sect. 6.2.1.3 are all bounded by the physical limits described in Chap. 2. For engines, motors, and human cyclists, these limits typically vary with the vehicle speed, whence the cross dependency on the state vector of the function $h(\cdot)$. In addition to these limits, drivability requirements might be applied to further constrain the powertrain force. The brake control input F_b, defined by (6.15) as a positive quantity, is generally bounded between zero and the maximal brake force that can be exerted by the braking system, $F_{b,max}$. When the gear also is to be controlled, $\gamma_{max}(t)$ and $\gamma_{min}(t)$ are actually functions of the gear engaged, thus requiring that the gear itself be included as a state variable.

State variables also are subject to inequality constraints. The vehicle speed is generally bounded in a position-dependent range

$$v_{min}(t, s(t)) \leq v(t) \leq v_{max}(t, s(t)) . \tag{6.20}$$

The quantity v_{max} is the most restrictive of several possible limitations, including the legal speed limit at position s, the average traffic-induced speed, the safe speed at turning v_{turn} defined in Chap. 4, and any subjective maximal speed allowed. A variable $v_{max}(t, s(t))$ can be also used to describe the constraints on speed imposed by infrastructure breakpoints such as traffic lights, stops, or intersections. As for the variable quantity v_{min}, it can be used to represent the minimal speed subjectively allowed or such that the traffic flow is not substantially disrupted.

A different class of state constraints concerns vehicle position,

$$s_{min}(t) \leq s(t) \leq s_{max}(t) . \tag{6.21}$$

where s_{max} might describe the presence of a leading vehicle that cannot be overtaken and s_{min} that of a trailing vehicle, whose positions obviously change with time.

When relevant, the state variable $\xi_b(t)$ is also confined in an admissible window $[\xi_{b,min}, \xi_{b,max}]$, whose width depends on the battery technology as described in Chap. 2.

6.2.1.6   Terminal Constraints

Terminal constraints concern the values of state variables at the end of the trip. In some scenarios, terminal position can be fixed with certainty, as it corresponds to the distance to the final destination or to a peculiar intermediate location. However, in other cases s_f is not specified (is free). In such scenarios, the ED-OCP can be solved by applying the transversality conditions or with an iterative procedure over s_f aimed at finding the best value that minimizes the objective function.

Similarly, terminal speed can be either free or constrained, more or less "rigidly", to a prescribed value v_f.

When an HEV is considered, an additional constraint concerns the final SoC. In charge-sustaining hybrids that cannot be recharged from external sources, the natural choice is to prescribe that the final SoC matches the initial value, $\xi_b(t_f) = \xi_i$. However, in plug-in hybrids, the terminal constraint rather targets a minimal value, $\xi_b(t_f) = \xi_{b,min}$, provided that after the trip the battery can be more efficiently recharged from the grid. This requirement is thus equivalent to the constraint (2.52) imposed on optimal energy management.

6.2.1.7 Interior Constraints

Equality constraints on both speed and position can be imposed point-wise along the trip, by traffic lights, stops, intersections, traffic queues, or other events that are typical of urban displacement scenarios. Formally these constraints read

$$s(t_{B,j}) = s_{B,j}, \quad v(t_{B,j}) = v_{B,j} , \tag{6.22}$$

where $t_{B,j}$ and $s_{B,j}, j = 1, \dots, n_B$ are the time instants and positions at which such constraints are imposed. All these event identifiers are called *breakpoints* in the rest of this book (Fig. 6.2).

Interior constraints can be enforced directly or by appropriately setting the state boundaries described in Sect. 6.2.1.5, for instance, $v_{min}(t_{b,j}, s_{b,j}) = v_{max}(t_{b,j}, s_{b,j}) = v_{b,j}$.

When speed and position are the sole state variables (i.e., excluding the HEV case) and both are fixed at each breakpoint, an alternative approach to enforcing interior constraints consists of separating the trip into $n_B + 1$ correlated sub-trips or *segments*. The original ED-OCP is then equivalent to $n_B + 1$ independent OCPs, each of which is characterized by its own temporal and spatial horizon, as well as by its own initial and terminal conditions

$$t_{f,j} = t_{B,j} - t_{B,j-1}, \quad s_{f,j} = s_{B,j} - s_{B,j-1}, \quad v_{f,j} = v_{i,j+1} = v_{B,j} , \tag{6.23}$$

with the boundary conditions $t_{B,0} = s_{B,0} = 0$, $v_{i,1} = v_i$, and $t_{f,n_B+1} = t_f, s_{f,n_B+1} = s_f$, $v_{f,n_B+1} = v_f$. The energy consumption of the whole trip is then the sum of the energy consumption of each segment.

Note that segments can be also used to separate the entire trip by (i) changes in road characteristics, such as legal speed limits, slope, etc., or by (ii) driver or traffic induced events, such as planned stops, intersections, traffic lights, or traffic queues. In the following, we shall consider the ED-OCP for a single segment between two breakpoints, unless when otherwise stated, e.g., for HEVs.

Fig. 6.2 Illustration of interior constraints and definition of segments or sub-trips

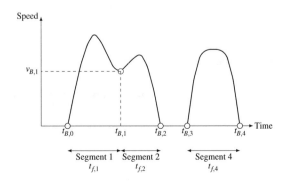

6.2.2   Solution Methods

Problems of the type (6.3)–(6.6) can be solved using various numerical techniques. Among them, *dynamic programming* (DP) and *Pontryagin's minimum principle* (PMP) are among the most used. In this section, these two solution methods are briefly introduced.

6.2.2.1   Pontryagin's Minimum Principle

This method is based on the definition of the Hamiltonian function that is formed as

$$H(\mathbf{x}, \mathbf{u}, t) = L(\mathbf{x}, \mathbf{u}, t) + \lambda f(\mathbf{x}, \mathbf{u}, t) , \tag{6.24}$$

where $\lambda \subset \mathbb{R}^n$ is a vector of costates, having the same dimension n as the state vector.

If state constraints are not present, the necessary conditions for the optimality of a control trajectory $\mathbf{u}(t)$, $t \in [0, t_f]$ include: the state dynamics (6.4)

$$\dot{\mathbf{x}}(t) = \frac{\partial H}{\partial \lambda}(\mathbf{x}(t), \mathbf{u}(t), t) \tag{6.25}$$

with the boundary conditions

$$\mathbf{x}(0) = \mathbf{x}_i, \quad x_j(t_f) = x_{j,f}, \quad j = 1, \ldots, q , \tag{6.26}$$

the costate dynamics (Euler-Lagrange equations)

$$\dot{\lambda}(t) = -\frac{\partial H}{\partial \mathbf{x}} \tag{6.27}$$

with the transversality conditions

$$\lambda_j(t_f) = \begin{cases} \text{free} & j = 1, \ldots, q \\ \dfrac{\partial \Phi}{\partial x_j}(\mathbf{x}) & j = q+1, \ldots, n \end{cases} , \tag{6.28}$$

and the Hamiltonian minimization condition of the minimum principle

$$\mathbf{u}(t) = \arg \min_{u \in U(\mathbf{x},t)} H(\mathbf{x}(t), u, t) . \tag{6.29}$$

This $2n$-dimensional system of coupled differential equations forms a *two-point boundary value problem* (TPBVP), since n boundary conditions are given at the initial time and the other n values at the final time. Of these latter values, q concern state variables and the remaining $n - q$ concern costate variables. This circumstance makes the TPBVP often difficult to solve.

Moreover, handling constraints (6.5) with PMP is not trivial. When the problem presents *pure state inequality constraints* of the form $g(\mathbf{x}(t), t) \le 0$, the indirect adjoining method [3, 4] can be used. Consider the case with $\ell = 1$ (just one such constraint). If $g(\mathbf{x}(t), t)$ is of the pth order, that is, it is differentiated p times with respect to time until the control variable $\mathbf{u}$ explicitly appears, then the term $g^{(p)}(\mathbf{x}, \mathbf{u}, t)$ is adjoined to the Hamiltonian with a multiplier η, to form the Lagrangian

$$L(\mathbf{x}(t), \mathbf{u}(t), t) \triangleq H(\mathbf{x}(t), \mathbf{u}(t), t) + \eta g^{(p)}(\mathbf{x}(t), \mathbf{u}(t), t) . \qquad (6.30)$$

In this case the necessary conditions for a control trajectory to be optimal are still (6.25)–(6.29) with the Lagrangian replacing the Hamiltonian, together with the jump conditions[5] at times at which state constraints become active (entry or contact times),

$$\lambda(\tau^-) = \lambda(\tau^+) + \sum_{j=0}^{p-1} \pi_j \frac{\partial g^{(j)}}{\partial \mathbf{x}}(\mathbf{x}(\tau), \tau) , \qquad (6.31)$$

$$H(\tau^-) = H(\tau^+) - \sum_{j=0}^{p-1} \pi_j \frac{\partial g^{(j)}}{\partial t}(\mathbf{x}(\tau), \tau) , \qquad (6.32)$$

as well as the complementary slackness conditions

$$\eta(t)g(\mathbf{x}(t), t) = 0, \quad (-1)^j \eta^{(j)}(t) \ge 0, \quad j = 0, \dots, p , \qquad (6.33)$$

and

$$\pi_j \ge 0, \quad \pi_j g(\mathbf{x}(\tau), \tau) = 0, \quad j = 0, \dots, p - 1 . \qquad (6.34)$$

For the special yet common case of first-order constraints ($p = 1$), (6.33) reduces to $\eta(t)g(\mathbf{x}(t), t) = 0$, $\eta(t) \ge 0$, $\dot{\eta}(t) \le 0$ with $\eta(t)$ as an additional unknown to be determined,[6] while (6.34) holds for the single unknown multiplier π_0. In summary, inequality constraints introduce additional unknowns that have to be determined as well, and additional conditions.

---

[5]These conditions derive from imposing that $[g^{(0)}(\tau), \dots, g^{(p-1)}(\tau)] = 0$ and treating these *tangency conditions* as interior point constraints that are adjoined to the Lagrangian through the additional multipliers π's.

[6]These conditions mean that when the constraint is not active ($g(\mathbf{x}(t), t) < 0$), then $\eta(t)$ is set to zero. When the constraint is active ($g(\mathbf{x}(t), t) = 0$), $\eta(t)$ must be positive [5] but unknown. The condition $g(\cdot) = 0$ provides the additional equation necessary in this case to determine the newly added unknown η.

Consequently, only rarely the TPBVP is solvable in closed form; generally, it is necessary to proceed iteratively. Most commonly used methods include *collocation methods*, such as the algorithm used in Matlab's `bvp4c` function, and (multiple) *shooting methods*. The latter approach works by iterating over the initial costates and checking the resulting values of the specified states at the final time. An example of PMP algorithm where the TPBVP is solved with shooting is shown in Algorithm 4.

Algorithm 4 PontryaginMinimumPrinciple

Require: λ_i
Ensure: $u_{opt}(1, \ldots, N), x_{opt}(0, \ldots, N)$
  repeat
    $\lambda_{opt}(0) = \lambda_i \in \Lambda$
    $x_{opt}(0) = x_i$

                                                ▷ Time loop
    for $k = 1, \ldots, N$ **do**
      $H_{opt} \leftarrow \infty$
      for $u \in U_k(x_{opt}(k-1))$ **do**

                                            ▷ Minimum Principle

        $\tilde{H} = L_k(x_{opt}(k-1), u) + \lambda(k-1) \cdot f_k(x_{opt}(k-1), u)$
        if $\tilde{H} < H_{opt}$ **then**
           $H_{opt} = \tilde{H}$
           $u_{opt} = u$
        end if
      end for

                                    ▷ Euler–Lagrange equations
      $\lambda_{opt}(k) = \lambda_{opt}(k-1) + \ell_k(x_{opt}(k-1), u_{opt}(k))\Delta$
      $x_{opt}(k) = x_{opt}(k-1) + f_k(x_{opt}(k-1), u_{opt}(k))\Delta$
    end for
  until $x_{opt}(N) = x_f$

6.2.2.2 Dynamic Programming

DP was developed during the 1950's by Richard Bellman [6] and has ever since been used as a tool to design optimal controllers for systems which are finite in their independent variable, with constraints on the state variables and the control inputs. Being a graph search method, it can be regarded as a generalization of Dijkstra's algorithm presented in Chap. 5 [7].

The method is generally based on a discretization of time as the independent variable,[7] state space, and control space. When the forward Euler scheme with time step Δ is applied to the problem (6.3)–(6.6), without terminal cost ($\Phi = 0$) for simplicity, the objective function reads

---

[7]In our problem, position could be alternatively chosen as the independent variable, as it will be shown later.

$$J = \sum_{k=1}^{N} L_k(\mathbf{x}_{k-1}, \mathbf{u}_k)\Delta , \tag{6.35}$$

where $N\Delta = t_f$, the state equations read

$$\mathbf{x}_k = \mathbf{x}_{k-1} + f_k(\mathbf{x}_{k-1}, \mathbf{u}_k)\Delta , \tag{6.36}$$

and the boundary conditions are

$$\mathbf{x}_0 = \mathbf{x}_i, \quad h(\mathbf{x}_N) = 0 . \tag{6.37}$$

A basic DP algorithm incorporating the main features of the method is shown as Algorithm 5. As this pseudocode shows, DP uses a discretized state space, $\mathbf{x}_k \in X_k$. The set X_k can vary with the position state to represent speed limits that are position-dependent (e.g., legal top speed) or additionally dependent on time (traffic-induced speed limit, passage speed at a traffic light, etc.). The control space is discretized as well, $\mathbf{u}_k \in U_k(\mathbf{x}_k)$, with the set U_k that can vary with the speed state to represent the force limits that are commonly speed-dependent. In this respect, the state and control constraints in (6.5) are naturally enforced by correspondingly selecting these subsets.

The procedure first initializes the *cost-to-go function* $\mathcal{J}_k(\mathbf{x})$ that represents the minimal cost to reach an admissible terminal state from a state $\mathbf{x}$ at time step k. At final time, an infinite cost is attributed to unfeasible states, e.g., any state for which $h(\mathbf{x}) \neq 0$. Then the algorithm proceeds backward in time, updating $\mathcal{J}$ taking advantage of Bellman's principle of optimality,

$$\mathcal{J}_k(\mathbf{x}) = \min_{\mathbf{u} \in U_{k+1}(\mathbf{x})} \{L_{k+1}(\mathbf{x}, \mathbf{u})\Delta + \mathcal{J}_{k+1}(\mathbf{x} + f_{k+1}(\mathbf{x}, \mathbf{u})\Delta)\}, \quad k = 0, \ldots, N-1 . \tag{6.38}$$

Correspondingly, the *feedback function* $\mathcal{U}_k(\mathbf{x})$, representing the optimal control input at any state $\mathbf{x}$ at time step k, is also evaluated. When these functions have been evaluated for the entire time-state grid, the algorithm proceeds forward in time from the initial state, using the functions $\mathcal{J}$ and $\mathcal{U}$ to compute the optimal trajectories of control input and state.

Despite the relatively simple structure of the Algorithm 5, there are several issues to consider when implementing dynamic programming. When applying Bellman's equation, the arguments of the term $\mathcal{J}_{k+1}$ are evaluated from the state equations, and may not match any of the discrete states in X_{k+1}, for which the cost-to-go function is defined. Therefore, this term must be approximated using a nearest-neighbor strategy or by interpolation. However, each of these methods have advantages and drawbacks in terms of computational speed and accuracy, and must be applied carefully. For instance, the common way to handle infeasible terminal states by assigning an infinite cost to those states becomes critical when used with an interpolation scheme. Readers interested in the DP algorithm are referred to standard textbooks [5, 8, 9].

Algorithm 5 DynamicProgramming

Require: $f(\cdot), L(\cdot), N, X, \Delta$
Ensure: $u_{opt}(1, \ldots, N), x_{opt}(0, \ldots, N)$

                                          ▷ End cost calculation step

  for $x \in X_N$ **do**
    $\mathcal{J}[x, N] \leftarrow \infty$
  end for
  $\mathcal{J}[x_f, N] \leftarrow 0$

                                    ▷ Intermediate calculation step

  for $k \in \{N-1, N-2, \ldots, 0\}$ **do**
    for $x \in X_k$ **do**
      $J_{opt} \leftarrow \infty$
      for $u \in U_{k+1}(x)$ **do**

                                  ▷ Bellman's equation

        $\tilde{J} \leftarrow L_{k+1}(x, u)\Delta + \mathcal{J}\left[x + f_{k+1}(x, u)\Delta, k+1\right]$
        if $\tilde{J} < J_{opt}$ **then**
          $J_{opt} \leftarrow \tilde{J}$
          $u_{opt} \leftarrow u$
        end if
      end for
      $\mathcal{J}[x, k] \leftarrow J_{opt}$
      $\mathcal{U}[x, k] \leftarrow u_{opt}$
    end for
  end for

                                    ▷ Forward calculation

  $x_{opt}(0) \leftarrow x_i$
  for $k = 1, \ldots, N$ **do**
    $u_{opt}(k) \leftarrow \mathcal{U}[x_{opt}(k-1), k-1]$
    $x_{opt}(k) \leftarrow x_{opt}(k-1) + f_k(x_{opt}(k-1), u_{opt}(k))\Delta$
  end for

Another issue of DP is that it suffers from "curse of dimensionality", as its computation time and memory grow exponentially with the number of states. It is thus desirable to reduce the size of the problem when possible. A state reduction can be achieved when perturbations and constraints (that is, the functions L, f, and g) depend on only one between time and position. In the former case (time-only-dependent problem), the position state would be irrelevant for the optimization except for the enforcement of the final state. Position can be therefore removed from the state vector, and its final value enforced with an additional tunable term $\beta' \sum_{k=1}^{N} v_{k-1}\Delta$ that is added to the objective function. To determine the right value of the tunable coefficient β', a root-finding method can be used to drive the final position error to zero.

More common is the case where the problem depends on position but not explicitly on time. This case applies for instance to scenarios with position-depending slopes and speed limits. In these cases, it could be more convenient to reformulate (6.35)–(6.36) with position as the independent variable and the terminal time constraint enforced with a tunable terminal cost. With a position step Δ', that yields

---

Algorithm 6 SingleStateEcoDrivingDynamicProgramming

Require: $f(\cdot), L(\cdot), N, N', X', \Delta', \text{TOL}, \beta_{min}, \beta_{max}$
Ensure: $u_{opt}(1, \ldots, N), x'_{opt}(0, \ldots, N)$
  $m = 1$
  while $m < M_{max}$ **do**
    $\beta \leftarrow \frac{\beta_{min} + \beta_{max}}{2}$
    $\{u_{opt}, x'_{opt}\} \leftarrow \text{DynamicProgramming}\left(\frac{f}{v}, \frac{L+\beta}{v}, N', X', \Delta'\right)$
    $t_f \leftarrow \sum \frac{\Delta'}{v}$
    if $|t_f - N| < \text{TOL}$ **then**
      return
    else
      $m \leftarrow m + 1$
      if $\text{sign}(\beta) = \text{sign}(\beta_{min})$ **then**
        $\beta_{min} \leftarrow \beta$
      else
        $\beta_{max} \leftarrow \beta$
      end if
    end if
  end while

$$J' = \beta \underbrace{\sum_{k=1}^{N'} \frac{\Delta'}{v_{k-1}}}_{t_f} + \sum_{k=1}^{N'} \frac{L_k(\mathbf{x}'_{k-1}, \mathbf{u}_k)}{v_{k-1}} \Delta' , \qquad (6.39)$$

where $N\Delta' = s_f$, and

$$\mathbf{x}'_k = \mathbf{x}'_{k-1} + \frac{f_k(\mathbf{x}_{k-1}, \mathbf{u}_k)}{v_{k-1}} \Delta' , \qquad (6.40)$$

respectively, where $\mathbf{x}'$ is the reformulation of the state vector that replaces position with time and subscript k now denotes discretized position. The constraint on the final position is now fulfilled by construction. Similarly to the previous case, to determine the right value of the tunable coefficient β, a root-finding method can be used to drive the final time error to zero [10–12], see Algorithm 6.

In the next sections the general eco-driving OCP will be particularized for various definitions of the running cost L.

6.3 Maximizing Wheel-to-Distance Efficiency

The simplest situation is when the objective function to be minimized is represented by the powertrain energy at the wheels. In this way, the ED-OCP and its solutions become independent from the particular type of powertrain used, except for the control constraints that necessarily vary according to the peak power and the regenerative braking capability.

6.3.1 Problem Formulation

The objective of this section is to find the velocity profile that minimizes the powertrain energy defined in (2.13), going from velocity of v_i to v_f over a time t_f and a distance s_f. As mentioned in Sect. 6.2.1.3, the control inputs are defined as $u_p \triangleq F_p/m$ and $u_b \triangleq F_b/m$. The optimal control problem thus reads

$$
\begin{aligned}
\underset{u_p(t),u_b(t)}{\text{minimize}} \quad & J = \int_0^{t_f} F_p(t)v(t)dt , \\
\text{subject to} \quad & \frac{ds(t)}{dt} = v(t) , \\
& \frac{dv(t)}{dt} = u_p(t) - \frac{C_1}{m}v(t) - \frac{C_2}{m}v^2(t) - h(s(t)) - u_b(t) , \\
& v(0) = v_i , \\
& v(t_f) = v_f , \\
& s(0) = 0 , \\
& s(t_f) = s_f , \\
& u_{p,min}(v(t), t) \leqslant u_p(t) \leqslant u_{p,max}(v(t), t) , \\
& 0 \leqslant u_b(t) \leqslant u_{b,max} , \\
& v_{min}(t, s(t)) \leq v(t) \leq v_{max}(t, s(t)) , \\
& s_{min}(t) \leq s(t) \leq s_{max}(t) ,
\end{aligned}
\tag{6.41}
$$

where $u_{p,min} \leqslant 0$ and $u_{p,max} > 0$ correspond to the extreme values of the powertrain force available, $u_{b,max}$ corresponds to the maximal braking force, and $h(s) \triangleq C_0/m + g\sin(\alpha(s))$.

6.3.2 Numerical Solutions

The solution of the optimal control problem (6.53) can be obtained with the dynamic programming methods introduced in Sect. 6.2.2. In particular, the single-state DP (Algorithm 6) can be used with a redefinition of the OCP as in (6.39)–(6.40) and $\mathbf{x}' = \{v\}$, $\mathbf{u} = \{u_p, u_b\}$.

This section presents some results obtained by varying the main parameters of the problem (6.41). The passenger car parameters considered are listed in Table 6.1. Trip time is fixed at $t_f = 60$ s, while the influence of initial and final velocities, trip distance, and road grade is investigated. The baseline situation is with $v_i = v_f = 0$, $\alpha = 0$ (flat road) and $s_f = 500$ m. This is a typical scenario in arterial driving when a vehicle travels a link between two stop signs in a desired time.

Figure 6.3a shows the velocity profiles that maximize wheel-to-distance energy efficiency for various traveling distances of $\{300, 500, 700, 1100\}$ m. The baseline scenario starts with a period of maximal acceleration leading to constant speed cruising, followed by a long period of coasting, and end with maximal braking. As the trip distance increases the top speed increases, the cruising stage gradually vanishes and gives way to longer coasting intervals.

Figure 6.3b shows the various optimal speed profiles obtained for road grades $\{-5,0,5\}$ %. While the general trend is kept, higher negative or positive grades enhance the duration of the coasting or cruising phase, respectively. However, in steep downhills, coasting may eventually lead to an acceleration as seen in the figure (-5% grade).

Figure 6.3c, d show the results for various initial and final speeds. From these figures and the ones above, it is apparent that in all situations the optimal speed profiles consist of the same four phases (maximal acceleration, cruising, coasting, and maximal deceleration), and that different boundary conditions translate into different timings of such phases. Note that in some cases, for instance in Fig. 6.3d, the same phase (maximal acceleration) can be repeated twice.

Table 6.1 Vehicle parameters used in Sect. 6.3.2

Vehicle	
m	1100 kg
ρ_a	1.184 kg/m^3
A_f	2.1 m^2
C_D	0.33
C_{rr}	0.015
g	9.81 m/s^2
Driveline	
$u_{p,max}$	3.0 m/s^2
$u_{p,min}$	0 (coasting)
$u_{b,max}$	2.0 m/s^2

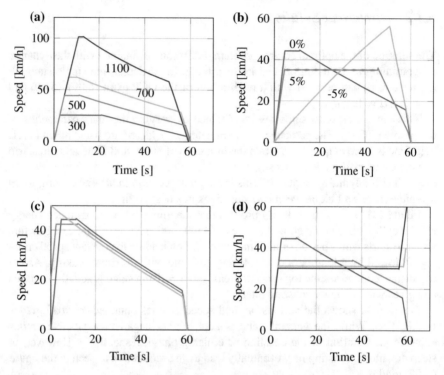

Fig. 6.3 Numerically computed speed profiles that minimize powertrain energy for a temporal horizon of 60 s and: varying spatial horizon (**a**), road grade (**b**), initial speed (**c**), final speed (**d**). Baseline values when not otherwise specified: $s_f = 500$ m, $v_i = v_f = 0$, $\alpha = 0$

6.3.3   Analytical Solutions

To have a better insight of the numerical results obtained, we derive in this section closed-form solutions of the ED-OCP using Pontryagin's minimum principle. To do that, we shall additionally assume that:

- the term C_1 of the resistance force is zero,[8]
- the term $h(s)$ is constant (constant slope),[9]
- there are no bounds on the states over the optimization horizon.

---

[8]This is not a very restrictive assumption since C_1 would model only second-order effects of speed on rolling resistance.

[9]This assumption can be removed by analyzing the problem with position as the independent variable, see [13].

Under these assumptions, the state equations read

$$\begin{aligned}
\dot{s}(t) &= v(t) \\
\dot{v}(t) &= u_p(t) - \beta v^2(t) - h - u_b(t)
\end{aligned} , \tag{6.42}$$

where $\beta \triangleq C_2/m$, while the running cost (powertrain power) is conveniently divided by the mass, yielding $u_p(t)v(t)$.

Following Pontryagin's Minimum Principle [14], the Hamiltonian H is formed as follows:

$$H = u_p(t)v(t) + \lambda(t)(u_p(t) - \beta v^2(t) - h - u_b(t)) + \mu(t)v(t) , \tag{6.43}$$

The variables λ and μ are the two costates, which have the following dynamics:

$$\begin{cases}
\dot{\mu}(t) = -\dfrac{\partial H}{\partial s} = 0 \Rightarrow \mu = \text{constant} \\[2mm]
\dot{\lambda}(t) = -\dfrac{\partial H}{\partial v} = -u_p(t) + 2\beta\lambda(t)v(t) - \mu
\end{cases} . \tag{6.44}$$

Boundary conditions for both λ and μ are free, since both states, s and v, are fixed at initial and final positions. We also note that μ is a constant over time, since its rate of change is zero, while the dynamics of λ is more complex.

The optimal control input should minimize the Hamiltonian. Since H is an affine function of u and therefore

$$\frac{\partial H}{\partial u_p} = v(t) + \lambda(t), \quad \frac{\partial H}{\partial u_b} = -\lambda(t) \tag{6.45}$$

are independent of the inputs, the Hamiltonian is minimized at extreme values of the inputs, except for when the partial derivative of H with respect to the input is zero, in which case a so-called singular interval may exist, see Fig. 6.4. Over a *singular interval* the input could take any value within its constraints. Therefore the optimal powertrain force, denoted here by u_p^*, is

$$u_p^* = \begin{cases}
u_{p,max}, & \text{if } v < -\lambda \text{ (maximal traction)} \\
u_{p,s}, & \text{if } v = \lambda \text{ (singular traction)} \\
u_{p,min}, & \text{if } v > -\lambda \text{ (overrun)}
\end{cases} , \tag{6.46}$$

where $u_{p,s}$ denotes the powertrain input during a possible singular interval.

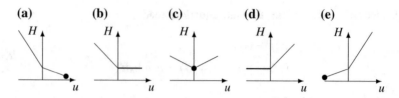

Fig. 6.4 Qualitative shape of the function $H(u)$ leading to **a** maximal acceleration, **b** singular traction, **c** coasting, **d** singular braking, **e** maximal braking. Black circles or segments indicate optimal u

The optimal braking force, denoted by u_b^* is

$$
u_b^* = \begin{cases} 0, & \text{if } \lambda < 0 \\ u_{b,s}, & \text{if } \lambda = 0 \\ u_{b,max}, & \text{if } \lambda > 0 \end{cases},
\tag{6.47}
$$

where $u_{b,s}$ denotes the braking input during a possible singular interval.

For a singular interval to exist during traction, the condition $\sigma \triangleq \lambda + v = 0$ must be valid for a time interval rather than just at one point. In other words, $d\sigma/dt$ must be zero over a singular interval. Upon substitution of state equations into (6.44),

$$
-h - \beta v^2 + 2\beta\lambda v - \mu = 0 ,
\tag{6.48}
$$

and, substituting $\lambda = -v$, we conclude that velocity during a singular traction interval must be

$$
v_{p,s} = \left(\frac{1}{3\beta}(-h - \mu) \right)^{1/2} ,
\tag{6.49}
$$

which is a constant, since μ was shown to be a constant. As a result, the singular traction force is $u_{p,s} = \beta v_{p,s}^2 + h = (2h - \mu)/3$. From these results, it should be clear that singular traction ($v_{p,s} > 0, u_{p,s} > 0$) can happen only for $\mu/2 < h < -\mu$. Since μ must be a negative quantity, singular traction is possible for both positive and negative slope.

For a singular interval to exist during braking, the condition $\lambda = 0$ should be valid for a position interval rather than just at one point. However, substituting $\lambda = 0$ and $u_p = u_{p,min}$ in (6.44) makes apparent that this condition cannot generally be sustained for any finite interval.[10] Thus singular braking is generally impossible.

---

[10]In principle, a whole trip could be singular if the boundary conditions would be satisfied with $\mu = -u_{p,min}$ and $\lambda \equiv 0$.

6.3.3.1 Control Modes

In summary, the optimal solution is comprised of four distinct modes:

- maximal traction, $u_p = u_{p,max}$, $u_b = 0$
- singular traction, $u_p = u_{p,s}$, $u_b = 0$
- overrun, $u_p = u_{p,min}$, $u_b = 0$
- maximal braking, $u_p = u_{p,min}$, $u_b = u_{b,max}$

that correspond to the phases found numerically in Sect. 6.3.2 since, for $u_{p,min} = 0$, overrunning becomes identical to coasting.

The particular sequence in which these modes occur is obtained by the solution of the two-point boundary value problem that reduces to find the two quantities λ_0, μ, such that:

$$
\begin{cases}
\dot{s} = v(t), \quad s(0) = 0, \quad s(t_f) = s_f \\
\dot{v} = u_p(t) - \beta v^2(t) - h - u_b(t), \quad v(0) = v_i, \quad v(t_f) = v_f \\
\dot{\lambda} = -u_p(t) + 2\beta\lambda(t)v(t) - \mu, \quad \lambda(0) = \lambda_0 \\
u_p(t) = (6.46), \quad u_b(t) = (6.47)
\end{cases}
\tag{6.50}
$$

It is relatively easy to solve (6.50) analytically on intervals where $u \triangleq u_p - u_b$ is piecewise constant. Note that the velocity dynamics is decoupled from the other two states, and the third equation is linear in λ. After lengthy but straightforward integrations, we obtain the following analytical expressions on intervals for which u is a constant:

$$
v(t) = \begin{cases}
a_1 \tanh\left(\tanh^{-1}\left(\frac{v_i}{a_1}\right) + a_1\beta(t - t_i)\right), & \text{if } u > h \\
\frac{v_i}{1 + v_i\beta(t - t_i)}, & \text{if } u = h \\
a_1 \tan\left(\tan^{-1}\left(\frac{v_i}{a_1}\right) + a_1\beta(t - t_i)\right), & \text{if } u < h
\end{cases}
\tag{6.51}
$$

where $a_1 \triangleq \sqrt{(u - h)/\beta}$ and the subscript i denotes the beginning of the interval. Note, however, that, despite the availability of analytical solutions (6.51), closed-form enforcement of boundary conditions and evaluation of powertrain energy consumption is generally not straightforward.

6.4 Maximizing Overall Efficiency of Combustion Engine Vehicles

The focus of the previous section was on increasing "wheel-to-distance" energy efficiency and did not address "tank-to-wheel" efficiency which is powertrain dependent. The two problems are not entirely decoupled: for instance we showed that low constant velocities improve "wheel-to-distance" energy efficiency due to lower drag. A gasoline engine on the other hand is not most efficient at low loads seen at low speeds. The engine sweet spot is typically at relatively large engine loads. To strike a balance (running the engine efficiently and maintaining a low average speed), during the cruising phase the engine could be periodically turned on at high load and then turned off, in a "pulse-and-glide" strategy. To investigate such behavior, in this section we expand on the analysis of Sect. 6.3 and include tank-to-wheel energy path for a combustion engine vehicle.

6.4.1 Problem Formulation

Instead of the powertrain energy, we choose the fuel energy defined in (2.16) as the objective function to be minimized. Therefore J represents the total fuel expended over the trip. The running cost is consequently given by the temporal rate at which fuel is injected into the engine, $\dot{m}_f(t)$. We use the representation given in Chap. 2 to relate $\dot{m}_f$ to the speed and the control input $u_p = F_p/m$. Assuming a fuel cut-off strategy and no penalty for resuming fuel injection, we obtain

$$
\dot{m}_f(t) =
\begin{cases}
\dfrac{1}{H_f}\left(\dfrac{k_{e,0}}{r_w}\gamma_e(t)v(t) + \dfrac{k_{e,1}}{r_w^2}\gamma_e^2(t)v^2(t) + \dfrac{k_{e,2}}{\eta_t}mv(t)u_p(t) + \right. \\
\left. \quad +\dfrac{k_{e,3}}{r_w\eta_t}\gamma_e(t)v^2(t)mu_p(t) + \dfrac{k_{e,4}}{r_w^2\eta_t}\gamma_e^2(t)v^3(t)mu_p(t)\right), & u_p(t) > 0 \\[2mm]
0, & u_p(t) \leqslant 0
\end{cases}
$$

$$(6.52)$$

The gear ratio can be in principle a control input to be optimized. However, as discussed in Sects. 6.1.2 and 6.2.1.3, we shall instead consider a predefined gear shift law of the type $\gamma_e(v(t), u_p(t))$.

With this assumption, the optimal control problem reads

$$
\begin{aligned}
\underset{u_p(t), u_b(t)}{\text{minimize}} \quad & J = \int_0^{t_f} \dot{m}_f(u_p(t), v(t))dt \\
\text{subject to} \quad & \frac{ds(t)}{dt} = v(t) , \\
& \frac{dv(t)}{dt} = u_p(t) - \frac{C_1}{m}v(t) - \frac{C_2}{m}v^2(t) - h(s(t)) - u_b(t) , \\
& v(0) = v_i , \\
& v(t_f) = v_f , \\
& s(0) = 0 , \\
& s(t_f) = s_f , \\
& u_{p,min}(v(t), t) \leqslant u_p(t) \leqslant u_{p,max}(v(t), t) , \\
& 0 \leqslant u_b(t) \leqslant u_{b,max} , \\
& v_{min}(t, s(t)) \leq v(t) \leq v_{max}(t, s(t)) , \\
& s_{min}(t) \leq s(t) \leq s_{max}(t) ,
\end{aligned}
$$

$$(6.53)$$

with the same definitions as in Sect. 6.3.

The maximal powertrain input $u_{p,max}$ is now an image of the maximal engine torque $T_{e,max}$ that, as described in Sect. 2.2.2, is generally a function of the engine rotational speed. The latter is related to the vehicle speed through a variable transmission ratio, whence the dependency of $u_{p,max}$ in (6.53) on both v and time. Similarly, the minimal powertrain input $u_{p,min}$ results from the minimal (braking) engine torque $T_{e,min}$.

6.4.2 Numerical Solutions

The solution of the optimal control problem (6.53) can be obtained with the dynamic programming methods introduced in Sect. 6.2.2. In particular, the single-state DP (Algorithm 6) can be used with a redefinition of the OCP as in (6.39)–(6.40) and $\mathbf{x}' = \{v\}$, $\mathbf{u} = \{u_p, u_b\}$.

This section presents results obtained with a single-state DP code by varying the main parameters of the problem (6.53). The vehicle considered is described in Table 6.2. Additionally, we consider a coasting strategy (see Sect. 2.2.2) with fuel cut-off, where the engine does not consume fuel for torques lower or equal to zero. Consequently, we enforce the system to operate with a $u_{p,min} = 0$.

The scenario considered is again a driving between two stops ($v_i = v_f = 0$) on a flat road ($\alpha = 0$), in a fixed time $t_f = 60$ s. Figure 6.5a shows the velocity profile

Table 6.2 ICEV parameters used in Sect. 6.4.2

Vehicle see Table 6.1	
Engine	
$k_{e,0}$	57.0
$k_{e,1}$	0.0697
$k_{e,2}$	2.80
$k_{e,3}$	−0.0032
$k_{e,4}$	$5.79 \cdot 10^{-6}$
$k_{e,8}$	50 N
$k_{e,9}$	0.1
$k_{e,10}$	300
$k_{e,11}$	521
$k_{e,12}$	−0.048
H_f	42.2 MJ/kg
Driveline	
r_w	0.2785 m
$u_{p,min}$	0
$u_{b,max}$	2.0 m/s^2

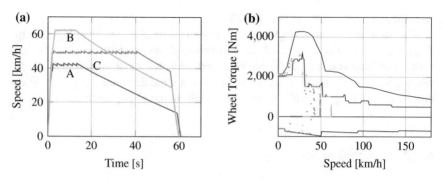

Fig. 6.5 Numerically computed speed profiles that minimize fuel energy for a temporal horizon of 60 s, $v_i = v_f = 0$, $\alpha = 0$, and variable s_f, v_{max}: A ($s_f = 500$ m), B ($s_f = 750$ m), C ($s_f = 750$ m, $v_{max} = 50$ km/h) (**a**). Corresponding powertrain operating points in the v-T_p plane, with curves $T_{p,max}$, $T_{p,\eta}$, $T_{p,min}$, and $T_{p,min} - T_{b,max}$ (**b**)

that minimizes fuel energy for three situations: A ($s_f = 500$ m), B ($s_f = 750$ m), and C (same as B with an additional speed constraint of $v_{max} = 50$ km/h). All profiles start with a phase with high powertrain torque, thus high acceleration. Then, speed reaches a cruising value that is kept by switching the engine on and off very fast. This peculiar mode has been already introduced in Sect. 6.1.2 and is known in the eco-driving literature as *Pulse and Glide* (P&G). In theory, when the engine can be rapidly turned on and off at no extra cost, the P&G can be done infinitely fast keeping the speed effectively constant.[11] After the P&G phase, a coasting phase with engine off decelerates the vehicle, which is then fully stopped by a braking phase. Note that adding a speed constraint (profile C) does not modify the nature of the speed phases but only their relative duration.

The optimal profiles A–C are represented as points in a speed-torque plane in Fig. 6.5b. The torque curves shown are the maximal powertrain torque, the powertrain torque corresponding to the engine OOL ($T_{e,\eta}$, see Sect. 2.2.2), the minimal powertrain torque, and the maximal braking torque ($T_{p,min} - T_{b,max}$). These values are expressed as a function of vehicle speed by combining the engine map and the gear shifting map. Clearly, acceleration is performed along the OOL, while braking points are adjacent to the maximal braking torque curve. Also apparent are the coasting points at zero torque. The intermediate points between the OOL and the coasting points are likely due to numerical effects and vary with the particular scenario considered.

---

[11]This behavior is similar to the process that *cylinder deactivation* exploits to save fuel. That is, instead of many cylinder firing events at lower manifold pressure and lower efficiency per event, the same average torque output is obtained with fewer cylinder firing events at higher manifold pressure and higher efficiency per event.

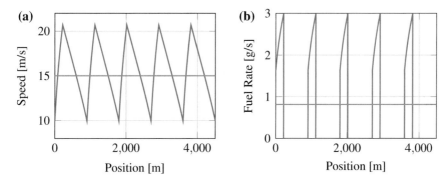

Fig. 6.6 Velocity (**a**) and fuel mass flow rate (**b**) variation of low-frequency (1/60 Hz) pulse and glide versus constant speed cruising at 15 m/s (54 km/h or 33.5 mph). The maximum traction acceleration (pulse acceleration) was set to 1 m/s^2

The evidence of a pulse and glide phase deserves a further analysis to understand its effectiveness. For this study simplified models of the engine and the gearbox are used. We compare P&G at a given engine on–off period with cruising at the corresponding constant average speed. For example, in Fig. 6.6a we depict a scenario where pulse and glide is exercised at a period of 60 s around the equivalent constant velocity of 15 m/s (representative of in city driving), with a pulse acceleration $u_{p,max} = 1$ m/s^2. The corresponding fuel consumption rates are shown in Fig. 6.6b. The fuel saving computed over an integer number of periods is 32%.

However, this fuel saving comes at the cost of the discomfort of a mild acceleration during a pulse stage. Moreover, the velocity varies between 10 and 20 m/s every minute, which could be disruptive to traffic. The velocity variation band can be narrowed by reducing the P&G period, however we are limited by how fast the engine can be turned on and off. We have not considered in this book the fuel cost of turning the engine on. This cost subtracts from fuel savings and is another constraint to consider when deciding on a viable P&G frequency. On the other hand, increasing the engine on/off period results in velocity fluctuation that increases aerodynamic drag losses and therefore a trade-off must be made. In theory, when the engine can be rapidly turned on and off at no extra cost, the Pulse and Glide (P&G) should be done infinitely fast keeping the speed effectively constant. Figure 6.7 summarizes fuel savings and velocity variation for different periods of P&G. Figure 6.7a confirms that the most effective energy savings are obtained when the P&G period virtually tends to zero (infinitely fast switching), a situation that is numerically approached by the results of Fig. 6.5. Figure 6.7a also shows the increase in aerodynamic drag losses as a result of velocity variations in P&G (illustrated by Fig. 6.7b). Beyond a certain P&G period, the drag losses offset the engine efficiency gain to the point that P&G is no longer efficient (negative fuel savings). The results shown are for an average speed of 15 m/s. The gains are expected to be less at higher average speeds.

The effectiveness of pulse and glide algorithms has been shown in [15], using theory of optimal control in [16], and experimentally in [17] but overall the existing

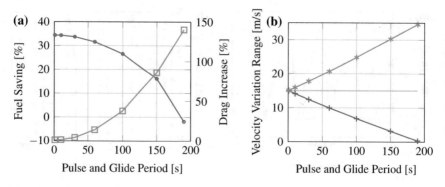

Fig. 6.7 Fuel saving, drag increase (**a**), and velocity variation (**b**) of pulse and glide versus constant speed cruising at 15 m/s (54 km/h or 33.5 mph). Results are shown for pulse acceleration of 1 m/s^2

literature presents mixed and sometimes conflicting results. Existence of similar *chattering arcs* have been shown before in flight control literature [18]. As a final remark, we have already noted that pulse and glide may not be a practical eco-driving strategy because velocity variations are uncomfortable to passengers and disruptive to traffic. Also according to [17] P&G may not be an effective approach in vehicles with automatic transmission due to torque converter losses.

6.4.3   Analytical Solutions

A theoretical justification of the chattering behavior illustrated in the previous section can be derived by analytically solving the ED-OCP (6.53) with PMP. We shall additionally assume that:

- the term C_1 of the resistance force is zero,
- the term $h(s)$ is constant (constant slope),
- the driveline efficiency η_t is constant,
- the gear shift law depends only on vehicle speed, $\gamma_e(t) = \gamma_e(v(t))$, that is, a piecewise-constant function of the speed similar to (2.18),
- the engine OOL torque coincides with the maximal torque, thus the model (2.23) is valid for the whole engine range,[12]
- the engine indicated efficiency is constant, $k_{e,3} = k_{e,4} = 0$,
- the engine friction torque is constant, $k_{e,1} = 0$,
- assuming coasting and fuel cut-off, engine brake is not used and $u_{p,min} = 0$,
- there are no bounds on the states over the optimization horizon.

---

[12] See discussion of this assumption in note 4 of Chap. 2.

Under these assumptions, the running cost reads

$$\dot{m}_f(t) = \begin{cases} p_0(v(t)) + p_1 u_p(t)v(t), & \text{if } u_p(t) > 0 \text{ (fueling)} \\ 0, & \text{if } u_p(t) = 0 \text{ (fuel cut-off)} \end{cases} \tag{6.54}$$

or, in a more compact way, $\dot{m}_f(t) = (p_0(v(t)) + p_1 u_p(t)v(t))\mathcal{H}(u_p)$, where $\mathcal{H}$ is the Heaviside's step function, $p_0 \triangleq k_{e,0}\gamma_e(v)v/(r_w H_f)$, and $p_1 \triangleq k_{e,2}m/(\eta_t H_f)$.

Utilizing the state dynamics in (6.42) and the above cost function, the Hamiltonian is

$$H = \left(p_0(v(t)) + p_1 u_p(t)v(t)\right)\mathcal{H}(u_p) + \lambda(t)\left(u_p(t) - \beta v^2(t) - h - u_b(t)\right) + \mu(t)v(t). \tag{6.55}$$

The costate dynamics read

$$\dot{\mu}(t) = -\frac{\partial H}{\partial s} = 0 \Rightarrow \mu = \text{constant},$$

$$\dot{\lambda}(t) = -\frac{\partial H}{\partial v} = -\left(\frac{\partial p_0}{\partial v} + p_1 u_p(t)\right)\mathcal{H}(u_p) + 2\beta\lambda(t)v(t) - \mu. \tag{6.56}$$

Boundary conditions for both λ and μ are free, since both states, s and v, are fixed at initial and final positions. We also note that μ is a constant over time, since its rate of change is zero, while dynamics of λ is more complex.

The Hamiltonian is piecewise-affine in the inputs[13] as in (6.24), with

$$\frac{\partial H}{\partial u_p} = p_1 v(t)\mathcal{H}(u_p) + \lambda(t), \qquad \frac{\partial H}{\partial u_b} = -\lambda(t), \tag{6.57}$$

but now it additionally presents a discontinuity for $u_p = 0$. Therefore, as Fig. 6.8 illustrates, its minimum is found by comparing the value at $u_{p,max}$ with the value at $u_p = 0$. The optimal input that minimizes the switching Hamiltonian is found as

$$u_p^* = \begin{cases} u_{p,max}, & \text{if } \lambda < \lambda_B \\ u_{p,max} \text{ or } 0, & \text{if } \lambda = \lambda_B \\ 0, & \text{if } \lambda > \lambda_B \end{cases} \tag{6.58}$$

where $\lambda_B \triangleq -p_0/(p_1 u_{p,max}) - p_1 v$. Note that there is no singular interval possible for positive u_p due to the switching Hamiltonian.[14] However, at the borderline when

---

[13]Note that, taking γ_e as an additional control input yields $\partial H/\partial\gamma_e = k_{e,0}v/(r_w H_f) > 0$, thus the optimal gear shifting law would be to use always the highest gear available (smallest value of γ_e), as discussed earlier in Sect. 6.1.2.

[14]Note that if the assumption of $u_{p,min} = 0$ and fuel cutoff is removed, there is no discontinuity in the Hamiltonian. Optimal driving modes include a singular arc similarly to Sect. 6.3 instead of the PnG mode.

Fig. 6.8 Qualitative shape of the function $H(u_p)$ leading to maximal traction (a), pulse and glide (b), coasting (c). Black circles indicate optimal u_p

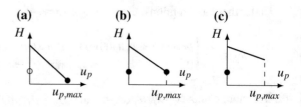

$\lambda = \lambda_B$, both $u_p = 0$ and $u_p = u_{p,max}$ are optimal and the optimal solution switches between the two values, whence the "pulse-and-glide" behavior already introduced.

Moreover, the switch between pulse and glide is infinitely fast. This assertion is proved by assuming a switch in finite time between the two solutions. Thus there is a first phase (I) where $u_p = u_{p,max}$ and $\lambda < \lambda_B$ and a second phase (II) where $u_p = 0$ and $\lambda > \lambda_B$. In order to avoid discontinuity in λ and ensure the periodicity of the solution, it must be that $\ddot{\lambda}^{(I)} > 0$ and $\ddot{\lambda}^{(II)} < 0$ so that λ crosses the value λ_B exactly at the switch between the two phases. Let us evaluate now the second derivative of λ in the two phases, by differentiating (6.56) for constant u_p:

$$\ddot{\lambda}(t) = 2\beta v(t)\dot{\lambda}(t) + 2\beta\lambda(t)\dot{v}(t) . \tag{6.59}$$

Assuming small variations of λ around λ_B and of v, (6.56) shows that $\dot{\lambda}^{(I)} \approx \dot{\lambda}^{(II)} - \left(\frac{\partial p_0}{\partial v} + p_1 u_{p,max}\right)$. Since $-\frac{\partial p_0}{\partial v}$ can be neglected if no gear change occurs during the switches, that yields

$$\ddot{\lambda}^{(I)} \approx \ddot{\lambda}^{(II)} + 2\beta\lambda_B(\dot{v}^{(I)} - \dot{v}^{(II)}) - 2\beta p_1 u_{p,max} v . \tag{6.60}$$

Since $\dot{v}^{(I)} > \dot{v}^{(II)}$ by construction, (6.60) shows that $\ddot{\lambda}^{(I)} < \ddot{\lambda}^{(II)}$, contrarily to what was required. This result means that the switch cannot occur in a finite time, but is infinitely fast.

As for the optimal braking input, we can easily show that it is still given by (6.47). As in the previous section, singular braking is not possible for a finite interval: for $\lambda = 0$, $u_p = 0$, the condition $\dot{\lambda} = 0$ cannot be generally sustained.

6.4.3.1   Control Modes

In summary, the optimal speed profiles of ICEVs comprise of at most of four distinct modes:

- maximal traction (A), $u_p = u_{p,max}$, $u_b = 0$
- pulse-and-glide (S), $u_p = u_{p,max}/0$, $u_b = 0$
- coasting (C), $u_p = 0$, $u_b = 0$
- maximal braking (B), $u_p = 0$, $u_b = u_{b,max}$.

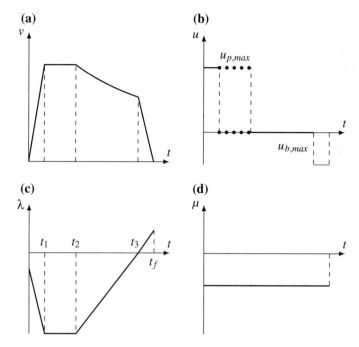

Fig. 6.9 Qualitative sketch of the analytical eco-driving solution for an ICEV, sequence A-S-C-B, in terms of speed (**a**), control inputs u_p (**b**, black) and u_b (**b**, gray), speed costate (**c**), and position costate (**d**)

The particular sequence and timing in which these modes occur depend on the boundary conditions and are obtained by the solution of the two-point boundary value problem (TPBVP) that reduces to finding the two quantities λ_0 and μ, such that:

$$
\begin{cases}
\dot{s} = v(t), \quad s(0) = 0, \quad s(t_f) = s_f, \\
\dot{v} = u_p(t) - \beta v^2(t) - h - u_b(t), \quad v(0) = v_i, \quad v(t_f) = v_f, \\
\dot{\lambda} = -\left(\dfrac{\partial p_0}{\partial v} + p_1 u_p(t)\right)\mathcal{H}(u_p) + 2\beta\lambda(t)v(t) - \mu, \quad \lambda(0) = \lambda_0, \\
u_p(t) = (6.58), \quad u_b(t) = (6.47).
\end{cases}
\tag{6.61}
$$

In practice, four basic sequences of control modes are possible: A-S-C-(B), A-S-A, C-S-C-(B), or C-S-A. For example, the sequence A-S-C-B is likely to be optimal when both v_i and v_f are small or zero. A qualitative sketch of a complete solution of this scenario, in terms of speed, control inputs, and costates is shown in Fig. 6.9.

6.4.3.2   Solution of the TPBVP

The sequential integration of the system (6.61), similarly to that of the corresponding system (6.50), as a function of the unknown initial values λ_0 and μ_0 is generally not possible since a PnG arc (phase S) always appears in the optimal sequence. However, the two-point boundary value problem can be reduced to a *parametric optimization* problem, where the correct sequence of control modes and the switching times between these modes, t_i, $i = 1, \ldots, 3$ are to be found.

For the complete, four-phase mode sequences, two switching times are determined by the terminal conditions $v(t_f) = v_f$, $s(t_f) = s_f$, while the third is left undetermined. This degree of freedom should be therefore found in such a way that the fuel consumption is minimized. Solving the TPBVP is thus reduced to a parametric MOOP of the type

$$\min_{t_1, t_2, t_3} \left(|v(t_f) - v_f|, |s(t_f) - s_f|, |m_f(t_f)| \right) , \tag{6.62}$$

where the three terms on the right-hand side clearly depend on the choice of the t_i's. For three-phase mode sequences, the two switching times t_1 and t_2 are fully defined by enforcement of the two terminal conditions; thus, only the first two terms appear at the right-hand side of (6.62). See Appendix A for a more detailed implementation of the method.

Instead of running a parametric optimization routine, a more practical online approach consists of directly evaluating the optimal switching times as a function of the boundary conditions, with the aid of a neural network [19]. For ICEVs, the latter would have an input vector $I = \{t_f, s_f, v_i, v_f, \alpha\}$ and an output vector $O = \{t_1, \ldots, t_3\}$.

6.5   Maximizing Overall Efficiency of Electric Vehicles

The two previous sections have shown that minimizing the powertrain energy leads to a bang-singular-bang behavior, while considering ICE-based powertrains and their affine-in-torque characteristic may lead to an optimal pulse-and-glide control. In this section, we extend our analysis to electric vehicles for which, due to their nonlinear characteristic, different types of speed profiles are optimal.

6.5.1   Problem Formulation

We choose now the battery energy defined in (2.30) as the objective function to be minimized. Therefore J represents the total electricity expended over the trip. The running cost is consequently given by the rate at which the battery is depleted, or the electrochemical power $P_b(t)$ defined in (2.39). We use the representation given in Chap. 2 to relate P_b to the speed and the control input $u_p = F_p/m$,

$$
P_b(t) =
\begin{cases}
\dfrac{1}{\eta_b}\left(k_{m,0} + k_{m,1}\dfrac{\gamma_m v(t)}{r_w} + k_{m,2}\dfrac{\gamma_m^2 v^2(t)}{r_w^2} + k_{m,3}\dfrac{mu_p(t)v(t)}{\eta_t}\right. + \\
\left. +k_{m,4}\dfrac{m^2 r_w^2 u_p^2(t)}{\gamma_m^2 \eta_t^2}\right), \quad u_p > 0 \\[2em]
\eta_b\left(k_{m,0} + k_{m,1}\dfrac{\gamma_m v(t)}{r_w} + k_{m,2}\dfrac{\gamma_m^2 v^2(t)}{r_w^2} + \right. \\
\left. +k_{m,3}mu_p(t)v(t)\eta_t + k_{m,4}\dfrac{\eta_t^2 m^2 r_w^2 u_p^2(t)}{\gamma_m^2}\right), \quad u_p \leqslant 0
\end{cases}
\tag{6.63}
$$

The optimal control problem reads

$$
\begin{aligned}
&\underset{u_p(t),u_b(t)}{\text{minimize}} \quad J = \int_0^{t_f} P_b(u_p(t), v(t))dt, \\
&\text{subject to} \quad \frac{ds(t)}{dt} = v(t), \\
&\frac{dv(t)}{dt} = u_p(t) - \frac{C_1}{m}v(t) - \frac{C_2}{m}v^2(t) - h(s(t)) - u_b(t), \\
&v(0) = v_i, \\
&v(t_f) = v_f, \\
&s(0) = 0, \\
&s(t_f) = s_f, \\
&u_{p,min}(v(t)) \leqslant u_p(t) \leqslant u_{p,max}(v(t)), \\
&0 \leqslant u_b(t) \leqslant u_{b,max}, \\
&v_{min}(t, s(t)) \leq v(t) \leq v_{max}(t, s(t)), \\
&s_{min}(t) \leq s(t) \leq s_{max}(t),
\end{aligned}
\tag{6.64}
$$

with the same definitions as in Sect. 6.3.

Similarly to what was discussed in the previous section, it should be clear that the control limits $u_{p,max}$ and $u_{p,min}$ result from the maximal and minimal (braking) motor torques, $T_{m,max}$ and $T_{m,min}$, respectively. Both of these quantities are functions of the motor speed and thus, assuming a fixed transmission ratio, of the vehicle speed.

6.5.2   Numerical Solutions

The solution of the optimal control problem (6.64) can be obtained with the dynamic programming methods introduced in Sect. 6.2.2. In particular, the single-state DP (Algorithm 6) can be used with a redefinition of the OCP as in (6.39)–(6.40) and $\mathbf{x}' = \{v\}$, $\mathbf{u} = \{u_p, u_b\}$.

This section presents some results obtained with DP by varying the main parameters of the problem (6.64). The electric vehicle considered is described in Table 6.3. The scenario considered is again a trip between two stops ($v_i = v_f = 0$) on a flat road ($\alpha = 0$), in a fixed time $t_f = 60$ s. Figure 6.10a shows the velocity profile that minimizes battery energy for the same three situations as in Sect. 6.4: A ($s_f = 500$ m), B ($s_f = 750$ m), and C (same as B with an additional speed constraint of $v_{max} = 50$ km/h). The optimal profiles A–C are represented as points in a speed-torque plane in Fig. 6.10b, which also shows the maximal and minimal powertrain torques and the minimal wheel torque ($T_{p,min} - T_{b,max}$). These values are expressed as a function of vehicle speed by combining the motor map and the driveline characteristics.

The optimal profiles are generally different from those obtained for an ICEV. Now the vehicle initially accelerates with decreasing torque rather than following the maximum or OOL torque curve. There is no cruising phase, unless induced by a speed limit, and the maximum speed is directly followed by a coasting phase. The

Table 6.3 EV parameters used in Sect. 6.5.2.

Motor	
$k_{m,0}$	96.2
$k_{m,1}$	-1.85
$k_{m,2}$	0.0034
$k_{m,3}$	1
$k_{m,4}$	-0.53
$k_{m,5}$	0.0123
$k_{m,6}$	$-1.98 \cdot 10^{-6}$
$k_{m,7}$	70 N
$\omega_{m,base}$	250 rad/s
Battery	
R_b	0.0255 Ω
V_{b0}	75.6 V
Driveline	
γ_m	9.91
η_t	0.95
r_w	0.2785 m
$u_{b,max}$	2.0 m/s^2

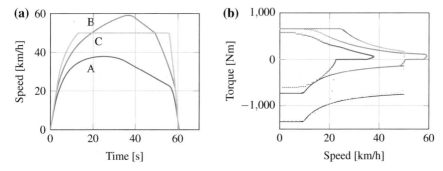

Fig. 6.10 Numerically computed speed profiles that minimize battery energy for a temporal horizon of 60 s, $v_i = v_f = 0$, $\alpha = 0$, and variable s_f, v_{max}: A ($s_f = 500$ m), B ($s_f = 750$ m), C ($s_f = 750$ m, $v_{max} = 50$ km/h) (**a**). Corresponding powertrain operating points in the v–T_p plane, with curves $T_{p,max}$, $T_{p,min}$, and $T_{p,min} - T_{b,max}$ (**b**)

vehicle also decelerates with decreasing torque. Note, however, that in scenario A the motor limits are never reached, neither in traction nor in braking, while that is the case in scenarios B and C, where additionally the braking limits are reached.

6.5.3 Analytical Solutions

Again we solve the ED-OCP (6.64) in a closed form using Pontryagin's minimum principle. Additional assumptions are:

- the term C_1 of the resistance force is zero,
- the term $h(s)$ is constant (constant slope),
- the driveline efficiency η_t is constant,
- in the motor model $b_0 = 0$, $k_{m,3} = 1$, and $k_{m,5} = k_{m,6} = 0$,
- there are no bounds on the states over the optimization horizon.

Under these assumptions, the running cost reads

$$P_b(t) = mu_p(t)v(t)\eta^{-\text{sign}(u_p)} + bu_p^2(t)(\eta\eta_t)^{-\text{sign}(u_p)}, \tag{6.65}$$

where $b \triangleq k_{m,4}m^2 r_w^2 \gamma_m^{-2}$, and $\eta \triangleq \eta_t\eta_b$.

Utilizing the state dynamics in (6.42) and the above cost function, the Hamiltonian is formed as

$$
\begin{aligned}
H = {}& mu_p(t)v(t)\eta^{-\text{sign}(u_p)} + bu_p^2(t)(\eta\eta_t)^{-\text{sign}(u_p)} + \\
& + \lambda(t)\left(u_p(t) - \beta v^2(t) - h - u_b(t)\right) + \mu(t)v(t).
\end{aligned} \tag{6.66}
$$

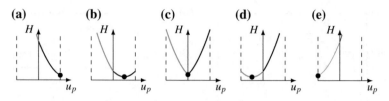

Fig. 6.11 Qualitative shape of the function $H(u_p)$ leading to maximal traction (**a**), optimal traction (**b**), coasting (**c**), optimal reg. braking (**d**), maximal reg (**e**). braking. Black circles indicate optimal u_p

The costate dynamics read

$$\dot{\mu}(t) = -\frac{\partial H}{\partial s} = 0 \Rightarrow \mu = \text{constant} ,$$

$$\dot{\lambda}(t) = -\frac{\partial H}{\partial v} = -mu_p(t)\eta^{-\text{sign}(u_p)} + 2\beta\lambda(t)v(t) - \mu . \tag{6.67}$$

Boundary conditions for both λ and μ are free, since both states, s and v, are fixed at initial and final positions. We also note that μ is a constant over time, since its rate of change is zero, while dynamics of λ is more complex.

This time the Hamiltonian, while still affine in the friction braking input, is quadratic in the powertrain input,

$$\frac{\partial H}{\partial u_p} = mv(t)\eta^{-\text{sign}(u_p)} + 2bu_p(t)(\eta\eta_t)^{-\text{sign}(u_p)} + \lambda(t), \quad \frac{\partial H}{\partial u_b} = -\lambda(t) , \tag{6.68}$$

leading to possible local minima

$$u_p^+(t) \triangleq -\frac{\eta_t}{2b}(\eta\lambda(t) + mv(t)) , \quad u_p^-(t) \triangleq -\frac{1}{2b\eta_t}\left(mv(t) + \frac{\lambda(t)}{\eta}\right) , \tag{6.69}$$

that are also global optima provided that $0 \leqslant u_p^+ \leqslant u_{p,max}$ or $u_{p,min} \leqslant u_p^- \leqslant 0$. Otherwise, according to the PMP, the optimal input lays at one of the boundaries, as Fig. 6.11 illustrates. Therefore, the optimal powertrain input that minimizes the Hamiltonian is found as

$$u_p^* = \begin{cases} u_{p,max} & \lambda \leqslant \lambda_{MT} & \text{(max. traction)} \\ u_p^+ & \lambda_{MT} < \lambda \leqslant \lambda_T & \text{(opt. traction)} \\ 0 & \lambda_T < \lambda \leqslant \lambda_B & \text{(coasting)} \\ u_p^- & \lambda_B < \lambda \leqslant \lambda_{MB} & \text{(opt. reg. braking)} \\ u_{p,min} & \lambda_{MB} < \lambda & \text{(max. reg. braking)} \end{cases} \tag{6.70}$$

where $\lambda_{MT} \triangleq -mv/\eta - 2bu_{p,max}/(\eta\eta_t)$, $\lambda_T \triangleq -mv/\eta$, $\lambda_B \triangleq -mv\eta$, λ_{MB} $\triangleq -mv\eta - 2b\eta\eta_t u_{p,min}$. As for the optimal braking input, we can easily show that it is still given by (6.47).

Note that coasting mode appears due the presence of the term $\eta^{-sign(u_p)}$ in the Hamiltonian function and, for the same reason, the optimal traction mode is distinguished from the optimal regenerative braking mode. Indeed, if $\eta = 1$, then $\lambda_B \equiv \lambda_T$ and $u_p^+ \equiv u_p^-$.

6.5.3.1 Control Modes

In summary, the optimal solutions comprise of up to six distinct modes:

- maximal traction (MA), $u_p = u_{p,max}$, $u_b = 0$
- optimal traction (A), $u_p = u_p^+$, $u_b = 0$
- coasting (C), $u_p = 0$, $u_b = 0$
- optimal powertrain braking (D), $u_p = u_p^-$, $u_b = 0$
- maximal powertrain braking (MD), $u_p = u_{p,min}$, $u_b = 0$
- maximal braking (B), $u_p = u_{p,min}$, $u_b = u_{b,max}$.

Clearly, this solution reflects the structure of the results obtained in Sect. 6.5.2, with the six modes corresponding to the various phases that are visible in Fig. 6.11.

The particular sequence in which these modes occur is obtained by the solution of the two-point boundary value problem that reduces to finding the two quantities λ_0, μ, such that:

$$\begin{cases} \dot{s} = v(t), \quad s(0) = 0, \quad s(t_f) = s_f \\ \dot{v} = u_p(t) - \beta v^2(t) - h - u_b(t), \quad v(0) = v_i, \quad v(t_f) = v_f \\ \dot{\lambda} = -mu_p(t)\eta^{-sign(u_p)} + 2\beta\lambda(t)v(t) - \mu, \quad \lambda(0) = \lambda_0 \\ u_p(t) = (6.58), \quad u_b(t) = (6.47) \end{cases} \tag{6.71}$$

A qualitative sketch of a complete solution with all six phases, in terms of speed, control inputs, and costates is shown in Fig. 6.12. Different boundary conditions may result in a different number of phases or a different ordering. For low distances, for example, the maximal traction and maximal braking phases may disappear, as it is the case in scenario A of Fig. 6.11.

6.5.3.2 Solution of the TPBVP

Assuming a full six-mode sequence as in Fig. 6.12, the calculation of the optimal velocity and position trajectories can be performed in a *sequential* fashion [20]. Start with unknown λ_0, μ, and integrate (6.71) for each phase of the sequence. Switching times between phases t_i, $i = 1, \ldots, 5$ are defined by the switching conditions $u_p^+(t_1) = u_{p,max}$, $u_p^+(t_2) = 0$, $u_p^-(t_3) = 0$, $u_p^-(t_4) = u_{p,min}$, $\lambda(t_5) = 0$. Then

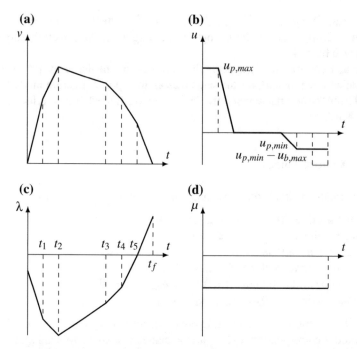

Fig. 6.12 Qualitative sketch of the analytical eco-driving solution, in terms of speed (**a**), control inputs u_p (**b**, black) and u_b (**b**, gray), speed costate (**c**), and position costate (**d**)

impose the two terminal conditions $v(t_f) = v_f$ and $s(t_f) = s_f$ and obtain two equations $f_1(\lambda_0, \mu) = 0$ and $f_2(\lambda_0, \mu) = 0$ in the two unknown initial conditions. It turns out that the former equation is affine in the variable λ_0 and can be solved in closed form. By replacing the resulting $\lambda_0(\mu)$ in the latter equation, a tenth-order polynomial equation $f_3(\mu) = 0$ is obtained. This equation has only one physically meaningful solution, although it is generally impossible to obtain in closed form and a numerical procedure is necessary.

Alternatively, the neural network approach mentioned in Sect. 6.4.3 is still a valid option for online implementation. While the input vector contains all the boundary conditions and is the same as in the ICEV case, $I = \{t_f, s_f, v_i, v_f, \alpha\}$, the output vector must include not only the switching times but also the values of the maximal or minimal control input (motor torque $\hat{u}_{p,max}$ or $\hat{u}_{p,min}$) during phase A (optimal traction) or D (optimal braking) when phase MA or MD, respectively, is not present. To avoid ambiguities, the output vector can be defined as $O = \{t_1, t_2, t_3, t_4, t_5, \hat{u}_{p,max}, \hat{u}_{p,min}\}$ for all cases. In the case of a full six-mode sequence, $\hat{u}_{p,max} = u_{p,max}$ and $\hat{u}_{p,min} = u_{p,min}$. Otherwise ($t_1 = 0$ or $t_4 = t_5$), $\hat{u}_{p,max}$ and $\hat{u}_{p,min}$ must be determined alongside with the switching times. In [19] the outputs have been divided between four feedforward smaller networks, with one hidden layer each.

6.5.3.3 Parabolic Speed Profile

If further simplifications are made to the model, namely:

- $C_2 = \beta = 0,$[15]
- $\eta_t = \eta_b = 1,$
- $u_{b,max} = 0,$
- $u_{p,max} = -u_{p,min} \to \infty,$

then the optimal solution would consist of one single phase, resulting from the merging of the optimal traction and optimal braking phases defined above ($u_p^+ \equiv u_p^-$). Given the first assumption, there is no coasting phase, and dissipative braking is also absent ($u_b \equiv 0$). The system (6.71) reduces to

$$
\begin{cases}
\dot{s} = v(t), & s(0) = 0, \quad s(t_f) = s_f \\
\dot{v} = u_p(t) - h, & v(0) = v_i, \quad v(t_f) = v_f \\
\dot{\lambda} = -mu_p(t) - \mu, & \lambda(0) = \lambda_0 \\
u_p(t) = -\dfrac{1}{2b}(\lambda(t) + mv(t))
\end{cases}
\tag{6.72}
$$

The optimal control trajectory $u_p(t)$ can be explicitly calculated and it is an affine function of time,

$$
u_p(t) = \left(h - \frac{4v_i}{t_f} - \frac{2v_f}{t_f} + \frac{6s_f}{t_f^2} \right) + \left(\frac{6v_i}{t_f^2} + \frac{6v_f}{t_f^2} - \frac{12s_f}{t_f^3} \right) t,
\tag{6.73}
$$

while the optimal trajectory $v(t)$ is a quadratic function of time [20, 21],

$$
v(t) = v_i + \left(-\frac{4v_i}{t_f} - \frac{2v_f}{t_f} + \frac{6s_f}{t_f^2} \right) t + \left(\frac{3v_i}{t_f^2} - \frac{6s_f}{t_f^3} + \frac{3v_f}{t_f^2} \right) t^2.
\tag{6.74}
$$

We will therefore refer to this approximated solution as the *"parabolic" speed profile* in the following. Note that the parabolic speed profile is completely determined by the boundary conditions t_f, s_f, v_i, and v_f and does not depend on the system's parameters.

The associated energy consumption, however, is a function of the vehicle parameters m, b, and h and is found to be

$$
\begin{aligned}
E_b = {}& mhs_f + m\frac{v_f^2 - v_i^2}{2} + bh^2 t_f + 2bh(v_f - v_i) + \\
& + 4b\left(\frac{3s_f^2}{t_f^3} - \frac{3s_f(v_i + v_f)}{t_f^2} + \frac{v_i^2 + v_i v_f + v_f^2}{t_f} \right).
\end{aligned}
\tag{6.75}
$$

---

[15]This is generally a strong assumption as it implies that the aerodynamic drag is zero. However, considering an electric city car that usually travels at low speeds, the error introduced may be limited.

Note that only the last term of (6.75) depends on the control law $u_p(t)$ used, while the other four depend only on the overall boundary conditions. We shall therefore introduce the following definition of *effective energy consumption* to be used later:

$$E_{b,e} \triangleq 4b \left(\frac{3s_f^2}{t_f^3} - \frac{3s_f(v_i + v_f)}{t_f^2} + \frac{v_i^2 + v_iv_f + v_f^2}{t_f} \right) . \tag{6.76}$$

Equation (6.74) represents an admissible speed profile only for certain combinations of the boundary conditions. In particular, conditions of the type $F(s_f, t_f) \geq 0$ for specific v_i and v_f define the domain of validity of the parabolic speed profile. These conditions impose that the speed is always positive, that its maximum does not exceed a given limit v_{max}, and that its maximum derivative (acceleration) or, alternatively, the control input, does not exceed the limit a_{max} or $u_{p,max}$, respectively. A general derivation of such conditions is presented in Appendix B. Specific cases will be discussed in Chap. 7.

6.6   Maximizing Overall Efficiency of Hybrid Vehicles

In hybrid-electric powertrains the minimization of the energy consumption is inherently related to the optimization of the energy management strategy, see (2.4.2). We refer in the following to a parallel hybrid configuration.

6.6.1   Problem Formulation

We use the "tank" energy defined in (2.46) as the objective function to be minimized but we additionally specify the value of the battery energy consumption by prescribing a final state of charge. This position is thus equivalent to minimizing the fuel energy under a terminal state constraint on the SoC. The control vector is now composed of the three inputs, corresponding to the wheel force coming from the engine, $u_{p,e}$, the wheel force coming from the motor, $u_{p,m}$, and the braking force, u_b. The fuel consumption rate is still given by (6.52) with $u_{p,e}$ replacing u_p, while the battery electrochemical power is given by (6.63) with $u_{p,m}$ replacing u_p. The state vector includes now the battery state of charge.

Due to the presence of SoC as a state variable, decoupling the OCP into several subproblems for each single subtrip as discussed in Sect. 6.2.1.7 is generally not possible, since the SoC is prescribed only at the end of the whole trip, and not at road discontinuities where the position and the speed can be reasonably prescribed. Interior constraints of the type (6.22) must be therefore taken into account explicitly.

The optimal control problem thus reads

$$\underset{u_{p,e}(t),u_{p,m}(t),u_b(t)}{\text{minimize}} \quad J = \int_0^{t_f} \dot{m}_f(u_{p,e}(t), v(t))dt,$$

$$\text{subject to} \quad \frac{ds(t)}{dt} = v(t),$$

$$\frac{dv(t)}{dt} = u_{p,e}(t) + u_{p,m}(t) - \frac{C_1}{m}v(t) - \frac{C_2}{m}v^2(t) - h(s(t)) - u_b(t),$$

$$\frac{d\xi_b(t)}{dt} = -\frac{P_b(u_{p,m}, v(t), t)}{V_{b0}Q_b},$$

$$v(0) = v_i,$$

$$v(t_f) = v_f,$$

$$s(0) = 0, \tag{6.77}$$

$$s(t_f) = s_f$$

$$\xi_b(0) = \xi_i,$$

$$\xi_b(t_f) = \xi_f,$$

$$s(t_{B,j}) = s_{B,j}, \quad v(t_{B,j}) = v_{B,j}, \quad j = 1, \ldots, n_B,$$

$$u_{p,e,min}(v(t), t) \leqslant u_{p,e}(t) \leqslant u_{p,e,max}(v(t), t),$$

$$u_{p,m,min}(v(t), t) \leqslant u_{p,m}(t) \leqslant u_{p,m,max}(v(t), t),$$

$$0 \leqslant u_b(t) \leqslant u_{b,max},$$

$$v_{min}(t, s(t)) \leq v(t) \leq v_{max}(t, s(t)),$$

$$s_{min}(t) \leq s(t) \leq s_{max}(t),$$

with the same definitions as in the previous sections, notably for what concerns the limits to the control inputs.

6.6.2 Numerical Solutions

The solution of the optimal control problem (6.77) can be obtained with the dynamic programming methods introduced in Sect. 6.2.2. In particular, the single-state DP (Algorithm 6) can be used with a redefinition of the OCP as in (6.39)–(6.40), and $\mathbf{x}' = \{v, \xi_b\}, \mathbf{u} = \{u_{p,e}, u_{p,m}, u_b\}$. This approach is referred to as *coupled optimization* of driving and energy management strategies.

To further reduce the computation time, a *bi-level* (or decoupled) approach can be used [22]. In this approach, the optimal control policy $\mathbf{u}$ is found with two nested loops by decoupling its components. In the outer loop the speed trajectory $(\mathbf{x}^{(1)} \triangleq v)$ is optimized with respect to the control vector $\mathbf{u}^{(1)} \triangleq \{u_p, u_b\}$, where $u_p = u_{p,e} + u_{p,m}$, thus in the same way as for an ICE or an EV. The running cost of this sub-problem is found by solving a second sub-problem (inner loop, with $\mathbf{x}^{(2)} \triangleq \xi_b$), where the power

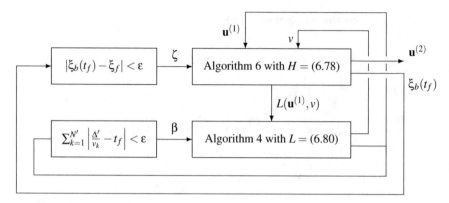

Fig. 6.13 Schematic of the bi-level solution of the ED-OCP for HEVs

split is optimized ($\mathbf{u}^{(2)} \triangleq \{u_{p,m}\}$) for a given speed state and wheel force, while $u_{p,e}$ is found by difference $u_p - u_{p,m}$. An overview of the algorithm is given in Fig. 6.13.

In a convenient embodiment of the method [22], the inner loop can be performed with the PMP-based technique known in the HEV control literature as Equivalent Consumption Minimization Strategy (ECMS) [9]. The Hamiltonian of this sub-problem is defined as

$$H(u_{p,m}, \mathbf{u}^{(1)}, v) = P_f(u_{p,m}, \mathbf{u}^{(1)}, v) + \zeta \cdot P_b(u_{p,m}, \mathbf{u}^{(1)}, v). \qquad (6.78)$$

The costate adjoint to the SoC, ζ, is found such to enforce the constraint over the final SoC (2.52). Hence, the powertrain torque provided by the motor is a function of vehicle speed and wheel force only,

$$u_{p,m}^*(t) = \arg\min_{u_{p,m}} H(u_{p,m}, \mathbf{u}^{(1)}(t), v(t)). \qquad (6.79)$$

As a consequence, also the fuel consumption is dependent on only those two variables and noted as $P_f^*(\mathbf{u}^{(1)}, v)$.

In the outer loop, DP is used with the position as the independent variable and the running cost

$$L(\mathbf{u}^{(1)}, v) = \frac{P_f^*(\mathbf{u}^{(1)}, v) + \beta}{v}. \qquad (6.80)$$

As described above, the coefficient β is tuned in an iterative process as to enforce the constraint over final time. In total, there are two root-finding processes to be performed, over ζ and β. Intermediate approaches between the fully coupled optimization and the bi-level optimization have been investigated in [23].

We present here some results obtained using the bi-level algorithm by varying the main parameters of the problem (6.77). The vehicle and powertrain parameters are a combination of Tables 6.1, 6.2 and 6.3. Figure 6.14a shows the optimal speed profile

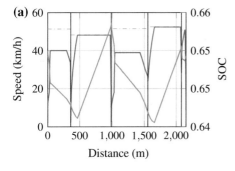
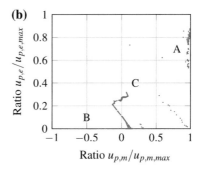

Fig. 6.14 Numerically computed speed and SoC profiles that minimize fuel energy for a given final SoC, for a temporal horizon of 200 s, $v_i = v_f = 0$, $s_f = 2154$ m and various interior and state constraints explicited in the main text **a**. Corresponding operating points in the adimensional engine torque versus motor torque plane **b**

and the SoC variation for a trip of $s_f = 2154$ m with $v_i = v_f = 0$ and $t_f = 200$ s. The trip has four intermediate breakpoints: $s_{B,1,...,4} = \{360, 988, 1555, 2080\}$ m, $v_{B,1,...,4} = \{2.2, 2.7, 24.0, 28.3\}\}$ km/h, $t_{B,1,...,4} = \{38, 92, 155, 193\}$ s. Each of the five resulting sub-trips has a different speed limit, $v_{max} = \{51.4, 48.3, 51.5, 52.5, 50.8\}$ km/h (resulting from posted speed limits and prevailing traffic speed), and slope, $\alpha = \{3.7, -7.4, 4.5, -7.8, -7.7\}$ %. The SoC at the end of the trip is imposed to be equal to the initial value of 65%. Following the procedure described in Sect. 6.2.1.7, five optimization problems have been solved for each sub-trip, coupled by the SoC variation.

The optimal profiles generally vary according to the sub-trip boundary conditions and constraints. The operating points plotted in Fig. 6.14b reveal the presence of strong acceleration phases, where both the engine torque and the motor torque are at or close to their maximal values (region denoted with A in the plot). Similarly, strong deceleration phases occur with the engine shut off and the motor close to its maximum regenerating capability (region B). Milder accelerations and decelerations are also present, where the engine is off and the motor torque is at an intermediate value. Additionally, constant speed phases (region C) are apparent from Fig. 6.14a, either due to the presence of a speed limit (2nd and 4th sub-trips) or as the result of a pulse-and-glide driving (1st and 3rd sub-trips). The latter mode corresponds to the cluster of operating points around the axis $u_{p,m} = 0$ evident in Fig. 6.14b. Similar trends are observed for other trips.

The optimal SoC trend generally follows the slope profile, with battery recharge during downhills and discharge during uphills. Indeed, the example trip suggests a certain correlation between the SoC variations per sub-trip and the corresponding altitude variations.

To investigate this aspect in more detail, a training data-set was created by generating 150 trips with randomly chosen parameters: number of sub-trips, boundary speeds, length, average speed (thus duration), top speed limit, and mean road grade of each sub-trip. The DP algorithm illustrated above was used to calculate the optimal

Fig. 6.15 Values of $\Delta\xi_b$ obtained for a database of 150 trips as a function of the useful energy variation per sub-trip (circles), with linear fitting (solid curve)

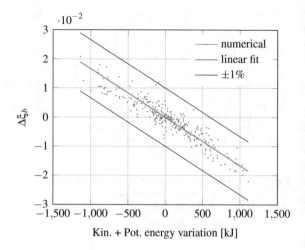

$E_{b,i}$ (or, equivalently, $\Delta\xi_{b,i}$) for each sub-trip $i \in [1, 150]$, subject to the constraint of invariant total SoC for the trip to which it belongs in the data-set.

Inspired by the optimal results obtained via DP, a deterministic model for the optimal SOC variation per road segment has been derived in [24], as already discussed in Sect. 4.4.3. Figure 6.15 shows the results obtained in terms of the optimal values of $\Delta\xi_b$ as a function of the kinetic and potential energy differences across the corresponding sub-trip. Also shown in the figure is a linear fit and its confidence intervals of $\pm1\%$, which matches quite well the numerical data. Consequently, a parametrization of the type

$$\Delta\xi_{b,i} \approx \rho \left(\frac{\frac{1}{2}m(v_i^2 - v_{i-1}^2) + mg(z_{i+1} - z_i)}{Q_b V_{b0}} \right) \tag{6.81}$$

where ρ is a parameter to be fitted, can be proposed to estimate a priori the optimal SoC variations from the sub-trip boundary conditions and further reduce the complexity of the eco-driving problem.

6.6.3   Analytical Solutions

It is possible to have a better insight of the optimal speed profiles of HEVs if we derive closed-form solutions of the ED-OCP (6.77) using PMP. We shall additionally assume that:

- the term C_2 of the resistance force is zero,
- the term $h(s)$ is constant (constant slope),
- the driveline efficiency η_t is constant,

- the gear shift law depends only on vehicle speed, $\gamma_e(t) = \gamma_e(v(t))$, that is, a piecewise-constant function of the speed similar to (2.18),
- the engine indicated efficiency is constant, $k_{e,2} = k_{e,3} = 0$,
- the engine friction torque is constant, $k_{e,1} = 0$,
- a fuel cutoff strategy is used, thus $u_{p,e,min} = 0$,
- in the motor model $k_{m,0} = k_{m,1} = k_{m,2} = 0$,
- the battery efficiency η_b is constant,
- there are no bounds on the states over the optimization horizon.

Under these assumptions, the running cost is given by (6.54), with $u_{p,e}$ replacing u_p, and the battery power by (6.65), with $u_{p,m}$ replacing u_p.

Utilizing the state dynamics in (6.42) and the above cost function, the Hamiltonian is formed as

$$
\begin{aligned}
H = & \left(p_0(v(t)) + p_1 u_{p,e}(t)v(t)\right) \mathcal{H}(u_{p,e}) + \\
& + \lambda(t)\left(u_{p,e}(t) + u_{p,m}(t) - \beta v^2(t) - h - u_b(t)\right) + \mu(t)v(t) + \\
& + \zeta(t)\frac{1}{V_{b0}Q_b}\left(mu_{p,m}(t)v(t)\eta^{-\mathrm{sign}(u_{p,m})} + bu_{p,m}^2(t)(\eta\eta_t)^{-\mathrm{sign}(u_{p,m})}\right),
\end{aligned}
\tag{6.82}
$$

where $\mathcal{H}$ is the Heaviside function.

The costate dynamics read

$$
\begin{aligned}
\dot{\mu}(t) &= -\frac{\partial H}{\partial s} = 0 \Rightarrow \mu = \text{constant}, \\
\dot{\lambda}(t) &= -\frac{\partial H}{\partial v} = -\left(\frac{\partial p_0}{\partial v}(t) + p_1 u_{p,e}(t)\right)\mathcal{H}(u_{p,e}) - \\
&\quad - \frac{\zeta(t)}{V_{b0}Q_b}\left(mu_{p,m}(t)\eta^{-\mathrm{sign}(u_{p,m})}\right) + 2\beta\lambda(t)v(t) - \mu, \\
\dot{\zeta}(t) &= -\frac{\partial H}{\partial \xi_b} = 0 \Rightarrow \zeta = \text{constant}.
\end{aligned}
\tag{6.83}
$$

Boundary conditions for both λ, μ, and ζ are free, since all states s, v, and ξ_b are fixed at initial and final positions. Note that the third costate ζ is constant since we have considered a constant battery efficiency that does not depend on the SoC. Clearly, this costate coincides with the equivalent factor introduced in Sect. 2.4.2 and denoted with the same symbol.

By differentiating the Hamiltonian with respect to the three control inputs, we obtain

$$
\begin{aligned}
\frac{\partial H}{\partial u_{p,e}} &= p_1 v(t)\mathcal{H}(u_{p,e}) + \lambda(t), \qquad \frac{\partial H}{\partial u_b} = -\lambda(t), \\
\frac{\partial H}{\partial u_{p,m}} &= \frac{\zeta}{V_{b0}Q_b}\left(mv(t)\eta^{-\mathrm{sign}(u_{p,m})} + 2bu_{p,m}(t)(\eta\eta_t)^{-\mathrm{sign}(u_{p,m})}\right) + \lambda(t).
\end{aligned}
\tag{6.84}
$$

Note that each of the partial derivatives depends either on $u_{p,e}$ or $u_{p,m}$, but not on both. The three inputs are thus independent functions of the costate λ. The engine input $u_{p,e}$ is given by (6.58), the motor input is given by (6.70) with λ replaced by $\lambda V_{b0} Q_b / \zeta$, and the braking input is given by (6.47).

6.6.3.1   Control Modes

In summary, the optimal solution comprises of several possible modes, namely,

- recharge modes (A-D), $u_b = 0$, $u_{p,e} = u_{p,e,max}$, and (i) $u_{p,m} = u_{p,m,min}$ or (ii) $u_{p,m} = u_{p,m}^-$,
- boost modes (A-A), $u_b = 0$, $u_{p,e} = u_{p,e,max}$, and (i) $u_{p,m} = u_{p,m,max}$ or (ii) $u_{p,m} = u_{p,m}^+$,
- purely electric modes (C-A), $u_b = 0$, $u_{p,e} = 0$, and (i) $u_{p,m} = u_{p,m,max}$ or (ii) $u_{p,m} = u_{p,m}^+$,
- purely ICE mode (A-C), $u_b = 0$, $u_{p,e} = u_{p,e,max}$, $u_{p,m} = 0$,
- regenerative braking modes (C-D), $u_{p,e} = 0$, and (i) $u_{p,m} = u_{p,m}^-$, $u_b = 0$, (ii) $u_{p,m} = u_{p,m,min}$, $u_b = 0$ or (iii) $u_{p,m} = u_{p,m,min}$, $u_b = u_{b,max}$,
- coasting mode (C-C), $u_{p,e} = 0$, $u_{p,m} = 0$, $u_b = 0$,
- pulse and glide or singular mode (S), $u_b = 0$, $v \approx$ const.

Figure 6.16 shows how some of these modes appear in the numerical results of Fig. 6.14. In addition, the scenario shown in the figure presents speed-constrained modes that will be discussed in the following chapter.

Fig. 6.16 Optimal modes identified for the numerical optimal profiles of Fig. 6.14. Dark blue: A-A, green: C-A, purple: C-D, orange: S, blue: constrained by v_{lim}

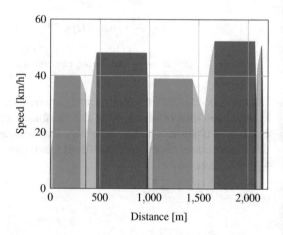

6.6.3.2 Solution of the TPBVP

Due to the large number of possible optimal modes, neither the sequential approach nor the parametric optimization approach proposed for ICEVs and EVs to evaluate online the optimal speed profiles are generally feasible for HEVs. Even the ANN-based approach requires special care, and two steps are necessary: the former to identify the optimal sequence of modes, and the latter to find the switching times) between each mode in the optimal control sequence.

In the mode sequence identification step, a *classification* ANN is required. The input vector contains the boundary conditions and the SoC variation in the sub-trip, $I = \{t_f, s_f, v_i, v_f, \alpha, \Delta\xi_b\}$, while the output describes the mode sequence. Although each combination of the modes of Sect. 6.6.3.1 is theoretically possible, for practical training sets only a small number of sequences actually appears in the output. In the switching-times identification step, a *regression* ANN can be conveniently used, with the same input vector I as above and the outputs depending on the mode sequence. In [25], one hidden layer with 5 neurons have been used for both types of neural networks.

References

1. Saerens B (2012) Optimal control based eco-driving. PhD thesis, Katholieke Universiteit Leuven
2. Hof T, Conde L, Garcia E, Iviglia A, Jamson S, Jopson A, Lai F, Merat N, Nyberg J, Rios S et al (2012) D 11.1: a state of the art review and users' expectations. EcoDriver project
3. Chachuat B (2007) Nonlinear and dynamic optimization: from theory to practice lecture ic-32: Spring term 2009. Technical report, EPFL
4. Sethi SP, Thompson GL(2000) Optimal control theory: applications to management science and economics. Springer, Berlin
5. Bryson AE, Ho Y-C (1975) Applied optimal control: optimization, estimation, and control. Routledge
6. Bellman R (2013) Dynamic programming. Courier corporation
7. Sniedovich M (2006) Dijkstra's algorithm revisited: the dynamic programming connexion. Control Cybern 35(3):599–620
8. Bertsekas DP (2005) Dynamic programming and optimal control. Athena Scientific, Belmont, MA
9. Guzzella L, Sciarretta A (2013) Vehicle propulsion system. Springer, Berlin
10. Hellström E, Ivarsson M, Åslund J, Nielsen L (2009) Look-ahead control for heavy trucks to minimize trip time and fuel consumption. Control Eng Pract 17(2):245–254
11. Dib W, Serrao L, Sciarretta A (2011) Optimal control to minimize trip time and energy consumption in electric vehicles. In: Proceedings of vehicle power and propulsion conference (VPPC), pp 1–8. IEEE
12. Mensing F, Bideaux E, Trigui R, Ribet J, Jeanneret B (2014) Eco-driving: an economic or ecologic driving style? Transp Res Part C: Emerg Technol 38:110–121
13. Han J, Vahidi A, Sciarretta A (2019). Fundamentals of energy efficient driving for combustion engine and electric vehicles: an optimal control perspective. Automatica 103:558–572
14. Kirk DE (2012) Optimal control theory: an introduction. Courier corporation
15. Gilbert EG (1976) Vehicle cruise: improved fuel economy by periodic control. Automatica 12(2):159–166

16. Li SE, Peng H (2011) Strategies to minimize fuel consumption of passenger cars during car-following scenarios. In: Proceedings of American control conference (ACC), pp 2107–2112. https://doi.org/10.1109/ACC.2011.5990786
17. Lee J (2009) Vehicle inertia impact on fuel consumption of conventional and hybrid electric vehicles using acceleration and coast driving strategy. PhD thesis, VirginiaTech
18. Marchal C (1973) Chattering arcs and chattering controls. J Optim Theory Appl 11(5):441–468
19. Thibault L, De Nunzio G, Sciarretta A (2018) A unified approach for electric vehicles range maximization via eco-routing, eco-driving, and energy consumption prediction. IEEE Trans Intell Veh 3(4):463–475
20. Dib W, Chasse A, Moulin P, Sciarretta A, Corde G (2014) Optimal energy management for an electric vehicle in eco-driving applications. Control Eng Pract 29:299–307
21. Petit N, Sciarretta A et al (2011) Optimal drive of electric vehicles using an inversion-based trajectory generation approach. In: Proceedings of IFAC world congress, vol 18, pp 14519–14526. IFAC
22. Ngo DV, Hofman T, Steinbuch M, Serrarens AFA (2010) An optimal control-based algorithm for hybrid electric vehicle using preview route information. In: Proceedings of American control conference (ACC), pp 5818–5823. IEEE
23. Maamria D, Gillet K, Colin G, Chamaillard Y, Nouillant C (2018) Computation of eco-driving cycles for hybrid electric vehicles: comparative analysis. Control Eng Pract 71:44–52
24. De Nunzio G, Sciarretta A, Gharbia IB, Ojeda LL (2018) A constrained eco-routing strategy for hybrid electric vehicles based on semi-analytical energy management. In: Proceedings of international conference on intelligent transportation systems (ITSC), pp 355–361. IEEE
25. Zhu J, Ngo C, Sciarretta A (2019) Real-time optimal eco-driving for hybrid-electric vehicles. In: Proceedings of international symposium on advances in automotive control, IFAC

Chapter 7
Specific Scenarios and Applications

In this chapter, the general methods for computing eco-driving strategies introduced in Chap. 6 are applied to several driving scenarios. These scenarios roughly reflect the list of Sect. 6.1.1 and comprise of: acceleration to a cruise speed (Sect. 7.1), deceleration to stop (Sect. 7.2), cruising with road slopes (Sect. 7.3), driving between stops with a speed limit (Sect. 7.5), approaching an intersection (Sect. 7.6), approaching a traffic light (Sect. 7.7), and car following (Sect. 7.8).

From the viewpoint of the driving optimization, the first two scenarios do not present external constraints. We shall study the influence of the optimization horizons, and compare the optimal speed profile to typical driving behavior as described by the models of Sect. 4.2.1. For the other scenarios, which are characterized by external constraints, we shall investigate the role of predictive information, which have already been broadly discussed in Sect. 6.1.2, and we shall evaluate the associated energy benefits. For all scenarios, we shall make use of the numerical methods of Chap. 6 to solve the corresponding ED-OCPs. To better illustrate the influence of the various parameters, we also solve several subproblems in an analytical fashion, using the simplified EV model of Sect. 6.5.3.3.

7.1 Acceleration

In this scenario, a target speed v_f is to be reached from rest in a given distance s_f and in a free time t_f, or with both distance and time being free parameters. A numerical analysis of this scenario is presented in Sect. 7.1.1. Then in Sect. 7.1.2 we use closed-form solutions of the ED-OCP (6.64) to corroborate the numerical results.

© Springer Nature Switzerland AG 2020
A. Sciarretta and A. Vahidi, *Energy-Efficient Driving of Road Vehicles*,
Lecture Notes in Intelligent Transportation and Infrastructure,
https://doi.org/10.1007/978-3-030-24127-8_7

7.1.1 Numerical Analysis

Figure 7.1 shows the numerical results obtained with the DP Algorithm 6 for the ICEV of Table 6.2 and the EV of Table 6.3, for varying end time and given the end position.

In the case of an ICEV, the speed profiles consist of two acceleration phases separated by a pulse-and-glide phase, whose duration increases with the required end time. In the case of an EV, the speed profiles are smoother. For both propulsion systems, when the required end time becomes shorter, the calculated speed profile might first cross the target final speed and then retrieve it with a deceleration phase. The energy consumption generally decreases with an increase of the final time.

As a basis of comparison, the output of Gipps' model (4.6) is also shown in the figures. The parameters of the latter are $a_{max} = 1.5$ m/s^2, $a_{min} = -1$ m/s^2, which yield an acceleration time, that is, the time at which the computed speed reaches within a band of ± 0.5 m/s around the target speed, of about 21 s. Calculation of energy consumption for both powertrains (datum not shown for ICEV due to scale) reveals that Gipps' driving is clearly far from being energy optimal.

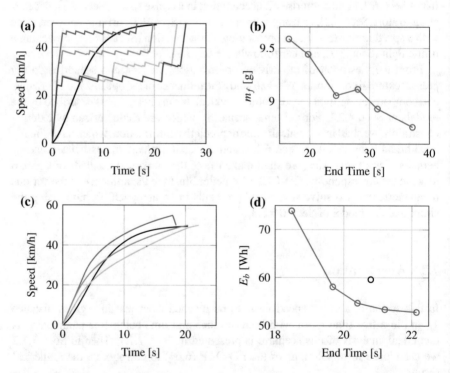

Fig. 7.1 Optimal speed profiles and Gipps' speed profile for an acceleration from 0 to 50 km/h with $s_f = 200$ m and varying final time t_f, for an ICEV (**a**) and an EV (**c**). Energy consumption as a function of final time, (**b**) and (**d**). The black curves and dots represent Gipps' model

7.1.2 Analytical Approach

In order to retrieve and explain the behavior shown by the numerical results of the previous section, we use the simplified EV model of Sect. 6.5.3.3, for which explicit solutions of the ED-OCP can be calculated.

7.1.2.1 Optimal Strategy

The solution of the ED-OCP for the considered EV model is given by (6.74) and (6.75), which we repeat here for $v_i = 0$,

$$v(t) = -\frac{2v_f t}{t_f} - \frac{6s_f t^2}{t_f^3} + \frac{6s_f t}{t_f^2} + \frac{3v_f t^2}{t_f^2}, \tag{7.1}$$

$$E_b = mhs_f + m\frac{v_f^2}{2} + bh^2 t_f + 2bhv_f + 4b\left(\frac{3s_f^2}{t_f^3} - \frac{3s_f v_f}{t_f^2} + \frac{v_f^2}{t_f}\right), \tag{7.2}$$

where only the last term represents the effective energy consumption $E_{b,e}$ as defined in Sect. 6.5.3.3.

First we study the domain of validity of (7.1) in terms of t_f and s_f, looking for conditions of the type $F(s_f, t_f) \geq 0$. We use the general results derived in Appendix B and particularize them for $v_i = 0$. A first condition (see (B.14)) is obtained by requiring that the speed profile never becomes negative, which is equivalent to imposing that $\dot{v}(0) \geq 0$, and reads

$$F_{UB1} \triangleq 3s_f - v_f t_f \geq 0. \tag{7.3}$$

Further, we can impose a maximum speed v_{max} not to exceed during the acceleration, which leads to the condition (see (B.16))

$$F_{LB1} \triangleq \frac{v_f + v_{max} + \sqrt{v_{max}^2 - v_f v_{max}}}{3} - \frac{s_f}{t_f} \geq 0. \tag{7.4}$$

We can also impose the maximum initial acceleration that is allowed. Denoting this value with a_{max}, we obtain (see (B.24) and (B.23))

$$F_{LB2} \triangleq -6s_f + 2v_f t_f + t_f^2 a_{max} \geq 0. \tag{7.5}$$

and

$$F_{UB2} \triangleq 6s_f - 4v_f t_f + t_f^2 a_{max} \geq 0. \tag{7.6}$$

The resulting domain of feasibility is illustrated in Fig. 7.2.

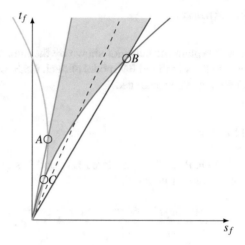

Fig. 7.2 Domain of feasibility (shaded gray area) of the parabolic speed profile in the plane t_f–s_f for the acceleration scenario. The curves shown are: F_{UB1} (orange), F_{LB1} (purple), F_{UB2} (green), and F_{LB2} (blue). Coordinates of the intersection points are A: $(2/3 \cdot \sigma, 2\tau)$, B: $(2/3 \cdot \sigma/\beta^2(1 + \sqrt{1-\beta})(\beta + 1 + \sqrt{1-\beta}), 2\tau/\beta(1 + \sqrt{1-\beta}))$, and C: $(\sigma/2, \tau)$, where $\tau \triangleq v_f/a_{max}$, $\beta \triangleq v_f/v_{max}$, and $\sigma \triangleq v_f^2/a_{max}$. Also shown (dashed) is the upper bound of the domain of feasibility of the simplified Gipps' profile

Now we can study the case of a free final time $t_f \in [t_{f,min}, t_{f,max}]$, where $t_{f,min}$ is given as a function of s_f by (7.4) or (7.5), while $t_{f,max}$ by (7.3) or (7.6), as depicted in Fig. 7.2. Local minima of E_b are found by setting

$$\frac{\partial E_b}{\partial t_f} = bh^2 + 4b\left(\frac{-9s_f^2}{t_f^4} + \frac{6s_f v_f}{t_f^3} - \frac{v_f^2}{t_f^2}\right) = 0 , \tag{7.7}$$

which has four solutions in t_f. For a realistic choice of the parameters, one solution is negative and must be discarded. Among the remaining three, one solution corresponds to a local maximum and two to local minima, with the energy becoming infinitely large for very large end times. The first local minimum is at

$$t_f^* = \frac{-v_f + \sqrt{v_f^2 + 6hs_f}}{h} . \tag{7.8}$$

For small h, this value may be larger than $t_{f,max}$. Therefore the energy consumption is a decreasing function of time and the optimal time is $t_{f,max}$. However, for large values of h (typically, positive slopes), the quantity (7.8) can be lower than the upper bound and thus represents the global minimum. Such a behavior is illustrated in Fig. 7.3, in terms of the normalized energy consumption E_b/E_W (see (2.5) for the definition of the energy at the wheels E_W) for two different road grades. The curve for $\alpha = 10\%$ has a minimum, while that for $\alpha = 0$ is continuously decreasing.

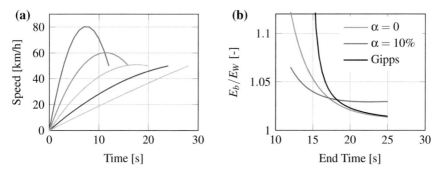

Fig. 7.3 Analytical optimal speed profiles for an acceleration from 0 to 50 km/h with varying final time, $s_f = 200$ m (**a**). Normalized energy consumption as a function of final time, for $\alpha = 0$ and $\alpha = 10\%$ (**b**). The simplified EV model parameters used are: $m = 1100$ kg, $h = 0.1472$ ($\alpha = 0$), $h = 1.1282$ ($\alpha = 10\%$), $b = 147.5$

7.1.2.2 Gipps' Driving

The full Gipps' model during an acceleration maneuver is not easily integrable to be used with the simplified EV energy consumption model. However, observing the results of its numerical simulation, it becomes apparent that a simple enough approximation of the speed profile is

$$v(t) = \begin{cases} a_1 t, & t \in [0, t_1] \\ v_f, & t \in [t_1, t_f] \end{cases}, \tag{7.9}$$

where $a_1 t_1 = v_f$. The fulfillment of final distance yields the further condition $v_f t_f - 1/2v_f^2/a_1 = s_f$, from which the relations

$$a_1 = \frac{v_f^2}{2(v_f t_f - s_f)}, \quad t_1 = \frac{2(v_f t_f - s_f)}{v_f} \tag{7.10}$$

hold. The latter equations bring the bounds for the final time, that is, $s_f \le v_f t_f \le 2s_f$, which are shown in Fig. 7.2.

The effective energy consumption is evaluated as

$$E_{b,e} = \frac{b v_f^3}{2t_f(v_f t_f - s_f)}, \tag{7.11}$$

which is shown to be always positive and larger than with the optimal strategy, see Fig. 7.3b.

7.2 Deceleration

In this scenario, the vehicle must decelerate from a given speed v_i to stop in a given distance s_f and free time t_f. A numerical analysis of this scenario is presented in Sect. 7.2.1. Then in Sect. 7.2.2 we use closed-form solutions of the ED-OCP (6.64) to corroborate the numerical results.

7.2.1 Numerical Analysis

Figure 7.4 shows the numerical results obtained with the DP Algorithm 6 for the ICEV of Table 6.2 and the EV of Table 6.3, for varying end time and given end position.

    In the case of an ICEV, the speed profiles may consist of an initial acceleration, followed by coasting and braking phases, as anticipated in the analysis of Sect. 6.4, although in principle slower decelerations could be achieved by interposing a coasting

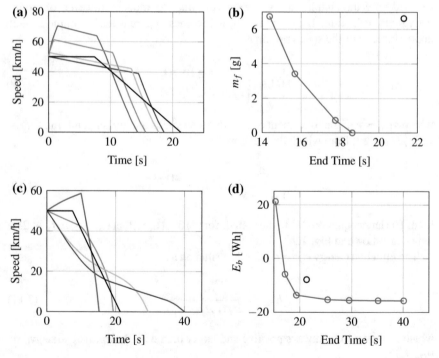

Fig. 7.4 Optimal speed profiles for a deceleration from 50 km/h to stop in 200 m with varying final time, for an ICEV (**a**) and an EV (**c**). Energy consumption as a function of final time, (**b**) and (**d**). The black curves and dots represent Gipps' model

phase between two braking phases. The maximum deceleration time considered is when only coasting and braking are used. In the case of an EV, the speed profiles are smoother and larger end times are allowed. The energy consumption generally decreases with the final time and becomes negative in the EV case due to larger proportion of regenerative braking.

The output of the Gipps' model (4.6) with $a_{min} = -1$ m/s^2 is also shown in the figures. Calculation of energy consumption for both powertrains again reveals that Gipps' driving is far from being energy optimal.

7.2.2 Analytical Approach

In order to retrieve and explain the behavior shown by the numerical results of the previous section, we use the simplified EV model of Sect. 6.5.3.3, for which explicit solutions of the ED-OCP can be calculated.

7.2.2.1 Optimal Strategy

The solution of the ED-OCP for the considered model is given by (6.74) and (6.75), which we repeat here for $v_f = 0$,

$$v(t) = v_i - \frac{4v_i t}{t_f} - \frac{6s_f t^2}{t_f^3} + \frac{6s_f t}{t_f^2} + \frac{3v_i t^2}{t_f^2} , \tag{7.12}$$

$$E_b = mhs_f - m\frac{v_i^2}{2} + bh^2 t_f - 2bhv_i + 4b\left(\frac{3s_f^2}{t_f^3} - \frac{3s_f v_i}{t_f^2} + \frac{v_i^2}{t_f}\right). \tag{7.13}$$

Similarly to the previous section, we find the conditions $F(t_f, s_f) \geq 0$ for which (7.12) is an admissible speed profile. We use the general results derived in Appendix B and particularize them for $v_f = 0$. Imposing that the speed profile never becomes negative yields

$$F_{UB1} \triangleq 3s_f - v_i t_f \geq 0 . \tag{7.14}$$

Imposing a maximum speed v_{max} yields

$$F_{LB1} \triangleq \frac{v_i + v_{max} + \sqrt{v_{max}^2 - v_i v_{max}}}{3} - \frac{s_f}{t_f} \geq 0 . \tag{7.15}$$

Imposing the maximum deceleration a_{min} yields

$$F_{LB2} \triangleq -6s_f + 2v_i t_f - t_f^2 a_{min} \geq 0 \tag{7.16}$$

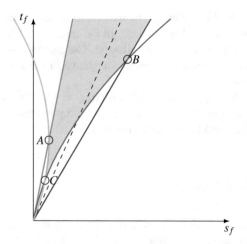

Fig. 7.5 Domain of feasibility (shaded gray area) of the parabolic speed profile in the plane t_f–s_f for the deceleration scenario. The curves shown are: F_{UB1} (orange), F_{LB1} (purple), F_{UB2} (green), and F_{LB2} (blue). The coordinates of the intersection points are A: $(2/3\sigma, 2\tau)$, B: $(2/3 \cdot \sigma/\beta^2(1 + \sqrt{1 - \beta})(\beta + 1 + \sqrt{1 - \beta}), 2\tau/\beta(1 + \sqrt{1 - \beta}))$, and C: $(\sigma/2, \tau)$, where $\tau \triangleq v_f/|a_{min}|, \beta \triangleq v_f/v_{max}$, and $\sigma \triangleq v_f^2/|a_{min}|$. Also shown (dashed) is the upper bound of the domain of feasibility of the simplified Gipps' profile

and

$$F_{UB2} \triangleq 6s_f - 4v_i t_f - t_f^2 a_{min} \geq 0 . \tag{7.17}$$

The resulting feasibility domain is illustrated in Fig. 7.2.

Now we study the case of a free final time $t_f \in [t_{f,min}, t_{f,max}]$, where $t_{f,max}$ is given as a function of s_f by (7.14), while $t_{f,min}$ is given by (7.15) or (7.16), as depicted in Fig. 7.5. In both cases, the same expressions as in the acceleration scenario are obtained, with $|a_{min}|$ replacing a_{max} and v_i replacing v_f. Therefore, (7.7) and (7.8) are still valid with the aforementioned replacements.

The consumed energy is now lower in absolute value than the wheel energy, which is a negative quantity. As shown in Fig. 7.6, the stationary point (7.8) is now a maximum. However, the latter belongs to the feasibility domain only for sufficiently low road grades (steep downhills).

7.2.2.2 Gipps' Driving

The Gipps' model is easily integrable in the deceleration case, at least for small step times Δt (see Eq. 4.6), yielding a speed profile

$$v(t) = \begin{cases} v_i, & t \in [0, t_1] \\ v_i + a_1 t, & t \in [t_1, t_f] \end{cases} , \tag{7.18}$$

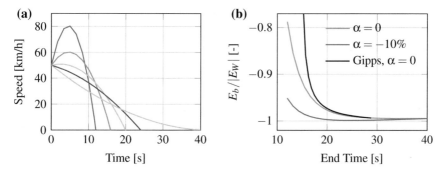

Fig. 7.6 Analytical optimal speed profiles for a deceleration from 50 km/h to stop, with varying final time, $s_f = 200$ m (**a**). Normalized energy consumption as a function of final time, for $\alpha = 0$ and $\alpha = -10\%$ (**b**). The simplified EV model parameters used are: $m = 1100$ kg, $h = 0.1472$ ($\alpha = 0$), $h = -0.834$ ($\alpha = -10\%$), $b = 147.5$

where $t_f - t_1 = v_i/|a_1|$. The fulfillment of the final distance yields the further condition $v_i t_f - v_f^2/(2|a_1|) = s_f$, from whence the relations

$$a_1 = -\frac{v_i^2}{2(v_i t_f - s_f)}, \quad t_1 = \frac{2s_f}{v_i} - t_f \tag{7.19}$$

hold. The latter equations impose the bounds on the final time, $s_f \le v_i t_f \le 2s_f$, which are shown in Fig. 7.5.

The effective energy is evaluated as

$$E_{b,e} = \frac{bv_i^3}{2(v_i t_f - s_f)}, \tag{7.20}$$

which is shown to be always positive and larger than with the optimal strategy, see Fig. 7.6.

7.3 Road Slopes

As anticipated in Chap. 1, a dominating factor in vehicle power demand is road grade, in particular on steep roads, and more so for heavier vehicles. Therefore, not surprisingly, energy-optimal speed profiles are strongly affected by road grade and its prior knowledge is highly beneficial in predictive eco-driving. For example, a vehicle can slow down in anticipation of a steep descent or speed up in preparation for a climb.

In this section we aim at illustrating the dependency between road grade and optimal speed. We consider the baseline situation of cruising at constant speed (e.g., on a highway) and we introduce a sinusoidal altitude profile of the type

$$z(s) = z_0 \sin(\Omega s) + z_1 , \tag{7.21}$$

from which grade is evaluated as

$$\alpha(s) = dz(s)/ds = \Omega z_0 \cos(\Omega s) . \tag{7.22}$$

Note that the absolute altitude level z_1 has no influence on α. A numerical analysis of this scenario is presented in Sect. 7.3.1. Then in Sect. 7.3.2 we use closed-form solutions of the ED-OCP (6.64) to corroborate the numerical results.

7.3.1 Numerical Analysis

We consider a baseline scenario ($z_0 = 0$) of cruising at a constant speed of 36 km/h. Figure 7.7a, b show the optimal speed profiles obtained with the DP (Algorithm 6) for the ICEV in Table 6.2 and the EV in Table 6.3, respectively for varying z_0. These optimization results confirm the intuitive rule that a vehicle should be slowed down before a downhill and accelerated before a climb. In other terms, one can observe an inverse correlation between optimal speed and altitude, that would be in principle useful to adapt speed when future altitude (grade) is estimated.

We can regard the optimal speed profile as the output of a predictive (P) eco-driving strategy that, using GPS and 3D GIS information (see Sect. 4.1), perfectly anticipates the upcoming road slope. To illustrate the effects of such preview, we compare the energy consumption of the P strategy with that of a non-predictive strategy (NP) that has no preview and would keep following a constant speed despite the slope changes. We define a measure of performance as

$$\varepsilon = \frac{E_T^{(NP)} - E_T^{(P)}}{E_T^{(P)}} , \tag{7.23}$$

where E_T is the tank energy, that is, the fuel energy for the ICEV and the battery energy for the EV, while the superscripts P, NP stand for the predictive and non-predictive strategies, respectively.

Figure 7.7c, d show the calculated values of ε as a function of z_0. Clearly, the higher z_0 in absolute value, the higher ε. The ICEV is affected more than the EV by the prediction of the road slope. The numbers reveal that keeping a constant speed (actually, the optimal speed for $z_0 = 0$, which is a pulse-and-glide profile) can consume two to three times more energy than following the optimal slope-sensitive speed profile.

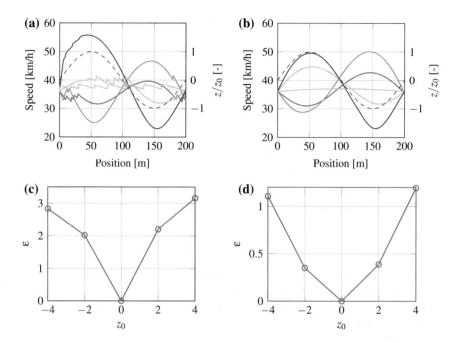

Fig. 7.7 Optimal speed profiles for sinusoidal slope profiles of the type (7.21), with $\Omega = \pi/100$ m^{-1} and $z_0 = \{-4, -2, 0, 2, 4\}$ m, $v_i = v_f = 36$ km/h, $s_f = 200$ m, $t_f = 20 \pm 1$ s, and measures of performance, for an ICEV (**a, c**) and an EV (**b, d**). Normalized altitude profile is shown as a dashed line

7.3.2 Analytical Approach

In order to retrieve and explain the behavior shown by the numerical results of the previous section, we use the simplified EV model of Sect. 6.5.3.3, for which explicit solutions of the ED-OCP can be calculated.

With respect to the assumptions in that section, here we relax the constancy of slope and make the resistance term $h(s)$ variable with the position. According to (7.21), the resistance term is evaluated as

$$h(s) = h_0 + g\alpha(s) = h_0 + g\Omega z_0 \cos(\Omega s) , \qquad (7.24)$$

where h_0 is the constant factor due to rolling resistance, g is gravity and Ω is chosen such that the net difference of altitude is zero, i.e., $\sin(\Omega s_f) = 0$.

To solve this problem in a closed form, we consider $v_i = v_f = s_f/t_f$ and perturb the solution for $z_0 = 0$, which is trivially $v(t) \equiv s_f/t_f$. We consider small altitude variations, i.e., $gz_0 \ll (s_f/t_f)^2$, and small resistance forces, i.e., $bh_0 \ll (s_f/t_f)$. With these positions, it is easy to show that the control input $u_p(t) \approx h_0$ and the speed profile

$$v(t) \approx \frac{s_f}{t_f} - 2bh_0 - \frac{gz_0t_f}{s_f} \sin(\Omega s(t)) \tag{7.25}$$

satisfies the necessary conditions for optimality and is thus the optimal solution sought. Note that for $z_0 = 0$, the profile (6.73)–(6.74) is retrieved, with $h = 0$.

The energy consumption can be evaluated by integrating the battery power over position and using the transformation $ds/dt = v$,

$$E_b = \int_0^{s_f} muds + \int_0^{s_f} b\frac{u^2}{v}ds = mh_0s_f + bh_0^2 \int_0^{s_f} \frac{ds}{v(s)} . \tag{7.26}$$

Under the assumptions above, the latter integral is equal to t_f. Thus the energy consumption of the optimal speed profile is

$$E_b^{(P)} = mh_0s_f + bh_0^2t_f , \tag{7.27}$$

the same value that would be obtained by (6.75) with $h \equiv h_0$ and a constant speed. These results confirm the observations made in the previous section that the optimal speed follows the altitude profile and that the optimal energy consumption is largely independent of the altitude.

The energy consumption of the non-predictive strategy that follows a constant speed despite the altitude variations is obtained by evaluating the control input as $u_p = h$ and then inserting it into the expression (7.26). The result is

$$E_b^{(NP)} = mh_0s_f + bh_0^2t_f + \frac{1}{2}bg^2\Omega^2z_0^2t_f . \tag{7.28}$$

Consequently, the measure of performance defined in (7.23) is evaluated as

$$\varepsilon = \frac{1}{2} \cdot \frac{bg^2\Omega^2z_0^2t_f}{mh_0s_f + bh_0^2t_f} , \tag{7.29}$$

whose quadratic dependency on z_0 clearly matches the numerical results of Fig. 7.7.

7.4 Constrained Eco-Driving

In the rest of this chapter (Sects. 7.5–7.8), we shall treat scenarios with trip constraints imposed on the optimization of the speed profile, under the form of state or interior constraints.

In particular, we shall present the constrained-optimal speed profiles for the scenarios considered. These solutions shall be regarded as the output of a *predictive* (P) eco-driving strategy that, assuming that the vehicle is equipped with dedicated sensors and/or communication technology, perfectly anticipates the trip constraints with

Fig. 7.8 Domain of
feasibility of the parabolic
speed profile (shaded gray
area) in the plane t_f–s_f for
the constrained scenarios.
The curves shown are: F_{LB1}
(purple) and F_{LB2} (blue).
The coordinates of the
intersection points are
$A \equiv C$: (0, 0) and B:
$(8/3 \cdot v_{max}^2/a_{max}$,
$4v_{max}/a_{max})$

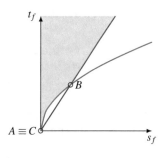

unlimited preview distance. The P speed profiles shall be compared with the respective unconstrained optimal speed profiles to illustrate the effects of the constraints on the energy consumption.

As discussed in Chap. 1, energy savings are highly dependent on the preview capability, and we shall study the effects of such anticipation. Therefore, we shall further compare the P speed profiles and energy consumption with those of *non-predictive* (NP) eco-driving strategies, which have no or limited preview of the trip constraints (but are supposed to know the trip duration).

Generally speaking, the NP strategy is composed of four phases for all the constrained scenarios considered: (i) in the first phase, it follows the unconstrained optimum until a constraint is detected, assuming visual preview only; (ii) upon detection, the NP strategy adjusts its trajectory to match the constraint and (iii) then it tracks the constraint perfectly; (iv) lastly, the NP strategy retrieves a sub-optimal behavior to complete the trip in the desired duration and distance. In some cases we shall assume the adjustment phase (ii) as instantaneous for simplicity.

In order to understand the differences between P and NP strategies, we use the simplified EV model of Sect. 6.5.3.3, for which explicit solutions of the ED-OCP can be calculated.

$$v^*(t) = \frac{6s_f}{t_f} \left(\frac{t}{t_f}\right)\left(1 - \frac{t}{t_f}\right). \tag{7.30}$$

In terms of t_f, s_f, both conditions F_{UB1} and F_{UB2} (see Appendix B) are always satisfied for $v_i = v_f = 0$, thus there is no upper bound on t_f for a given s_f, see Fig. 7.8. Condition F_{LB2} reads $t_f \geq \sqrt{6s_f/a_{max}}$, while condition F_{LB1} reads $t_f \geq 3s_f/(2v_{max})$. The unconstrained minimal energy consumption is found as

$$E_b^* = mhs_f + bh^2 t_f + \frac{12bs_f^2}{t_f^3}. \tag{7.31}$$

7.5 Speed Limit

In this section, we discuss the eco-driving along a route in the presence of a maximum speed limit,

$$v(t) - v_{max}(t) \leq 0, \quad t \in [0, t_f], \tag{7.32}$$

which is a pure state constraint of the form (6.5). The key factor for effective eco-driving in such a situation is the ability to anticipate the limit and its variability along the route.

A numerical analysis of this scenario is presented in Sect. 7.5.1. Then in Sect. 7.5.2 we use closed-form solutions of ED-OCP (6.64) to corroborate the numerical results and highlight the influence of the preview ability and other parameters of the problem.

7.5.1 Numerical Analysis

We consider a trip having fixed distance s_f and duration t_f, with a constant speed limit v_{max}. The unconstrained optimal speed profile, used as a baseline, is denoted as $v^*(t)$. We further compare predictive (P) and non-predictive (NP) constrained optimal speed profiles, as they have been defined in Sect. 7.4. The latter has no information about the upcoming limit, while predictive eco-driving perfectly anticipates the speed constraint.

Speed profiles P and NP can be obtained numerically using the methods introduced in Chap. 6. With dynamic programming, a speed constraint is easily enforced by making unfeasible all the points of the grid that exceed the limit value. The predictive speed profile is thus obtained by solving the constrained OCP for the original boundary conditions, see Table 7.1. The non-predictive speed profile is obtained by concatenating (i) the unconstrained solution until a time t_N, such that $v^*(t_N) = v_{max}$, with (ii) the solution of the constrained OCP from t_N to the end time, with a distance to cover $s_f - s^*(t_N)$, an initial speed equal to v_{max} and a final speed of zero.

Table 7.1 Boundary conditions for the speed-limit-constrained scenario

		Constrained optimization?	Duration	Distance	Initial speed	End speed
Unconstrained		N	t_f	s_f	0	0
Predictive		Y	t_f	s_f	0	0
Non-predictive	#1	N	t_N	$s^*(t_N)$	0	v_{max}
	#2	Y	$t_f - t_N$	$s_f - s^*(t_N)$	v_{max}	0

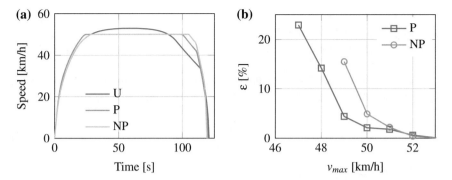

Fig. 7.9 Unconstrained optimal (U), predictive (P), and non-predictive (NP) strategies for a trip with $s_f = 1500$ m, $t_f = 120$ s, $v_i = v_f = 0$, $v_{max} = 50$ km/h. Speed profiles (**a**) and performance indexes (**b**)

Figure 7.9a shows the three speed profiles for $s_f = 1500$ m, $t_f = 120$ s, $v_i = v_f = 0$, $v_{max} = 50$ km/h, computed with the DP Algorithm 6 and the EV model of Table 6.3. Note how the non-predictive speed profile is forced to keep the top speed longer than the P profile in order to cover the prescribed distance in the prescribed time.

As a measure of performance of the eco-driving strategies, the relative energy loss with respect to the unconstrained optimum is defined as

$$\varepsilon^{(\{P,NP\})} \triangleq \frac{E_b^{(\{P,NP\})} - E_b^*}{E_b^*}, \tag{7.33}$$

where E_b^* is the energy consumption for the unconstrained optimal strategy and $E_b^{(\{P,NP\})}$ are those for the predictive and non-predictive strategy, respectively.

The values obtained for $\varepsilon^{(P)}$ and $\varepsilon^{(NP)}$ for the aforementioned scenario are plotted in Fig. 7.9b as a function of the speed limit. The latter has an influence only if it is smaller than the maximum speed reached in the unconstrained case (53 km/h for this scenario). Clearly, the energy losses increase dramatically as the speed constraint becomes more aggressive, because the vehicle does not have enough time to take advantage of coasting.

7.5.2 Analytical Approach

In order to retrieve and explain the behavior shown by the numerical results of the previous section, we use the simplified EV model of Sect. 6.5.3.3, for which explicit solutions of the ED-OCP can be calculated. We assume that the maximum road speed limit is held constant over the sub-trip.

This model, with $v_i = v_f = 0$, yields an unconstrained optimal speed profile v^* given by (7.30) and an overall energy consumption E_b^* given by (7.31), and will be now compared with the values achieved by the predictive and the non-predictive strategies as a function of v_{max}.

For later use, we introduce here the ratio between the speed limit and the maximum speed in the unconstrained case (larger speed limits would have no effect),

$$r_s \triangleq \frac{2v_{max}t_f}{3s_f} \, , \tag{7.34}$$

with $2/3 < r_s < 1$ by virtue of condition F_{LB1} (see Sect. 7.4).

7.5.2.1 Predictive Strategy

Anticipating the presence of the speed limit allows the implementation of a constrained-optimal speed profile. The necessary conditions for optimality are obtained from (6.72) in the Lagrangian form with the addition of the conditions (6.31)–(6.34). Using the nomenclature introduced in Sect. 6.2.2.1, the speed constraint $g(\mathbf{x}(t), t) = v(t) - v_{max} \leq 0$ is of the first order ($p = 1$), since $\dot{g} = g^{(1)} = \dot{v} = u_p - h$. Additionally, we have $\partial g/\partial v = 1$, $\partial g/\partial s = 0$ and, from the assumption of constant v_{max}, $\partial g/\partial t = 0$. Therefore, the system of equations to be solved reads

$$\begin{cases} \dot{s} = v(t), \quad s(0) = 0, \quad s(t_f) = s_f \\ \dot{v} = u_p(t) - h, \quad v(0) = v_i, \quad v(t_f) = v_f \\ \dot{\lambda} = -mu_p(t) - \mu, \quad \lambda(0) = \lambda_0 \\ u_p(t) = -\dfrac{1}{2b}(\lambda(t) + mv(t) + \eta(t)) \\ \eta(t)g(v(t)) = 0, \quad \eta(t) \geq 0, \quad \dot{\eta} \leq 0 \\ \lambda(\tau^-) = \lambda(\tau^+) + \pi_0, \quad \pi_0 \geq 0, \quad \pi_0 g(v(t)) = 0 \\ H(\tau^-) = H(\tau^+) \end{cases} \tag{7.35}$$

where τ is a junction time (entry or contact time) for the constraint and the Hamiltonian is $H = mu_p v + bu_p^2 + \lambda(u_p - h) + \mu v$.

When the speed constraint is not active (interior intervals), $\eta(t) \equiv 0$ and the optimal control input is linear in time with a constant derivative $\dot{u}_p = (\mu + mh)/(2b)$. Thus the speed is quadratic in time as in the unconstrained case, see (6.73)–(6.74). The jump conditions together with the algebraic relation between u_p and λ imply the continuity of the control input.[1] Since inside the boundary interval (when the constraint is active), it must be $\dot{v} \equiv 0$, which is ensured by $u_p \equiv h$, then $u_p(\tau^-) = u_p(\tau^+) = h$.

Once the solution has left the first boundary interval, the control input keeps varying with time with the same constant derivative as in the first interior interval. That means that the trajectory $u_p(t)$ cannot reach the boundary value h more than once. In other terms, the speed constraint can be active only on a single boundary interval, between an entry time t_1 and an exit time t_2. The constrained-optimal control law is thus made up of just three phases, see Fig. 7.11a.

Given the constancy of $\ddot{v} = (\mu + mh)/(2b) - h$ in the two unconstrained (parabolic) phases and the symmetric boundary conditions $v_i = v_f = 0$, the two unconstrained phases must be symmetrical, thus $t_2 = t_f - t_1$. Further imposing the continuity of the control input at the junction times ($\dot{v}(t_1) = \dot{v}(t_2) = 0$), the speed profile is completely characterized by the boundary conditions and the unknown parameter t_1. Explicitly, it reads[2]

$$
v(t) = \begin{cases}
\dfrac{2v_{max}}{t_1}t - \dfrac{v_{max}}{t_1^2}t^2, & t \in [0, t_1) \\[2mm]
v_{max}, & t \in [t_1, t_2] \\[2mm]
\dfrac{2v_{max}}{t_1}(t_f - t) - \dfrac{v_{max}}{t_1^2}(t_f - t)^2, & t \in (t_2, t_f]
\end{cases}
\tag{7.36}
$$

The time t_1 is found by imposing the overall distance,

$$
t_1 = \frac{3}{2} \cdot \frac{v_{max}t_f - s_f}{v_{max}}.
\tag{7.37}
$$

We find now the conditions $F(t_f, s_f) \geq 0$ for which (7.36) is an admissible speed profile. As discussed earlier, the general conditions on the unconstrained speed profile F_{UB1} and F_{UB2} are both inactive for $v_i = v_f = 0$, while F_{LB1} and F_{LB2} fix the unconstrained lower bound.

This lower bound can be exceeded by the constrained speed profile, as discussed in detail in Appendix B. A first obvious constraint is

$$
v_{max} - \frac{s_f}{t_f} \geq 0.
\tag{7.38}
$$

---

[1]Denoting $H(\tau^-) - H(\tau^+)$ with ΔH and analogously for the other jumps at a junction time, we can compute $\Delta H = mv\Delta u_p + b\Delta u_p^2 + \Delta(\lambda u_p) - h\Delta\lambda$ from the Hamiltonian definition and the fact that $\Delta v = 0$. The algebraic relation between λ and u_p provides $\Delta\lambda = -2b\Delta u_p - \Delta\eta$, which equals $-2b\Delta u_p + \eta$ for the slack conditions, and $\Delta(\lambda u_p) = -2b\Delta u_p^2 - mv\Delta u_p - \Delta(\eta u_p)$, which equals $-2b\Delta u_p^2 - mv\Delta u_p - \eta h$ by construction. Combining these equations, we obtain $\Delta H = -b\Delta u_p^2 + 2bh\Delta u_p$. Since this quantity must be zero for the jump conditions, $\Delta u_p = 0$ results.

[2]See Appendix B for the general solution when $v_i \neq 0$, $v_f \neq 0$.

Fig. 7.10 Domain of
feasibility of the
unconstrained speed profile
(shaded gray area) and of the
constrained speed profile
(dark gray area) in the plane
t_f–s_f for the limit speed
scenario. The curves shown
are: F_{LB1} (purple), F_{LB2}
(blue), $F_{LB2'}$ (black), and the
inactive constraint (7.38)
(dashed)

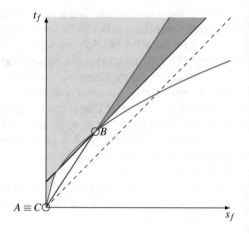

On the other hand, limiting the acceleration/deceleration of the constrained speed
profile to a maximum value a_{max}, that is, imposing $\dot{v}(0) = -\dot{v}(t_f) = a_{max}$ yields

$$F_{LB2'}(t_f, s_f) \triangleq v_{max}t_f - s_f - \frac{4v_{max}^2}{3a_{max}} \geq 0 . \qquad (7.39)$$

Note that the latter condition implicitly ensures that t_1 given by (7.37) is a positive
quantity. It also is intrinsically more restrictive than (7.38), which is thus never active.
The domain of feasibility is illustrated in Fig. 7.10.

Finally, the effective energy consumption, as defined in (6.76) for the unconstrained profile, is evaluated by particularizing and summing (6.76) for the three
phases, with phase 2 that has no contribution. After tedious calculations, we obtain

$$E_{b,e}^{(P)} = \frac{12bs_f^2}{t_f^3} \left(\frac{r_s^3}{3r_s - 2} \right) . \qquad (7.40)$$

7.5.2.2 Non-predictive Strategy

The non-predictive strategy initially follows the unconstrained optimal speed profile
$v^*(t)$. Then, at a time t_N, the speed limit is reached and the strategy is forced to keep
the speed constant. As the numerical results of the previous section have shown, there
is a time, denoted here as $t_N < t_T < t_f$, when the NP strategy retrieves a sub-optimal
(parabolic, within the assumption of the model) speed profile until stop. Therefore the
NP speed profile is made of three phases and, imposing the continuity of the control
input ($\dot{v}(t_T) = 0$), it is completely characterized by the boundary conditions and the
unknown parameters t_N, t_T. The latter are found by imposing that $v^*(t_N) = v_{max}$ and
the overall distance.

The first condition, using (7.30) and (7.34), yields

$$t_N = \frac{t_f}{2} \left(1 - \sqrt{1 - r_s} \right) ,$$ (7.41)

while the second condition yields

$$t_T = t_f \left(-\frac{1}{2} + \frac{1}{r_s} - \left(1 - \frac{1}{r_s} \right) \sqrt{1 - r_s} \right) .$$ (7.42)

The effective energy consumption is finally found as

$$E_{b,e}^{(NP)} = \frac{6bs_f^2}{t_f^3} \left(\frac{3r_s^2 + 2r_s - 3 + (2r_s^2 - 5r_s + 3)\sqrt{1 - r_s}}{4r_s - 3} \right) ,$$ (7.43)

where $s_N \triangleq s^*(t_N)$ is easily evaluated from (7.30).

7.5.2.3 Analysis

As a measure of performance of eco-driving strategies, we use here the relative effective energy loss with respect to the unconstrained optimum, defined as

$$\varepsilon_e^{(\{P,NP\})} \triangleq \frac{E_{b,e}^{(\{P,NP\})} - E_{b,e}^*}{E_{b,e}^*} .$$ (7.44)

With this definition, the influence of vehicle parameters vanishes and (7.44) is a function of the scenario parameter r_s only.

For the predictive strategy,

$$\varepsilon_e^{(P)} = \frac{r_s^3 - 3r_s + 2}{3r_s - 2} ,$$ (7.45)

while for the non-predictive strategy

$$\varepsilon_e^{(NP)} = \frac{3r_s^2 - 6r_s + 3 + (2r_s^2 - 5r_s + 3)\sqrt{1 - r_s}}{2(4r_s - 3)} .$$ (7.46)

Functions (7.45)–(7.46) are plotted in Fig. 7.11b. Although they cannot be compared quantitatively, the analytical predictions show the same trend of the numerical results of Fig. 7.9. Note that the energy losses become unrealistically high as r_s approaches its lower bound of 2/3 (when $s_f = v_{max}t_f$ the acceleration and deceleration phases become infinitely fast).

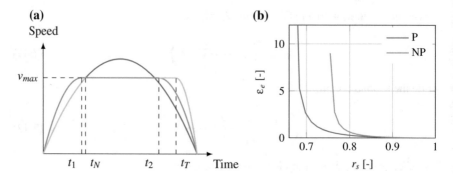

Fig. 7.11 Schematic strategies for the limit speed scenario: unconstrained optimal, non-predictive sub-optimal, and predictive (**a**); performance index as a function of the parameter r_s (**b**)

7.6 Intersection

This section discusses the effect of an intersection or a stop sign in the middle of a route that is otherwise unconstrained. This circumstance imposes the interior constraints

$$s(t_t) = s_t, \quad v(t_t) = v_t \, , \tag{7.47}$$

with s_t denoting the position of the intersection, v_t the prescribed crossing speed, while the crossing time t_t is free.

A numerical analysis of this scenario is presented in Sect. 7.6.1. Then in Sect. 7.6.2 we use closed-form solutions of the ED-OCP (6.64) to corroborate the numerical results and highlight the influence of the preview ability and other parameters of the problem.

7.6.1 Numerical Analysis

As in the previous sections, we consider a trip having a fixed distance s_f and duration t_f. The unconstrained optimal speed profile, used as a baseline, is denoted as $v^*(t)$. We further compare predictive (P) and non-predictive (NP) constrained optimal speed profiles, as they have been defined in Sect. 7.4. The latter has just some limited visual preview of the intersection, quantified by the *preview distance* $r_t s_t$ ($0 \le r_t < 1$), while predictive eco-driving perfectly anticipates the intersection constraint with unlimited preview distance ($r_t = 1$).

Speed profiles P and NP can be obtained using the methods introduced in Chap. 6. In particular, if the intersection crossing time t_t, is fixed, the optimization of the trip ahead of the intersection becomes independent from that after it. The predictive speed profile is thus obtained by solving a first unconstrained OCP from the initial time to

Table 7.2 Boundary conditions for the intersection-constrained scenario

		Duration	Distance	Initial speed	End speed
Unconstrained		t_f	s_f	0	0
Predictive	#1	t_t	s_t	0	v_t
	#2	$t_f - t_t$	$s_f - s_t$	v_t	0
Non-predictive	#1	t_N	$s_t(1 - r_t)$	0	$v^*(t_N)$
	#2	$t_t - t_N$	$r_t s_t$	$v^*(t_N)$	v_t
	#3	$t_f - t_t$	$s_f - s_t$	v_t	0

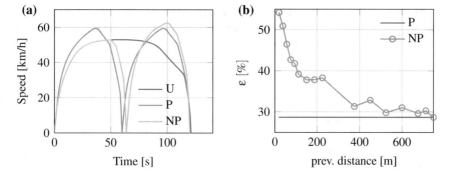

Fig. 7.12 Unconstrained optimal (U), predictive (P), and non-predictive (NP) strategies for a trip with $s_f = 1500$ m, $t_f = 120$ s, $v_i = v_f = 0$, $s_t = s_f/2 = 750$ m, $v_t = 0$. Speed profiles (**a**) and performance indexes (**b**)

the crossing time, then a second one from t_t to the end time. As for the NP speed profile, it results from the concatenation of: (i) the unconstrained optimal speed profile until the time t_N at which the intersection is detected, such that $s^*(t_N) = s_t(1 - r_t)$, (ii) the solution of an unconstrained OCP from time t_N to the crossing time t_t, and (iii) downstream of the intersection, the solution of an unconstrained OCP from t_t to the end time. In both cases, the crossing time t_t is further optimized to have minimal energy consumption. See Table 7.2 for a more detailed list of the respective boundary conditions.

Figure 7.12a shows the three speed profiles for $s_f = 1500$ m, $t_f = 120$ s, $v_i = v_f = 0$, $s_t = s_f/2 = 750$ m, $v_t = 0$ (stop), $r_t = 0.2$ (preview distance of 150 m), computed with the DP Algorithm 6 and the EV model of Table 6.3. The optimal crossing time for the predictive strategy is found as $t_t = t_f/2 = 60$ s, which yields two equal speed profiles before and after the intersection, while for the NP strategy it is found to be around 64 s.

The values of the measures of performance (7.45) are plotted in Fig. 7.12b as a function of the preview distance. While $\varepsilon^{(P)}$ does not depend on the preview distance, $\varepsilon^{(NP)}$ generally decreases with it. Note that the optimal crossing time in the NP strategy varies with the preview distance, ranging from 65 s at small preview distances, while decreasing to 60 s as r_t increases.

7.6.2 Analytical Approach

In order to retrieve and explain the behavior shown by the numerical results of the previous section, we use the simplified EV model of Sect. 6.5.3.3, for which explicit solutions of the ED-OCP can be calculated.

In particular, the benefits of preview and anticipation are analyzed as a function of two parameters, namely the preview distance and the crossing speed, while setting the intersection position at $s_t = s_f/2$. We define for later use the ratio between the crossing speed v_t and the average speed along the route,

$$r_v \triangleq \frac{v_t t_f}{s_f}, \tag{7.48}$$

with $0 \le r_v \le 3/2$ since v_t cannot exceed the top speed of the unconstrained scenario, $2/3(s_f/t_f)$. As for the preview distance, we convert it into a more easily treatable preview time, by using the average speed of the unconstrained profile. In such a way, the ratio r_t takes the new formulation

$$r_t = \frac{t_t - t_N}{t_t}, \tag{7.49}$$

with $r_t \in [0, 1]$.

7.6.2.1 Predictive Strategy

The predictive speed profile is composed of two separate parabolic phases, before and after the intersection. Imposing the crossing speed and the distance of both phases, the P profile is completely defined by the crossing time t_t, see Fig. 7.13. The optimal value for this parameter is found by minimizing the consumption. Using (7.30), the effective energy consumption as a function of t_t is

$$E_{b,e}^{(P)}(t_t) = bs_f^2 \left\{ 3 \left(\frac{1}{t_t^3} + \frac{1}{(t_f - t_t)^3} \right) - 6\frac{r_v}{t_f} \left(\frac{1}{t_t^2} + \frac{1}{(t_f - t_t)^2} \right) + \right.$$
$$\left. + 4\frac{r_v^2}{t_f^2} \left(\frac{1}{t_t} + \frac{1}{(t_f - t_t)^2} \right) \right\}, \tag{7.50}$$

Fig. 7.13 Schematic
strategies for the intersection
scenario: unconstrained
optimal, non-predictive
sub-optimal, and predictive,
with $r_v = 0$

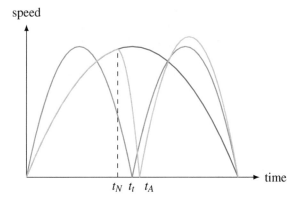

The minimum of this function is obtained for

$$\frac{\partial E_{b,e}^{(P)}}{\partial t_t} = 0 \rightarrow t_t = \frac{t_f}{2} \, , \tag{7.51}$$

meaning that the predictive speed profile is made of two equal phases, the result numerically found in Sect. 7.6.1. With this optimal crossing time, the effective consumption is derived as

$$E_{b,e}^{(P)} = \frac{16bs_f^2}{t_f^3} \left(3 - 3r_v + r_v^2\right) \, . \tag{7.52}$$

7.6.2.2 Non-predictive Strategy

The non-predictive strategy initially follows the unconstrained optimal speed profile v^*. This phase ends at time $t_N = t_t(1 - r_t)$ when the presence of the intersection is detected. The corresponding distance and speed are easily evaluated as

$$s^*(t_N) = \frac{s_f}{4}(2 - 3r_t + r_t^3) \, , \tag{7.53}$$

$$v^*(t_N) = \frac{3s_f}{2t_f}(1 - r_t^2) \, , \tag{7.54}$$

respectively. The intersection is then approached with a second parabolic profile to be reached at time t_t. After that, a third parabolic speed profile is performed until stop at the final distance. Imposing the crossing speed and the distance of the two latter phases, the NP speed profile is completely characterized by the boundary conditions and the unknown parameter t_t. The latter should in principle be found by minimizing the energy consumption, similarly to the predictive scenario. Instead of attempting such an optimization that would lead to formulas difficult to manage, we shall consider for simplicity the same value as in the predictive scenario, i.e., $t_t = t_f/2$, for which the overall consumption is derived as

$$E_{b,e}^{(NP)} = \frac{4bs_f^2}{t_f^3} \cdot \frac{(9 - 12r_v + 4r_v^2) + r_t(15 - 12r_v + 4r_v^2)}{2r_t}. \tag{7.55}$$

Note that for $r_t = 1$, the fully predictive result (7.52) is retrieved.

7.6.2.3 Analysis

The performance indexes for the predictive and the two non-predictive strategies are obtained from (7.52), (7.55) as

$$\varepsilon_e^{(P)} = \frac{(3 - 2r_v)^2}{3} \tag{7.56}$$

and

$$\varepsilon_e^{(NP)} = \frac{(1 + r_t)(3 - 2r_v)^2}{6r_t}. \tag{7.57}$$

Functions (7.56)–(7.57) are plotted in Fig. 7.14a. As in the previous section, these results are not quantitatively comparable with the numerical ones shown in Fig. 7.12. However, the qualitative trend shows that, as the preview horizon increases, energy loss of the non-predictive strategy monotonically decreases.

On the other hand, the energy loss decreases with an increase of the desired speed at the intersection, as shown in Fig. 7.14b for a preview distance corresponding to $r_t = 0.25$. If the vehicle must stop at the traffic signal ($r_v = 0$), the resulting energy loss with the NP strategy is several times higher than that of the P strategy, which is due to higher electric losses during the regenerative braking process.

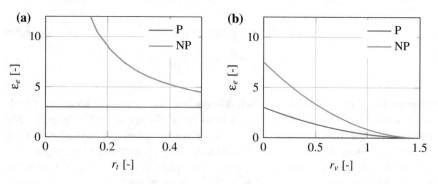

Fig. 7.14 performance indexes for the intersection scenario as a function of the parameters r_t (**a**) and r_v (**b**)

These results are largely independent of the choice of the boundary conditions. For a slightly different scenario where the vehicle would cruise at a constant speed if the intersection was not present ($v_i = v_f = s_f/t_f$), the effective energy consumption of the three strategies are

$$E_{b,e}^* = 0, \quad E_{b,e}^{(P)} = \frac{16bs_f^2}{t_f^3}(1 - r_v)^2, \quad E_{b,e}^{(NP)} = \frac{8bs_f^2}{t_f^3}(1 - r_v)^2 \left(1 + \frac{1}{r_t}\right),$$

(7.58)

showing a similar dependency on r_v, r_t than (7.52)–(7.55).

7.7 Traffic Light

The scenario studied in this section is the presence of a signalized intersection (traffic light) in the middle of the route (at $s_t = s_f/2$), allowing the crossing only during fixed time slots that correspond to the green light periods. In particular, the light is set to be red at time $t_f/2$. Denoting the crossing time with t_t (such that $s(t_t) = s_t$), the constraint reads

$$t_t \leq \frac{t_f}{2}(1 - r_x) \quad \vee \quad t_t \geq \frac{t_f}{2}(1 + r_x),$$

(7.59)

where r_x is the parameter that defines the duty cycle of the traffic light (see Sect. 4.1.3) and Fig. 7.15.

A numerical analysis of this scenario is presented in Sect. 7.7.1. Then, in Sect. 7.7.2 we use closed-form solutions of ED-OCP (6.64) to corroborate the numerical results and highlight the influence of the preview ability and other parameters of the problem.

Fig. 7.15 Schematic definition of the traffic light scenario

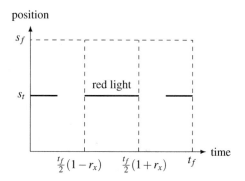

Table 7.3 Boundary conditions for the traffic-light-constrained scenario

		Duration	Distance	Initial speed	End speed
Unconstrained		t_f	s_f	0	0
Predictive	#1	t_t	s_t	0	v_t
	#2	$t_f - t_t$	$s_f - s_t$	v_t	0
Non-predictive	#1	t_N	$s^*(t_N)$	0	$v^*(t_N)$
	#2	$t_A - t_N$	$s_t - s^*(t_N)$	$v^*(t_N)$	0
	#3	$t_f - t_A$	$s_f - s_t$	0	0

7.7.1 Numerical Analysis

As in the previous sections, we consider a trip having a fixed distance s_f and duration t_f. The unconstrained optimal speed profile, used as a baseline, is denoted as $v^*(t)$. We further compare predictive (P) and non-predictive (NP) eco-driving, as they have been defined in Sect. 7.4. The latter has just visual information of the signal phase and color, while predictive eco-driving perfectly anticipates the traffic light constraint.

The enforcement of the interior constraint (7.59) requires that the optimization of the speed profiles P and NP is made with a DP algorithm having speed and time as state variables. To speed up the calculation, a single-state DP code can still be used, provided that time t_t is considered a free parameter to be further optimized.

For the predictive strategy, we can reasonably assume from physical consider-ations that the optimal choice for t_t is to equal the last useful green time, that is, $t_t = t_f/2(1 - r_x)$.[3] The original OCP can thus be split into two partial OCPs, before and after the traffic light crossing. However, these two OCPs are not independent as in the intersection case of Sect. 7.6, but coupled by the crossing speed v_t that plays the role of final speed in the former OCP and initial speed in the latter. The parameter v_t must therefore be further optimized to have the minimal energy consumption.

As for the NP speed profile, it first follows the unconstrained solution until the time at which the traffic light is assumed to be detected, which we also set at $t_N = t_f/2(1 - r_x) = t_t$. Then, having no preview about the red phase duration, the NP strategy should decelerate and stop in an arbitrary time, and wait for the next green phase. The faster the deceleration, the larger is the energy consumption. We consider the best-case scenario (from the NP perspective), where the strategy exactly chooses to stop at the last time on red, $t_A = t_f/2(1 + r_x)$, only to start again soon after. Thus, the rest of the NP profile is found by concatenating (i) the solution of an unconstrained OCP from time t_t to the time t_A, and (ii) after the traffic light, the solution of an unconstrained OCP from t_A to the end time. See Table 7.3 for a more detailed list of the respective boundary conditions of the three strategies.

---

[3] Due to the symmetry of the problem in this particular case, the same result would be obtained with t_t as the first useful green time after the central red phase.

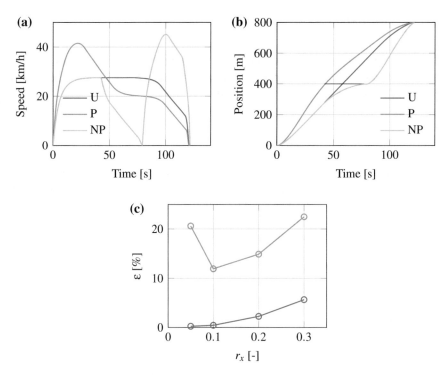

Fig. 7.16 Unconstrained optimal (U), predictive (P), and non-predictive (NP) strategies for a traffic light scenario with $s_f = 800$ m, $t_f = 120$ s, $v_i = v_f = 0$, $s_t = s_f/2 = 400$ m, $r_x = 0.3$. Speed (**a**) and position (**b**) profiles and performance indexes (**c**)

Figure 7.16a shows the three speed profiles for $s_f = 800$ m, $t_f = 120$ s, $v_i = v_f = 0$, $s_t = s_f/2 = 400$ m, $r_x = 0.3$, computed with the DP Algorithm 6 and the EV model of Table 6.3. Figure 7.16b shows the respective position trajectories, together with the forbidden window with the traffic light being red. The optimal crossing speed of the predictive strategy is found to be around 30 km/h in this case.

The values of the measures of performance (7.33) are plotted in Fig. 7.16c as a function of the duty cycle parameter r_x. While $\varepsilon^{(P)}$ increases with an increase of r_x and tends to zero as r_x tends to zero, $\varepsilon^{(NP)}$ results from two opposite trends and shows a minimum. On the one hand, small values of r_x mean small preview times and strong decelerations/accelerations to catch the new green period. On the other hand, large values of r_x mean longer red intervals and thus stronger accelerations in the second green period to complete the trip.

7.7.2   Analytical Approach

In order to retrieve and explain the behavior shown by the numerical results of the previous section, we use the simplified EV model of Sect. 6.5.3.3, for which explicit solutions of the ED-OCP can be calculated.

7.7.2.1   Predictive Strategy

The predictive strategy perfectly anticipates the presence of the traffic light and chooses the optimal crossing time and speed satisfying the constraint imposed by its timing. As discussed above, we assume the optimal timing choice as $t_t = t_f/2(1 - r_x)$. However, the crossing speed v_t is free and can be optimized. Similarly to (7.50), we evaluate the overall consumption as a function of v_t,

$$E_{b,e}^{(P)}(v_t) = 4b \left(\frac{12s_f^2}{t_f^3} \frac{1 + 3r_x^2}{(1 - r_x^2)^3} - \frac{12s_f v_t}{t_f^2} \frac{1 + r_x^2}{(1 - r_x^2)^2} + \frac{4v_t^2}{t_f} \frac{1}{(1 - r_x^2)} \right) . \quad (7.60)$$

The minimum of function (7.60) is obtained for

$$\frac{\partial E_{b,e}^{(P)}}{\partial v_t} = 0 \rightarrow v_t = \frac{3s_f}{2t_f} \cdot \frac{1 + r_x^2}{1 - r_x^2} . \quad (7.61)$$

Note that for $r_x = 0$ the top speed of the unconstrained optimal profile is retrieved. With this choice of crossing speed, the effective consumption is derived as

$$E_{b,e}^{(P)} = \frac{12bs_f^2}{t_f^3} \cdot \frac{1 + 6r_x^2 - 3r_x^4}{(1 - r_x^2)^3} \quad (7.62)$$

and equals the unconstrained optimum for $r_x = 0$.

7.7.2.2   Non-predictive Strategy

For the non-predictive strategy, we set $t_N = t_f/2(1 - r_x)$ and we impose a stop ($v_t = 0$) at the traffic light at time $t_A = t_f/2(1 + r_x)$, as in the numerical analysis. Thus the effective consumption is derived as

$$E_{b,e}^{(NP)} = \frac{12bs_f^2}{t_f^3} \cdot \frac{(1 + r_x^2)(3 + r_x(31 + (-3 + r_x)r_x(8 + r_x^2)))}{16(1 - r_x)^3 r_x} . \quad (7.63)$$

Fig. 7.17 Performance index for the traffic light scenario as a function of the parameter r_x

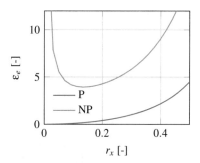

7.7.2.3 Analysis

The performance indexes (7.44) are evaluated as

$$\varepsilon_e^{(P)} = \frac{r_x^2(3 - r_x^2)^2}{(1 - r_x^2)^3} \tag{7.64}$$

and

$$\varepsilon_e^{(NP)} = \frac{3 + 15r_x + 27r_x^2 - 9r_x^3 - 11r_x^4 + 9r_x^5 - 3r_x^6 + r_x^7}{16(1 - r_x)^3 r_x}. \tag{7.65}$$

Figure 7.17 shows the two functions (7.64)–(7.65) as a function of the parameter r_x. Clearly, the trend numerically obtained and shown in Fig. 7.16c is retrieved, with $\varepsilon^{(P)}$ increasing with r_x and $\varepsilon^{(NP)}$ showing a minimum.

These results are largely independent of the particular boundary conditions chosen. For a slightly different scenario where the vehicle is cruising at a constant speed ($v_i = v_f = s_f/t_f$) except for the traffic light, the effective energy consumption of the three strategies are

$$E_{b,e}^* = 0, \quad E_{b,e}^{(P)} = \frac{48bs_f^2}{t_f^3} \cdot \frac{r_x^2}{(1 - r_x)^3}, \quad E_{b,e}^{(NP)} = \frac{8bs_f^2}{t_f^3} \cdot \frac{1 + r_x + r_x^2}{(1 - r_x)^3}, \tag{7.66}$$

showing a similar dependency on r_x to (7.62)–(7.63).

7.8 Car Following

In the scenario of this section, the host vehicle is subject to avoiding rear-end collisions with a preceding vehicle, whose motion is described by a temporal law $s_p(t)$.

That imposes a constraint of the type

$$s(t) - (s_p(t) - s_{min}(v(t), v_p(t))) \leq 0, \quad t \in [0, t_f], \tag{7.67}$$

where s_{min} is the minimum inter-vehicle safe distance. The latter quantity can actually describe a distance headway or a time headway enforced by the road law or other considerations. While the minimal safe distance headway is usually a constant, the minimal safe time headway depends on the relative speed of the two vehicles. In this section, we shall assume that the safe distance at most depends on the preceding vehicle's speed v_p only (not on the host vehicle's speed), and thus it will be considered as a prescribed function of time, $s_{min}(t)$.

A numerical analysis of this scenario is presented in Sect. 7.8.1. Then in Sect. 7.8.2 we use closed-form solutions of ED-OCP (6.64) to corroborate the numerical results and highlight the influence of the preview ability and other parameters of the problem.

7.8.1 Numerical Analysis

Position constraint (7.67) plays the role of a state constraint in the ED-OCP. Additionally, there might be speed constraints such as those studied in Sect. 7.5. The key factor for performing eco-driving in such a situation is the ability to detect the position and the velocity of the preceding vehicle. That is usually feasible with current ADAS-type sensors (see Sect. 3.2.1), which nevertheless have a finite range. An extension of the preview distance provided by standard ADAS sensors could be made possible by the use of dedicated vehicle-to-vehicle or vehicle-to-infrastructure communication (see Sect. 3.1).

As in the previous sections, we consider a trip having a fixed distance s_f and duration t_f. The preceding vehicle is assumed to advance at constant acceleration, that is, we assume $s_p(t) = s_{p,0} + v_{p,0}t + a_p t^2/2$.[4] The unconstrained optimal speed profile, used as a baseline, is denoted as $v^*(t)$. We further compare predictive (P) and non-predictive (NP) constrained optimal speed profiles, as they have been defined in Sect. 7.4. The latter has just a limited preview of the preceding vehicle, defined by the preview distance $r_d s_f$, with $0 \leq r_d \leq 1$. Here we consider only the case $r_d = 0$. The predictive eco-driving perfectly anticipates the preceding vehicle constraint with unlimited preview distance ($r_d = 1$).

The speed profile P is obtained using the methods introduced in Chap. 6. In particular, a full two-state DP code can be used, with time as the independent variable and velocity and position as state variables. In this way it is possible to enforce the position constraint directly, by making unfeasible all the points of the position grid that

---

[4]In Chap. 9, we present a case study where motion of the preceding vehicle is modeled as a Markov chain and predicted probabilistically.

Table 7.4 Boundary conditions for the position-constrained scenario

		Constrained optimization?	Duration	Distance	Initial speed	End speed
Unconstrained		N	t_f	s_f	0	0
Predictive		Y	t_f	s_f	0	0
Non-predictive	#1	N	t_N	$s_p^*(t_N)$	0	$v^*(t_N)$
	#2	Y	$t_f - t_N$	$s_f - s^*(t_N)$	$v_p(t_N)$	0

exceed the maximal allowed position $s_p(t) - s_{min}(t)$. The predictive speed profile is thus obtained by solving the constrained OCP for the original boundary conditions, see Table 7.4. The non-predictive speed profile is obtained by concatenating (i) the unconstrained solution until a time t_N such that $s^*(t_N) = s_p(t_N) - s_{min}(t_N)$ with (ii) the solution of the constrained OCP from t_N to the end time, with a distance to cover $s_f - s^*(t_N)$, an initial speed equal to $v^*(t_N)$ and a final speed of zero. Note that this definition might imply an unrealistic speed discontinuity at t_N if $v_p(t_N) < v^*(t_N)$.

Figure 7.18a, b shows the U and P profiles in terms of speed and position computed with the two-state DP and the EV model of Table 6.3 for $s_f = 500$ m, $t_f = 60$ s, $v_i = v_f = 0$, and two leader's scenarios: $s_{p,0} = 20$ m, $v_{p,0} = 15$ km/h, $a_p = 0.14$ m/s² (P1) and $s_{p,0} = 50$ m, $v_{p,0} = 15$ km/h, $a_p = 0.17$ m/s² (P2). Clearly, the P profiles present milder initial accelerations to allow a rendez-vous with the leader when the two vehicles have the same speed. For the sake of figure readability, these figures do not show the NP profiles.

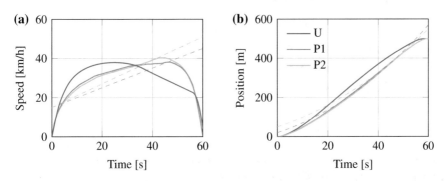

Fig. 7.18 Unconstrained optimal (U) and predictive (P) strategy for a trip with $s_f = 500$ m, $t_f = 60$ s, $v_i = v_f = 0$, $s_{p,0} = \{20, 50\}$ m, $v_{p,0} = 15$ km/h, and $a_p = \{0.14, 0.17\}$ m/s²

7.8.2 Analytical Approach

In order to retrieve and explain the behavior shown by the numerical results of the previous section, we use the simplified EV model of Sect. 6.5.3.3, for which explicit solutions of the ED-OCP can be calculated.

We further assume that the preceding vehicle departs from the same position as the host vehicle ($s_{min} = 0$) and drives at a constant speed, denoted by v_p. A more general case, including initial separation and relative speed between the two vehicles, and a constant leader acceleration, is treated in Appendix B. The unconstrained solution in terms of speed is still given by (7.30).

7.8.2.1 Predictive Strategy

The predictive speed profile is the result of a constrained optimization. The necessary conditions for optimality derive from (6.72) in the Lagrangian form with the addition of the conditions (6.31)–(6.34). Using the nomenclature introduced in Sect. 6.2.2.1, the position constraint $g(\mathbf{x}(t), t) = s(t) - v_p t \leq 0$ is of the second order ($p = 2$), since $\ddot{g} = g^{(2)} = \dot{v} = u_p - h$. Additionally, we have $\partial g/\partial s = 1$, $\partial g^{(1)}/\partial v = 1$, $\partial g/\partial t = -v_p$, and $\partial g^{(1)}/\partial t = \partial g^{(1)}/\partial s = \partial g/\partial v = 0$. Therefore, the system of equations to be solved reads

$$\begin{cases} \dot{s} = v(t), \quad s(0) = 0, \quad s(t_f) = s_f \\ \dot{v} = u_p(t) - h, \quad v(0) = v_i, \quad v(t_f) = v_f \\ \dot{\lambda} = -mu_p(t) - \mu, \quad \lambda(0) = \lambda_0 \\ \dot{\mu} = 0, \quad \mu(0) = \mu_0 \\ u_p(t) = -\dfrac{1}{2b}(\lambda(t) + mv(t) + \eta(t)) \\ \eta(t)g(s(t), t) = 0, \quad \eta(t) \geq 0, \quad \dot{\eta} \leq 0, \quad \ddot{\eta} \geq 0 \\ \lambda(\tau^-) = \lambda(\tau^+) + \pi_1 \\ \mu(\tau^-) = \mu(\tau^+) + \pi_0 \\ \pi_0 \geq 0, \quad \pi_1 \geq 0, \quad \pi_0 g(s(t), t) = 0, \quad \pi_1 g(s(t), t) = 0 \\ H(\tau^-) = H(\tau^+) + \pi_0 v_p \end{cases} \tag{7.68}$$

where τ is a junction time (entry or contact time) for the constraint and the Hamiltonian is $H = mu_p v + bu_p^2 + \lambda(u_p - h) + \mu v$.

When the position constraint is not active, $\eta(t) \equiv 0$ and the optimal control input is linear in time with a constant derivative $\dot{u}_p = (\mu_0 + mh)/(2b)$. Thus the speed is quadratic in time as in the unconstrained case, see (6.73)–(6.74). The position constraint can be active either on a boundary interval or at a contact point. Here only the case with a contact point t_1 is considered. The tangency conditions impose that $s(t_1) = s_p(t_1) = v_p t_1$ and $v(t_1) = v_p$. The jump conditions together with the algebraic relation between u_p and λ imply the continuity of the control input at the contact point.[5]

The constrained-optimal trajectory is thus made up of two parabolic phases separated by the contact point, see Fig. 7.20a, b. Imposing the tangency conditions, the speed profile is completely characterized by the boundary conditions and the unknown parameter t_1. Explicitly, it reads

$$
v(t) = \begin{cases}
\dfrac{4v_p}{t_1}t - \dfrac{3v_p}{t_1^2}t^2, & t \in [0, t_1) \\[2ex]
3v_p - \dfrac{2v_p}{t_1}t + \dfrac{v_p}{t_1}\dfrac{2t_f - 3t_1}{(t_f - t_1)^2}(t - t_1)^2, & t \in [t_1, t_f]
\end{cases}
\tag{7.69}
$$

Note that the first of (7.69) implies that the first phase has a maximum at $2/3t_1$. The maximum speed reached is $4/3v_p$. Here we shall assume that this speed is lower than any top speed limit.

The contact time is found by imposing the overall distance, yielding

$$
t_1 = \frac{t_f^2 v_p}{4t_f v_p - 3s_f}.
\tag{7.70}
$$

We find now the conditions $F(t_f, s_f) \geq 0$ for which (7.69) is an admissible speed profile. As discussed earlier, the general conditions on the unconstrained speed profile F_{UB1}, F_{UB2}, and F_{LB1} are always satisfied when $v_i = v_f = 0$ and $v_{max} \to \infty$. Thus the only conditions to be fulfilled, which defines the limits of feasibility of the unconstrained speed profile, are F_{LB2} (maximum acceleration/deceleration) and F_{LB3} that is imposed by the position constraint. The latter condition can be expressed by the requirement that the quadratic-in-time equation $s^*(t) = v_p t$ has at most one degenerate real root. As shown in Appendix B, that reads

$$
F_{LB3}(t_f, s_f) \triangleq 8v_p t_f - 9s_f \geq 0.
\tag{7.71}
$$

The domain of feasibility of the unconstrained speed profile can be exceeded by the constrained speed profile. However, we must obviously require that the final position does not exceed the final leader position, that is,

$$
F_{LB3''}(t_f, s_f) \triangleq v_p t_f - s_f \geq 0.
\tag{7.72}
$$

Note that this condition implicitly ensures that t_1 given by (7.70) is a positive quantity.

---

[5]Denoting $H(\tau^-) - H(\tau^+)$ with ΔH and analogously for the other jumps at a junction time, we can compute $\Delta H = mv_p\Delta u_p + b\Delta u_p^2 + \Delta(\lambda u_p) - h\Delta\lambda + v\Delta\mu$ from the Hamiltonian definition and the fact that $\Delta v = 0$. The algebraic relation between λ and u_p provides $\Delta\lambda = -2b\Delta u_p - \Delta\eta$, which equals $-2b\Delta u_p$ for the slack conditions, and $\Delta(\lambda u_p) = -2b\Delta u_p^2 - mv_p\Delta u_p - \Delta(\eta u_p)$, which equals $-2b\Delta u_p^2 - mv_p\Delta u_p$ by construction. Combining these equations, we obtain $\Delta H = -b\Delta u_p^2 + 2bh\Delta u_p + v_p\Delta\mu$. Since this quantity must equal $v_p\pi_0 = v_p\Delta\mu$ for the jump conditions, $\Delta u_p = 0$ results. In addition, $\pi_1 = \Delta\lambda = 0$.

In the position-constrained profile's domain of feasibility, the lower bound F_{LB2} is replaced by the condition that the maximum acceleration and deceleration of the position-constrained profile, $a(0)$, are lower than a_{max} and $|a_{min}| = a_{max}$, respectively. That yields

$$F_{LB2''a} \triangleq a_{max} - \frac{4}{t_f^2}(4t_f v_p - 3s_f) \geq 0, \qquad (7.73)$$

which is always satisfied in the region of interest, and

$$F_{LB2''d} \triangleq a_{max} - \frac{2}{3t_f^2}\frac{(4t_f v_p - 3s_f)(2t_f v_p - 3s_f)}{t_f v_p - s_f} \geq 0, \qquad (7.74)$$

which in turn limits the domain of feasibility of the position-constrained profile.

The domain of feasibility in terms of s_f, t_f is illustrated in Fig. 7.19. For later use, we introduce here a speed ratio

$$r_p \triangleq \frac{v_p}{s_f/t_f} - 1, \qquad (7.75)$$

with $0 \leq r_p \leq 1/8$ as a consequence of (7.71).

Finally, the effective energy consumption is derived as

$$E_{b,e}^{(P)} = \frac{4bs_f^2}{9t_f^3} \cdot \frac{(4r_p + 1)^3}{r_p}. \qquad (7.76)$$

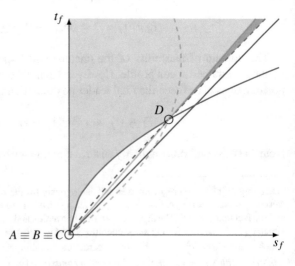

Fig. 7.19 Domain of feasibility of the unconstrained speed profile (shaded gray area) and the constrained speed profile (dark gray area) in the plane t_f–s_f for the car following scenario. The curves shown are: F_{LB2} (blue), F_{LB3} (yellow), $F_{LB2''a}$ (dashed green), $F_{LB2''d}$ (dashed blue), and $F_{LB3''}$ (red). The coordinates of the intersection points are D: $(128/27\sigma, 16/3\tau)$, where $\sigma \triangleq v_p^2/a_{max}, \tau \triangleq v_p/a_{max}$

7.8.2.2 Non-predictive Strategy

The non-predictive speed profile initially follows the unconstrained optimal speed profile v^*, see Fig. 7.20. This phase ends at time t_N when the preceding vehicle is reached (assuming $s_{min} = 0$). The preceding vehicle is then followed, assuming for simplicity instantaneous adaptation (i.e., infinitely fast deceleration) of the NP speed to v_p. This phase lasts until time t_T, at which the non-predictive speed profile retrieves a parabolic sub-optimal profile until stop.

Therefore, the speed profile is made of three phases and, imposing continuity of the control input at t_T ($\dot{v}(t_T) = 0$), it is completely determined by the boundary conditions and the unknown parameters t_N, t_T. Imposing the condition $s^*(t_N) = s_p(t_N) = v_p t_N$, the contact time is evaluated as

$$t_N = \frac{3t_f}{4}\left(1 - \sqrt{1 - \frac{8t_f v_p}{9s_f}}\right). \tag{7.77}$$

On the other hand, the switching time t_T is found by imposing the overall distance, yielding

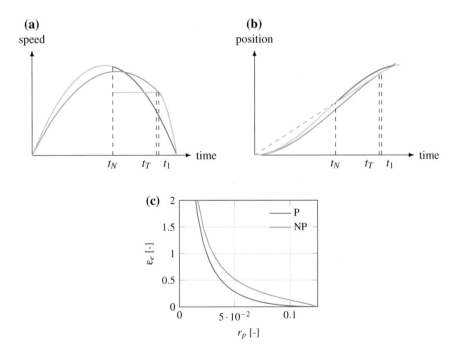

Fig. 7.20 Schematic strategies for the car following scenario: unconstrained optimal (blue), predictive (orange), and non-predictive (green) speed (**a**) and position (**b**) profiles; performance indexes as a function of the parameter r_p (**c**)

$$t_A = \frac{3s_f}{v_p} - 2t_f .$$
(7.78)

Finally, the effective energy consumption is evaluated as

$$E_{b,e}^{(NP)} = \frac{4bs_f^2}{t_f^3} \cdot \frac{(1+r_p)^2((1+r_p)\sqrt{1-8r_p} - 3(1+13r_p))}{9r_p(\sqrt{1-8r_p} - 3)} .$$
(7.79)

7.8.2.3 Analysis

The performance indexes (7.44) are evaluated as

$$\varepsilon_e^{(P)} = \frac{(1-8r_p)^2(1+r_p)}{27r_p}$$
(7.80)

and

$$\varepsilon_e^{(NP)} = \frac{9r_p(1+r_p)\sqrt{1-8r_p} + (1-2r_p)(2-17r_p-r_p^2)}{54r_p} .$$
(7.81)

Note that $\varepsilon^{(P)}$ goes to zero for $r_p = 1/8$, where $\varepsilon^{(NP)}$ would be negative. It is thus clear that the domain of feasibility of the NP profile in terms of r_p is smaller than that of the P profile. These trends are shown in Fig. 7.20c where the two performance indexes are plotted as a function of the speed ratio r_p.

*

* *

The scenarios presented in this chapter are obviously oversimplified and represent real-life situations only in an idealized way. However, their analysis has allowed us to identify for each of them the key parameters that induce energy inefficiencies and how to alleviate them with predictive optimal driving control. While the existing literature has often addressed these scenarios separately with ad hoc strategies, the next chapter, grounded on the theory of Chap. 6, will discuss their implementation in a unified eco-driving system. More realistic conclusions can be drawn after several of these scenarios are treated in the detailed case studies of Chap. 9.

Chapter 8
Eco-Driving Practical Implementation

The aim of this section is to present a few practical implementation issues of eco-driving. In Sect. 8.1 we shall discuss the various eco-driving systems that implement partly or fully the concepts described in the previous chapters. All of these systems use the localization, perception, and planning/control functions that have been treated throughout this book. A few additional and specific algorithmic and implementation issues are further detailed in Sect. 8.2. Then, we will discuss in Sect. 8.3 the issues related to advising a human driver about the optimal speed to follow and the direct implementation of eco-driving via an autonomous driving system.

8.1 Implementation of Eco-Driving Concepts

In this section we present, in order of increasing complexity and comprehensiveness, eco-coaching (Sect. 8.1.1), PCC (8.1.2), eco-ACC (Sect. 8.1.3), and the most general predictive eco-driving schemes (Sect. 8.1.4).

8.1.1 Eco-Coaching

Eco-coaching is the online assessment of the driving style of human drivers by comparing their actual speed traces to the energy-optimal speed profiles. An example of eco-coaching system that has been demonstrated for EVs and ICEs is Geco [1, 2]. Its practical implementation follows a number of steps as illustrated by the flowchart in Fig. 8.1.

© Springer Nature Switzerland AG 2020   215
A. Sciarretta and A. Vahidi, *Energy-Efficient Driving of Road Vehicles*,
Lecture Notes in Intelligent Transportation and Infrastructure,
https://doi.org/10.1007/978-3-030-24127-8_8

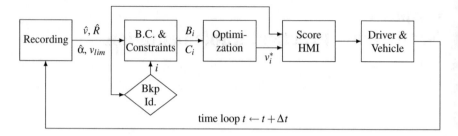

Fig. 8.1 Conceptual sketch of an eco-coaching system

The system is based on the real-time recording of the actual vehicle speed $\hat{v}$. This measure can be obtained from several sources, including integration of GPS coordinates and on-board measurements (odometry) available on the vehicle communication bus. Relying only on GPS-based measurement makes the system free from in-vehicle connections and able to run on a mobile device or as a web service, but suffers from inaccuracies due to satellite signal quality. However, the current precision of GNSS is considered sufficient for this type of application, and accuracy is expected to further improve with the anticipated trend in localization precision. In addition to speed, also curvature $\hat{R}$, slope $\hat{\alpha}$, and legal speed limit v_{lim} need to be extracted from GPS and map data, e.g., using algorithms introduced in Sect. 8.2.1.

These speed traces are continuously analyzed to identify when a *breakpoint* occurs. As explained in Sect. 8.2.2, such identification is based on the definition of breakpoints either as the locations where the characteristics of the road change, or the points where surrounding traffic induces strong decelerations or stops.

When a breakpoint is detected, a sub-trip is defined encompassing the last two breakpoints, see Fig. 8.2. For this sub-trip, the distance covered, the travel time, and the average speed are evaluated as

$$t_f^{(i)} = \tau_{i+1} - \tau_i; \quad s_f^{(i)} = \int_{\tau_i}^{\tau_{i+1}} \hat{v}(\tau)d\tau; \quad v_i^{(i)} = \hat{v}(\tau_i); \quad v_f^{(i)} = \hat{v}(\tau_{i+1}), \quad (8.1)$$

where τ is time of record, τ_{i+1} is current time and τ_i is the time of last breakpoint. These quantities form the boundary condition vector $B_i \triangleq \{t_f^{(i)}, s_f^{(i)}, v_i^{(i)}, v_f^{(i)}\}$, while the road parameter vector $C_i \triangleq \{\alpha^{(i)}, R^{(i)}, v_{lim}^{(i)}\}$ is composed of the slope, curvature, and speed limit in the sub-trip.

A speed optimization is then performed using the elements of B_i and C_i as the optimization parameters. The methods described in Chap. 6 can be generally used to optimize the speed profile and compute the minimal energy E_T^*. For EVs, the simpler approach is to use the parabolic speed profile (6.74). Alternatively, optimal speed profiles can be obtained by training neural networks to evaluate the optimal sequence of control modes and their switching times, that is, provide the vector $O_i = \{t_1, \ldots\}$ as a function of the input vector I_i as also described in Chap. 6.

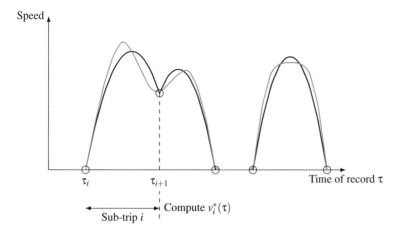

Fig. 8.2 Illustration of the eco-coaching approach, with the recorded speed (black), the breakpoints (circles), a sub-trip, the time instant at which an optimization is performed, and the optimized speed profile (gray)

The output of such a process is ultimately the optimal speed profile $v_i^*(\tau)$, $\tau \in [\tau_i, \tau_{i+1}]$ that the driver should have followed during the last sub-trip, with the same constraints in terms of duration, distance, initial and final speed, maximal speed, than the actual profile performed.

The optimal speed profile is typically displayed to the drivers by an HMI (see Sect. 8.3.1), in order to make them aware of the best driving practices. Additionally, the system can provide an "eco-driving" score, comparing a model-based estimation of the actual (tank) energy consumed during the sub-trip, with the optimum calculated. The drivers are expected to learn from these scores and adapt their driving style toward the optimum.

8.1.2 Predictive Cruise Control

In this book we adopt the terminology of Predictive Cruise Control (PCC) to denote those cruise control systems that, in addition to the standard feature of tracking a reference speed, are mainly focused on anticipation of road slopes. A few such systems are actually on the market as of this book's writing,[1] particularly for heavy-duty truck applications. As shown in Sect. 7.3, a speed profile that anticipates and reacts to road slopes in a certain way (reach a minimum speed at the top of a hill, maximum speed at the bottom of a downhill) may actually coincide with the minimal-energy speed profile.

---

[1] Introduced by Daimler Trucks in 2009, now PCC systems are available on trucks by DAF, Kenworth, Volvo Trucks, as well some Mercedes and Volkswagen cars, among others.

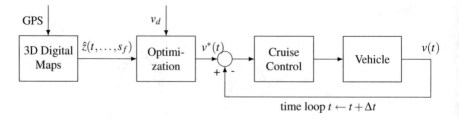

Fig. 8.3 Conceptual sketch of a predictive cruise control system

In a typical embodiment of PCC, see Fig. 8.3, the GPS signal is an input. Using stored or retrieved 3D digital road maps, PCC estimates the altitude profile of the road ahead, $\hat{z}(\sigma)$, $\sigma \in [s, s_f]$, up to the selected horizon s_f. In order to limit the computing resources, this horizon might be shorter than the whole trip. An algorithm then calculates the optimal speed profile[2] $v^*(\tau)$, $\tau \in [t, t_f]$, for the temporal horizon t_f that is linked to s_f through the desired average speed v_d. The first value of the optimal speed reference is sent to a speed controller (which is in fact a conventional cruise controller), which ultimately pilots the vehicle's actuators to track the reference speed as a function of the measured actual speed.

8.1.3 Eco-ACC

As briefly explained in Sect. 4.2.2, adaptive cruise controllers (ACC) are cruise control systems that automatically adjust the vehicle speed to keep a safe gap from the vehicle ahead. Besides the simple PI control schemes described in Sects. 3.3.4.1 and 4.2.2, ACC could in principle be implemented by solving, at each time, an optimal control problem that minimizes some safety-related cost. Those systems that are known as eco-ACC use in addition an energy-efficiency-related cost, whence comes their name.

Figure 8.4 shows a flowchart illustrating the concept. Unlike in PCC where a single optimization is performed at the beginning of an entire trip or of a long section, in an eco-ACC system a new optimization is performed at every time step for a prediction or look-ahead horizon. This approach is often generically referred to as Model Predictive Control (MPC), and is described more extensively in Sect. 8.2.5.

Typically, in eco-ACC systems the prediction horizon t_f is fixed, in an embodiment that is called a *receding horizon* approach. Over this horizon, the disturbances and constraints acting on the vehicle must be predicted. Considering for simplicity only road grade and curvature, top speed limit and preceding vehicle, the road parameter vector is $C(t) \triangleq \{\alpha(s(\tau)), R(s(\tau)), v_{lim}(s(\tau)), s_p(\tau)\}$, $\tau \in [t, t + t_f]$.

---

[2]For ICEVs, optimal gear profile is usually also calculated.

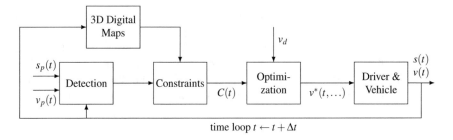

Fig. 8.4 Conceptual sketch of an Eco-ACC system

The road characteristics within the prediction horizon are retrieved from detailed 3D maps that are either stored within the system or accessed remotely. The presence of a preceding vehicle (moving or stationary obstacle), together with its relative position and speed to the host vehicle, is assumed to be detected thanks to ADAS sensors such those introduced in Chap. 3. Based on the current relative speed and headway, a dedicated algorithm estimates the leader position s_p, for the upcoming horizon. Such estimation is very critical since it strongly affects the effectiveness of the optimization results, and is described in some detail in Sect. 8.2.3. In addition, V2I communication or camera sensors can provide information on the state of upcoming traffic lights, which also affects the optimization process by introducing additional constraints. A few simple algorithms for traffic light detection are presented in Sect. 8.2.4.

Once the constraints are set, optimization of the speed profile of the host vehicle can be run for the entire prediction horizon. The cost function considers several factors. It generally includes the energy-based cost related to the tank energy, the safety cost related to the headway distance with the preceding vehicle, but might also include a regulation cost that penalizes the variations of the control inputs, and a speed smoothness cost that penalizes deviations from the desired average speed. The OCP is completed by the state equations and by the control and speed constraints, and can be summarized as

$$\underset{\mathbf{u}(t)}{\text{minimize}} \quad J = \int_{t}^{t+t_f} \left[w_1 P_T(v(\tau), \mathbf{u}(\tau), \alpha(s(\tau))) + w_2(s(\tau) - s_p(\tau))^2 + \right.$$
$$\left. + w_3(v(\tau) - v_d)^2 + w_4(\mathbf{u}(\tau) - \mathbf{u}_d)^2 \right] d\tau$$

$$\text{subject to} \quad \dot{s}(\tau) = v(\tau) \tag{8.2}$$
$$\dot{v}(\tau) = f(v(\tau), \mathbf{u}(\tau), \alpha(s(\tau)))$$
$$\mathbf{u}_{min}(v(\tau), \tau) \leqslant \mathbf{u} \leqslant \mathbf{u}_{max}(v(\tau), \tau)$$
$$0 \leq v(\tau) \leq \min(v_{lim}(s(\tau)), v_{turn}(R(s(\tau))))$$

where $\mathbf{u}$ is the control vector, $\mathbf{u}_d$ is the steady-state control corresponding to cruising at constant speed v_d, f is the speed dynamics equation (6.15), and the w's are weighting factors chosen opportunely. The output of this optimization block is the optimal speed profile for the entire horizon $v^*(\tau)$, $\tau \in [t, t + t_f]$.

Finally, this speed profile is either advised to the human driver through an HMI (see Sect. 8.3.1) or directly realized through a vehicle controller in the case of an automated speed controller. The actual position and speed of the host vehicle serve as the inputs to update the boundary conditions. A new iteration is performed at each time step.

8.1.4 Predictive Eco-Driving

The eco-ACC concept presented in the last section is not the most general embodiment of eco-driving techniques. Particularly in urban driving, the receding fixed horizon approach must be replaced by a "shrinking" horizon based on a predictive breakpoint detection. In addition, a reference speed to track might not necessarily be available, and in the most general case the speed must be fully optimized.

The fundamentals of such predictive eco-driving have been presented and discussed in the previous chapters. Figure 8.5 shows a flowchart illustrating the practical implementation of an eco-driving function. Similarly to an eco-ACC system, the approach is intrinsically iterative, with a new optimization that is performed every time step for the whole extent of the prediction horizon.

Typically, this horizon is set as the remaining trip to the *next infrastructure* breakpoint (e.g., a signalized intersection) and must be defined by the distance to be covered $s_f(t)$ and an estimation of the travel time $t_f(t)$. These pieces of information can be obtained, e.g., from an eco-routing system or directly from real-time traffic and infrastructural data. Together with the initial speed and desired final speed, they form the boundary conditions vector $B(t) \triangleq \{t_f(t), s_f(t), v_i(t), v_f(t)\}$ to the optimization. The process of setting the boundary conditions is described in more detail in Sect. 8.2.6.

Other elements that have to be predicted are the disturbances and constraints acting on the vehicle during the just-defined horizon. A road parameter vector $C(t) \triangleq$

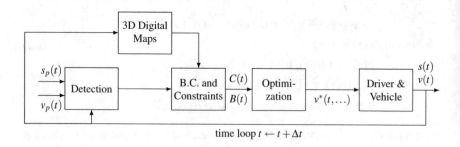

Fig. 8.5 Conceptual sketch of a predictive eco-driving system

$\{\alpha(s(\tau)), R(s(\tau)), v_{max}(\tau), s_p(\tau)\}$, $\tau \in [t, t + t_f(t)]$, can be obtained similarly to what was discussed for the eco-ACC approach.

Once the boundary conditions and the leader and infrastructure constraints are set, an optimization of the speed profile of the host vehicle can be run for the entire horizon. The OCP to be solved is generally that described in Chap. 6, which we rewrite here in a form that is valid for ICEVs and EVs:

$$
\begin{aligned}
\underset{\mathbf{u}(t)}{\text{minimize}} \quad & J = \int_{\tau=t}^{t+t_f(t)} P_T(v(\tau), \mathbf{u}(\tau), \alpha(s(\tau)))d\tau \\
\text{subject to} \quad & \dot{s}(\tau) = v(\tau) \\
& \dot{v}(\tau) = f(v(\tau), \mathbf{u}(\tau), \alpha(s(\tau))) \\
& v(t) \triangleq v_i(t), \\
& v(t + t_f(t)) = v_f(t), \\
& s(t) = 0, \\
& s(t + t_f(t)) = s_f(t) \\
& \mathbf{u}_{min}(v(\tau), \tau) \leqslant \mathbf{u} \leqslant \mathbf{u}_{max}(v(\tau), \tau) \\
& 0 \leq v(\tau) \leq \min(v_{lim}(s(\tau)), v_{turn}(R(s(\tau)))) \\
& 0 \leq s(\tau) \leq s_p(\tau)
\end{aligned}
\tag{8.3}
$$

where $\mathbf{u}$ is the control vector and f is the speed dynamics equation (6.15). For HEVs, the dynamics of the state of charge ξ_b must be additionally taken into account, as well as the additional boundary condition $\xi_b(t + f_f(t)) = \xi_f$. The quantity ξ_f is not known a priori, as it is the result of the SoC trajectory optimization over the whole trip, and not just one sub-trip (see Sect. 6.6). However, it can be approximated by heuristic expressions such as that described by (4.47). Overall, the output of this optimization block is the optimal speed profile for the entire horizon $v^*(\tau)$, $\tau \in [t, t + t_f(t)]$.

Similarly to an eco-ACC system, the optimal speed profile is either advised to the human driver through an HMI (see Sect. 8.3.1) or directly realized through a vehicle controller in the case of an automated speed controller. The actual position and speed of the host vehicle serve as the feedback input to update the boundary conditions for a new iteration after each time step.

8.2 Practical Issues

Many of the building blocks appearing in the flowcharts of the previous sections have been extensively treated in this book. For instance, the use of 3D digital maps is the subject of Sect. 4.1. Also, the characteristics of the various sensors used, not explicitly shown in the flowcharts, can be found in Sect. 3.2.1. However, many corollary functions and sub-systems have not been treated yet with an adequate detail and thus shall be discussed in this section. Among these are, the speed recording and the breakpoint detection of eco-coaching (Sects. 8.2.1 and 8.2.2), the leader position prediction and the MPC algorithms of both eco-ACC and predictive eco-driving (Sects. 8.2.3–8.2.5), as well as predictive boundary condition setting (Sect. 8.2.6).

8.2.1   Speed and Path Recording

As discussed in Sect. 3.2, accurate measurement of vehicle velocity often needs multiple sensors (GPS, IMU, wheel odometer) and the fusion of their output through extended Kalman filtering. Simple and fast-running eco-coaching systems like the one shown in Fig. 8.1 might nevertheless rely only on relatively noisy GPS data beside 3D map data, from which longitudinal velocity (speed), road slope, and road curvature must be extracted. Raw data (latitude, longitude, elevation) on a sub-trip need first to be transformed into local Cartesian coordinates (x_c, y_c, z_c) based on the UTM (Universal Transverse Mercator) system, to which record times t_c, $c = 1, 2, \ldots, C$ are associated. If s is the curvilinear position and θ is the heading angle, numerical differentiation of the data set should in principle reveal the speed,

$$v(s) = \frac{ds}{dt} , \tag{8.4}$$

the curvature

$$\frac{1}{R(s)} = \frac{d\theta(s)}{ds} , \tag{8.5}$$

and the slope

$$\alpha(s) = \arctan\left(\frac{dz(s)}{ds}\right) , \tag{8.6}$$

with

$$dx = \cos\theta ds , \tag{8.7}$$
$$dy = \sin\theta ds . \tag{8.8}$$

To differentiate the available data numerically, a smoothing filter such as the *Savitzky–Golay filter* [3] can be used given its ability to increase the signal-to-noise ratio without greatly distorting the signal [4]. With this approach, the curvature can be obtained directly from the first- and second-order position derivatives, without the intermediate evaluation of the heading.

Another approach consists of setting an ad-hoc optimal control problem to find the time profiles of speed, curvature, and slope that minimize the error between GPS data (x_c, y_c, z_c) and reconstructed positions evaluated at times t_c [5]. To do so, time is chosen as the independent variable,[3] while the "control" vector consists of the derivatives of the sought functions $\tilde{u}(t) \triangleq [dv/dt, d(1/R)/dt, d\alpha/dt]$. The performance index to be minimized can be defined for the i-th sub-trip as

$$\int_0^{t_f} r\tilde{u}^2 dt + \sum_{c=1}^{C} \left((x_c - x(t_c))^2 + (y_c - y(t_c))^2 + (z_c - z(t_c))^2\right), \tag{8.9}$$

with (8.4)–(8.8) as dynamic equations and t_f given by the first of (8.1). The weighting factor r trades off the variations in the reconstructed curvature profile with the reconstruction accuracy. Appropriate boundary conditions and constraints complete the OCP.

More accurate estimation might be obtained using IMU data and/or sensor fusion, particularly for on-board systems that are integrated with the in-vehicle data bus.

8.2.2 Breakpoint Detection

In analyzing a recorded speed and path profile, breakpoints can be defined with two types of considerations. On one hand, fixed or stationary breakpoints can be defined at locations where: (i) slope α, (ii) road curvature R, or (iii) top speed limit v_{lim} change more than a predefined threshold with respect to their value in the previous sub-trip, or (iv) the trip is likely to undergo a discontinuity, like in the presence of signalized intersections, stops, or other types of intersections.

On the other hand, the analysis of the actual speed profile $\hat{v}$ generally reveals additional breakpoints that were likely induced by surrounding traffic, which is not predictable only using static road information. The detection of such breakpoints can be done, in principle, by identifying the speed minima (including stops) subsequent to decelerations that are not already accounted for by the static breakpoints.

Minima detection is, however, not always a simple task as the recorded speed profile can be very noisy and spurious minima can appear. The latter could be eliminated by keeping, e.g., only minima that have a sufficient prominence[4] or that have a minimum time separation from other minima. These features can be imposed while using functions that are available in commercial software, for example Matlab's *findpeak*.

However, often it is not sufficient to isolate the representative breakpoints. Digital filters such as the aforementioned Savitzki–Golay filter [3] can smooth the data and reduce the noise, but they are an option only if they preserve the main minima that can be attributed to breakpoints. Ramer–Douglas–Peucker algorithm [6] can be used

---

[3]Curvilinear position is the independent variable in the original formulation of [5], although this choice generates some difficulties.

[4]Analogously to the prominence of a peak, the prominence of a valley measures how much the valley stands out due to its intrinsic depth and its location relative to other valleys. A shallow isolated valley can be more prominent than one that is deeper but is an otherwise unremarkable member of a deep sequence.

to decrease the number of points in the recorded speed trace, which can also reduce the noise. Mathematical morphology can also be used, directly on the speed trace or rather on a filtered signal, to isolate the valleys from the original signal. However, the position of these valleys might be displaced when their shape is strongly asymmetric.

8.2.3 Leader Position Prediction

The optimality of the iterative schemes of Figs. 8.4 and 8.5 is strongly affected by the quality of prediction of the preceding vehicle's states. The simplest approach to predicting the leader's future speed is to assume that the vehicle continues to move with the same acceleration until it stops or exceeds a maximum velocity [7]. But this method can be unrealistic because it may lead to a very high or a zero predicted velocity at the end of a long horizon. Therefore, this model is often [8] improved by considering a speed-dependent acceleration value.

To this end, the car-following models described in Sect. 4.2.1 could in principle be applied to the leader and the generic equation (4.2) integrated over the prediction horizon to yield $v_p(\tau)$, $\tau \in [t, t + t_f(t)]$. Normally, the information about the speed of the vehicle preceding the leader and its gap from the leader are not available, so that (4.2) is integrated assuming either free flow,

$$\dot{v}_p(\tau) = F(v_p(\tau), \delta \to \infty) , \tag{8.10}$$

or approaching a fixed obstacle, e.g., a red light or an intersection, whose position is known,

$$\dot{v}_p(\tau) = F(v_p(\tau), \delta(\tau)) \tag{8.11}$$
$$\dot{\delta}(\tau) = -v_p(\tau) . \tag{8.12}$$

The integration of these differential equations in closed form is only possible for very simple car following models. For more realistic ones, such as the Gipps' model or the IDM, numerical integration is required, which is not always practical in an onboard system.

Therefore, more heuristic approaches have been often used in the literature. For example, introducing a multiplier to weigh the current acceleration as

$$a_p(\tau) = a_p(t) \cdot \zeta(v_p(\tau)) , \tag{8.13}$$

where $\zeta(v_p)$ is defined as

$$\zeta(v_p) \triangleq \frac{1}{(1 + e^{-\beta_1(v_p - w_1)})(1 + e^{\beta_2(v_p - w_2)})} . \tag{8.14}$$

Fig. 8.6 Function (8.14)

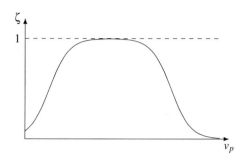

The parameters w_1 and w_2 define an approximate range of velocities. The function (8.14) states that $a_p(\tau) \approx a_p(t)$ if $v_p(\tau)$ is within that range, otherwise $a_p(\tau) \approx 0$, and $\beta_1 > 0$ and $\beta_2 > 0$ express the sharpness of the sigmoid function, see Fig. 8.6. That means that the acceleration of the preceding vehicle approaches zero when it reaches a maximum velocity or stops completely.

In the case the preceding vehicle is approaching a red signal, prediction using (8.14) is not appropriate, since it is certain that the vehicle will stop at the end of the section after a known distance $\delta(t) = s_f - s_p(t)$. For this special case, a prediction model of the preceding vehicle based on experimental driving data has been proposed in [7] as

$$v_p(\delta(\tau)) = \sum_{i=0}^{5} c_i \delta^i(\tau) = -\dot{\delta}(\tau) , \tag{8.15}$$

with the coefficients c_i experimentally determined. The corresponding stopping rate is obtained as $a_p(\tau) = -v_p(\delta(\tau)) \cdot dv_p(\delta)/d\delta$.

With V2V connectivity, the future intent of the preceding vehicle can be communicated to the follower, which reduces the uncertainty. A vehicle cannot know its future moves with certainty but if it employs an MPC approach for longitudinal control it generates, as a by product, a future sequence of its optimal control moves conditioned on its current state. This "intended" sequence of its future moves can be communicated to following vehicles as a best estimate of what it is going to do. The ego following vehicle can replace, in its cost function and constraints, the intended position of the preceding vehicle over the horizon. The MPC problems can be solved sequentially by each participating CAV from the front to the tail of a string resulting in a distributed MPC solution of the car following problem as also described in [9].

Even in absence of V2V connectivity, latest sensor observations can be used to estimate the probable position of the preceding vehicle along a future planning horizon. In [10–12], different flavors of a Markov chain approach are proposed to predict the likely position of the preceding vehicle over ego car planning horizon. By counting historical occurrences, probabilities are assigned to the transition between two states (i.e., vehicle velocity) in consecutive time steps. In a Markov chain model the assumption is that a state at any point in time is only dependent on the states of one prior step and not earlier steps. In [11] only velocity transitions of the preceding vehicle were modeled, but more precise predictions can be made if enough data is available to include the acceleration state as well. To reduce the size of the transition probability matrix[5] and the need for training data it is possible to use only a categorical variable describing whether the preceding vehicle is accelerating, decelerating, or cruising. For instance, the velocity can be quantized at 1 m/s steps between 0 and maximum allowable velocity while the acceleration state can be chosen from {accelerating, decelerating, cruising} as shown in [12]. A higher resolution quantization of acceleration is used in [13, 14].

Once the velocity of preceding vehicle is estimated and given the current position of the preceding vehicle $s_p(0)$, future positions can be determined using kinematics. Because a model is used to generate the final output, this approach always gives physically realistic trajectories. To each sequence of velocities and resulting positions a probability calculated from the constructed Markov transition matrix can be assigned. With a sufficient number of samples, a distribution can be fit to the generated position histogram to be used in anticipative car following as described in a case study in Sect. 9.3.

8.2.4 Probabilistic Traffic Light Prediction

There can be much uncertainty in the phase and timing of a traffic signal which makes predicting its future state quite challenging. For fixed-time traffic signals which do not respond to traffic conditions (see Sect. 4.1.3) and operate only on a timing table, the traffic signal clock may drift during a 24 h period (this is potentially due to variations in the electric grid frequency, see [15] for more details). Therefore, it is not possible to know with certainty the start of greens and reds, even for fixed-time signals. The level of uncertainty is higher for actuated and adaptive traffic signals which do respond to traffic conditions. Although they have a base timing table, the timings of actuated and adaptive lights may change according to traffic conditions, rendering not only the start of reds and greens but also the phase lengths uncertain.

Due to the aforementioned uncertainties, it is difficult to always determine the start and duration of greens deterministically. One can employ a probabilistic prediction framework to handle the case with partial or uncertain information. Here we describe the approach that first appeared in [16].

Assume that the current phase (color) and the average red and green lengths for a signal are known. This information can be used to predict the probability of a green over the planning horizon. The state of a light is denoted by $\ell(t)$ which can assume two values, g and r, representing green and red respectively. The goal is to determine the probability of a light being green at time $t + t_p$ conditioned on its current color at time t. To form this conditional probability function, the durations of green and red are assumed to be t_g and t_r respectively on average. The traffic signal is assumed

---

[5]A matrix $P_{\alpha\beta} \triangleq P\left(v_p(\tau + \Delta\tau) = \beta | v_p(\tau) = \alpha\right)$ indicating the probability to go from velocity state α to state β.

to operate cyclically, which is true for many traffic signals, and as a result the total cycle time is fixed and equal to $t_g + t_r$. Here yellow time is included in red.

Using relatively straight-forward probabilistic reasoning, the chance of a green light in t_p seconds, given a green at current time t can be found to be:

$$P[\ell(t + t_p) = g | \ell(t) = g] = \begin{cases} \dfrac{t_g - t_m}{t_g}, & \text{if } t_m \leq t_r \wedge t_m \leq t_g \\ \dfrac{t_g - t_r}{t_g}, & \text{if } t_r \leq t_m \leq t_g \\ 0, & \text{if } t_g \leq t_m \leq t_r \\ \dfrac{t_m - t_r}{t_g}, & \text{if } t_g \leq t_m \wedge t_r \leq t_m \end{cases}, \qquad (8.16)$$

where $t_m \triangleq \mathrm{mod}(t_p, t_g + t_r)$ is the residue of division of t_p by $t_g + t_r$ and its role is to remove whole cycle times from t_p. In other words because the signal clock is assumed to be periodic, the resulting conditional probability is also going to be a periodic function of time so the probability of green (or red) can be described in terms of t_m instead of t_p.

Now for instance when $t_m \leq \min(t_g, t_r)$, then $\ell(t + t_m) = g$ for $0 \leq t < t_g - t_m$ and $\ell(t + t_m) = r$ for $t_g - t_m \leq t \leq t_g$, so that the probability of light being green now and in t_m seconds is $\frac{t_g - t_m}{t_g}$, which explains the first line of Eq. 8.16. But if $t_g \leq t_m \leq t_r$ and the light is green now, it is impossible for it to be green in t_m seconds as in third line of Eq. 8.16. The details may be best understood graphically and therefore they are not explained here.

Similarly, the chance of a green light in t_p seconds, given a red at time t is:

$$P[\ell(t + t_p) = g | \ell(t) = r] = \begin{cases} \dfrac{t_m}{t_r}, & \text{if } t_m \leq t_r \wedge t_m \leq t_g \\ 1, & \text{if } t_r \leq t_m \leq t_g \\ \dfrac{t_g}{t_r}, & \text{if } t_g \leq t_m \leq t_r \\ \dfrac{t_g + t_r - t_m}{t_r}, & \text{if } t_g \leq t_m \wedge t_r \leq t_m \end{cases}. \qquad (8.17)$$

Figures 8.7 and 8.8 show several probabilistic prediction examples with different splits between red and green but with the same cycle length. These are visualizations of the probabilities which we use in probabilistic simulation cases described later in Sect. 9.1.

8.2.5 MPC Schemes

In contrast to the *offline* optimal control framework described in Sects. 6 and 7, MPC needs to solve an optimization problem *online* at each control time step to compute the optimal control input. This fact has prevented the application of MPC in several

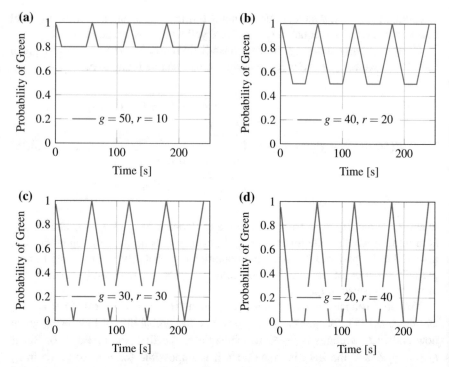

Fig. 8.7 Conditional future probability of green given that the light is currently green, for four different light timing patterns. In all patterns the total cycle time is 60 s, with the lengths of green and red indicated in the legends. The time axis is t_p as described in (8.16)–(8.17)

contexts, either because the processor technology needed to solve the OCP within the sampling time is too expensive or simply infeasible for onboard automotive-grade controllers, or because the implementation of the numerical solver causes software certification and/or explicability concerns, especially in safety-critical applications. At the same time, such limitations have motivated the development of *explicit MPC* approaches, where the optimal control inputs are pre-computed offline as a function of the states [17], or even the analytical and semi-analytical approaches presented throughout this book. However, in non-critical and research applications, numerical MPC remains the most common option.

Practical methods to solve problem (8.3) online generally require its discretization, that is, replacing the integral J with a sum

$$\sum_{\tau=t}^{t+t_f(t)} P_T(\cdot)\Delta\tau \tag{8.18}$$

and the dynamic equations with algebraic equations $v(\tau + \Delta\tau) = v(\tau) + f(\cdot)\Delta\tau$, $s(\tau + \Delta\tau) = s(\tau) + v(\tau)\Delta\tau$, with the prediction time step $\Delta\tau$ not necessarily equal

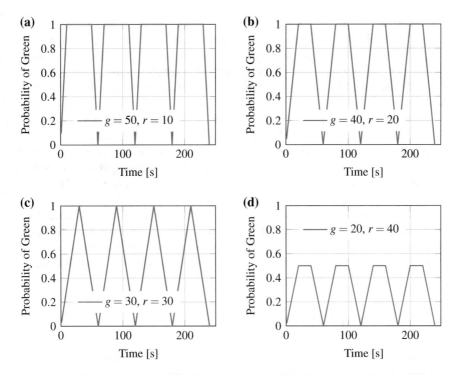

Fig. 8.8 Conditional future probability of green given that the light is currently red, for four different light timing patterns. In all patterns the total cycle time is 60 s, with the lengths of green and red indicated in the legends. The time axis is t_p as described in Eqs. 8.16 and 8.17

to the control time step Δt. The stability of MPC has been proven at least for OCP without terminal state constraints and terminal cost [18]. An excellent survey on stability in MPC is given in [19].

For linear systems with linear constraints and a quadratic cost, the discretized problem can be rewritten into a standard quadratic programming (QP) form, as it is the case, for instance, of the quadratic EV model (6.74). A variety of methods (gradient, interior-point, etc.) are commonly used to solve QP problems exactly and many fast and efficient solvers are available, including toolboxes in the main programming languages [20].

More generally, nonlinear MPC (NMPC) problems must be solved iteratively and involve approximations of the problem equations. As far as optimal control is concerned, the two solution approaches can be recognized, those of indirect and direct methods.

Indirect methods, based on the PMP (see Sect. 6.2.2) and resulting in a two-point boundary value problem, suffer from a general lack of efficient and robust generic software packages to be implemented online. In contrast, the use of *direct methods* is nowadays much more common. With this approach, the NPMC must be converted into a nonlinear programming (NLP) problem. To do that, pseudospectral (orthogonal collocation) algorithms are commonly employed, which use polynomials to approximate states and control inputs at some collocation points. Depending on the choice of the polynomial basis and of the collocation points, several pseudospectral methods are available, such as those named after Gauss (GPM), Chebyshev (CPM), and Legendre (LPM).

Once cast into an NLP, the latter can be solved using the several existing Newton-type methods. These are roughly classified into sequential quadratic programming (SQP) and interior-point (IP) methods [21, 22]. For both categories, several solvers and software packages are available [22]. Energy-efficient driving systems have been actually implemented with SQP solvers such as NPSOL [23] and SNOPT [24], as well as with an IP-type solver such as C/GMRES [25] and IPOPT.

The use of such methods needs, however, special treatments of the original OCP to, e.g., regularize constrained or singular arcs, smooth discontinuous functions, etc. The use, when available, of intrinsically robust, analytical or semi-analytical solution procedures, such as those presented in this book, could overcome these difficulties.

8.2.6 Setting of Boundary Conditions

As discussed in Sect. 8.1.4, the optimization block in the predictive eco-driving scheme of Fig. 8.5 requires proper boundary conditions

$$B(t) = \{t_f(t), s_f(t), v_i(t), v_f(t)\} , \tag{8.19}$$

where the argument t explicitly denotes that such setting is made at each new optimization run.

The initial speed of the new optimization horizon is naturally set to the current speed,

$$v_i(t) = v(t) .$$

As for the other boundary conditions, their definition results from the knowledge of the road network ahead and the estimated leader position. We shall assume good predictive capability of both these elements, which form the road parameter vector $C(t) = \{\alpha(s(\tau)), v_{max}(\tau), s_p(\tau)\}, \tau \in [t, t + t_f(t)]$.

The spatial horizon can be either constant or variable. The former option, $s_f(t) \equiv s_f$, is also known as *sliding horizon* and is sometimes appropriate for highway trips, in the absence of any sensible breakpoints. This is the preferred option for PCC and eco-ACC implementations, as discussed in Sects. 8.1.2 and 8.1.3.

In most driving situations, however, road topology naturally fixes the next breakpoint. Referring to the scenarios of Sect. 7, at least four situations may occur, where the next breakpoint is the position (i) where road slope and/or curvature change sensibly with respect to the current value, (ii) where a different speed limit than the current is posted, (iii) of a signalized intersection, or (iv) of a stop. Denoting the position of the next breakpoint as $\hat{s}_{NB}(t)$, where the hat denotes an estimation and the argument t the fact that such estimation can vary with time during the trip, the spatial horizon is naturally set as

$$s_f(t) = \hat{s}_{NB}(t) - s(t) . \tag{8.20}$$

As for the temporal horizon, it also is rarely fixed and sliding. More often, it is estimated based on $\hat{s}_{NB}$ and some desired or average speed (e.g., traffic-induced) $V(t)$ that is needed to reach it,

$$t_f(t) = \frac{s_f(t)}{V(t)} . \tag{8.21}$$

Though, in case (iii) of a signalized intersection, it is more natural to set a breakpoint time $\hat{t}_{NB}(t)$ during the green phase and, similarly to (8.20), evaluate the time horizon as the remaining time to catch the green,

$$t_f(t) = \hat{t}_{NB}(t) - t . \tag{8.22}$$

The terminal speed $v_f(t)$ also is to be estimated. In case (i) above, that is a typical highway situation, often $v_f(t) = v_i$ (cruising). In case (ii) to avoid having to abruptly decelerate or brake at the beginning of the new segment, the obvious choice is to set $v_f = \hat{v}_{max,NB}(t) - \Delta v$, where $\hat{v}_{max,NB}$ is the speed limit in the next sub-trip and Δv is a safety margin to allow further acceleration in the next sub-trip. In case (iii), $v_f(t)$ should equal the crossing speed v_t at the traffic light, and could in principle be the result of a further optimization as shown in Sect. 7.7.[6] In case (iv), $v_f(t) \equiv 0$.

The feasibility of the sub-trip horizons defined in one of the ways discussed above must be checked. For that, the feasible range of durations and distances satisfying all constraints, $F(t_f(t), s_f(t)) \geq 0$, must be constructed as shown in Chap. 7 using the available information contained in the road parameter vector $C(t)$. Additionally, the vehicle and powertrain characteristics can be used to set the maximum acceleration or deceleration.

If the pair $(s_f(t), t_f(t))$ is not feasible, a correction must occur. Two cases may be identified. In case, e.g., of a preceding vehicle that is slower than thought ($V(t)$ overestimated in (8.21)), then likely an increase of $t_f(t)$ suffices to bring the boundary conditions back in the feasible range. If, however, a non-estimated obstacle appears, for instance, the preceding vehicle will stop before the estimated position $\hat{s}_{NB}$, then the most natural option is to decrease $s_f(t)$ and possibly $t_f(t)$ according to (8.21). These two situations are illustrated in Fig. 8.9. In both cases, a safety margin can be conveniently applied with respect to the closest boundary of the feasibility domain.

---

[6] In this OCP, $v_f(t)$ is open so the corresponding costate is fixed.

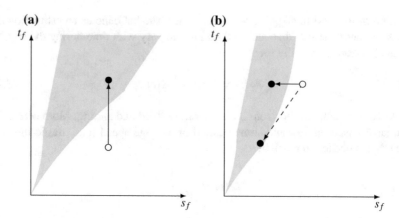

Fig. 8.9 Correction of boundary conditions that are infeasible for an underestimation of t_f (left) and an overestimation of s_f (right)

8.3 On-Board Implementation

The effective implementation of the concepts presented in the previous sections requires several technical arrangements. Here we discuss the ergonomy of the HMIs aimed at advising drivers about eco-practices (Sect. 8.3.1) and of their response to these advice (Sect. 8.3.2). Finally some issues related to the automated implementation are discussed in Sect. 8.3.3.

8.3.1 Human-Machine Interfaces

The energy-efficiency systems described in the previous sections and chapters must be complemented with human-machine interfaces to communicate with a human driver. As discussed above, the role of an eco-coaching HMI is to effectively assess the driver's behavior in-trip or post-trip; that of an eco-driving HMI is to advise a speed to follow, and possibly other actions. In addition, eco-routing HMIs advise the route to follow.

The main types of driver assistance systems are: (i) visual interfaces, (ii) visual-auditory assistance systems, and (iii) haptic assistance systems.

8.3.1.1 Visual and Acoustic Assistance Systems

Visual and visual–acoustic assistance systems are based on dedicated on-board multi-modal displays or personal mobile devices (smartphones, tablets). The system architecture may include several tabs, related to different functions or different drive stages (e.g., pre-trip eco-routing, in-trip eco-driving, post-trip eco-coaching).

Eco-routing output is usually presented similarly to current navigation systems, with the user providing origin and destination by entering physical addresses or

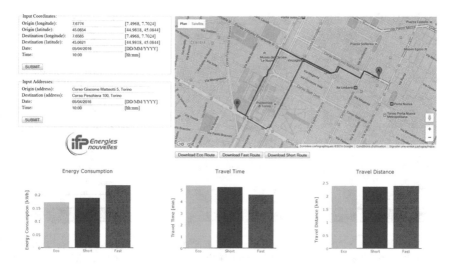

Fig. 8.10 Example screenshot of an eco-routing interface (http://isntsv-optem:8080/EcoRouting/), showing the user's inputs addresses and coordinates, the routes displayed on a map, and the corresponding energy, time, and distance attributes

clicking on a map, and possibly time of departure (if not automatically detected). The system then displays the computed eco-route on the map. Information on predicted trip energy consumption, duration, and distance as compared with standard route choices (typically, the fastest and the shortest route) can be also useful for the driver. A screenshot of an example eco-routing interface is shown in Fig. 8.10.

In-trip or post-trip eco-coaching HMIs must provide the user with useful feedback on his/her driving style. For this purpose, scores are commonly used. These scores can be indirect or direct. An example of the first type is the time fraction spent with a vehicle speed different from the optimal. On the other hand, direct scores compare the actual (tank) energy consumed[7] during the sub-trip, E_T, with the calculated optimum, E_T^*, for instance using the definition

$$\text{EDS} \triangleq 10 \left(2 - \frac{E_T}{E_T^*} \right) . \tag{8.23}$$

The numerical coefficients of (8.23) ensure that, when $E_T = E_T^*$, the score is equal to 10 while, if the actual consumption is twice the minimum, then the score is zero.

Scores can be presented by exact values, horizontal bars, or various *visual metaphors*, see the screenshots in Fig. 8.11. A history with scores for recent sub-trips or entire trips is also often useful, particularly if the system allows calculating analytics, e.g., the average score of all the saved trips. That allows the driver to compare his/her current performance with recent ones and to get an impression about

---

[7]Note that the means to actually measure the energy consumption on board is a complex subject that is beyond the scope of this book.

Fig. 8.11 Example screenshots of an eco-coaching interface (GECO [26]) showing a presentation of the EDS scoring as a numerical value and of the energy consumption (in CO_2 g/km) as a metaphor (left); three sliders rating accelerations, decelerations, and smoothness (middle); a visual comparison of the optimal vs. the actual velocity profiles in the last sub-trip (right)

his/her progress. This learning process supports the long-term motivation for an energy-efficient behavior of the driver.

Since they must induce a desired behavior of the driver, predictive eco-driving HMIs might be more complex. They typically include several kinds of icons, representative of different events during the drive, visual advices, feedback on fulfillment of the eco-driving behavior through scores, as well as game (intrinsic reward) elements [27].

Visual advice can be provided by presenting the current speed calculated by the eco-driving function as a numerical value, on a tachometer-like icon, or as a horizontal bar in comparison with the current actual speed. Moreover, the recommended speed can be conveniently visualized as a range, besides its exact computed value [28]. Second-order information on the required change in pedal depression has been demonstrated to be more effective than first-order information on the current speed error [29].

In addition to visual advise, feedforward information of upcoming events (traffic light with distance and possibly status, traffic sign, curves, roundabout, etc.) is often also presented in eco-driving assistance interfaces [27].[8]

The use of game elements in the driver assistance context is also increasingly popular [30]. The objective of such gamified interfaces is to facilitate intrinsically-motivated energy-saving behaviors that are less amenable to change than with simple prescriptions. According to the *self-determination theory*, a psychology theory of human personality and motivation, the factors to be considered when developing such interfaces include: (i) avoiding external rewards (e.g., prizes, cash incentives);

---

[8]See a few examples in the case studies of Chap. 9.

(ii) providing positive feedback to drivers regarding their eco-driving behavior; and, (iii) fostering social interactions (e.g., through cooperative systems) [30].

Game elements can be provided to drivers through challenges based on different levels of difficulty, to facilitate feelings of enhanced competence as they progress through levels. Research in this field has highlighted the need to balance difficulty and user skill, in order to avoid boredom (when challenges are too easy) and/or frustration (when challenges are too difficult).

Social-interaction features can be designed to let the users share their scores with friends or other drivers, in order to socially engage and cooperate toward the common goal of eco-driving. Competitive challenges for users or teams (of friends, co-workers) are also organized by providers of eco-driving systems and local authorities to promote the adoption of energy-saving driving styles [31].

8.3.1.2 Haptic Assistance Systems

In a typical vehicle, there is a proportional relationship between the force applied to the accelerator pedal and the pedal depression produced. In a haptic system, this relationship is changed by increasing or decreasing the accelerator pedal resistance. The variable force-depression profile should somehow reflect the acceleration that is advised to the drivers and thus expected from them. Two methods have been demonstrated in recent years [29], see Fig. 8.12.

The *haptic force method* requires the driver to produce a significant extra force to increase the pedal depression beyond that calculated by the eco-driving function to be optimum for energy efficiency at the current instant. In contrast, to discourage under-accelerating, a weakening of the pedal resistance below that of a standard accelerator pedal is enforced for pedal depressions below the advised one, so that the pedal is easier to push than in a vehicle not equipped with the haptic assistance system.

The *haptic stiffness method* discourages an over-acceleration through a change in pedal stiffness rather than a step-change in force for the drivers to overcome. Again, this method is able to encourage harsher acceleration when energy efficiency can be improved by increasing pedal force, through a reduction in the resistance of the accelerator pedal relative to a non-haptic standard pedal.

The effectiveness of these two methods and in comparison to visual assistance systems is still a matter of debate, with the few experimental results being somehow contradictory [29].

8.3.2 Human Response

The effectiveness of energy-efficient driving assistance systems depends not only on the algorithms and the HMI used, but also strongly on how the human drivers react to the advice and adopt it.

Effectiveness of the system as a whole can be evaluated through field tests, driving simulators, or traffic simulations. Commonly, multiple participants are selected for

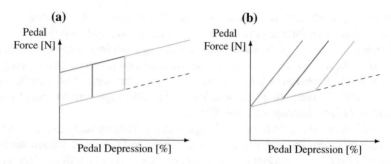

Fig. 8.12 Haptic force (**a**) and haptic stiffness (**b**) system profiles for three setpoints of pedal depression. Dashed: standard pedal; color curves: increasing acceleration setpoints

such tests, and the same driving scenarios are completed at several separate occasions. To limit the influence of external factors, such repetitions are aiming at performing or reproducing, e.g., conditions at the same day of the week and departure time of the day. Though often professional drivers are involved in such tests for practical reasons, the use of non-instructed participants should be favored.

The outcomes of these tests can be of a different nature. In particular, the human response to HMIs can be assessed by several measures. Questionnaires filled by participants to an experimental campaign may inform about the user perception of the interface design (how engaging it is, how easily it is understood, its perceived usability), their self-perception while using the system (e.g., self-evaluating their level of anxiety, annoyance, curiosity), or about the impact of using the interface on driving.

Quantitative indexes are also used in the ergonomic science, such as the percentage of success in correctly reacting and the reaction time to feedback. To evaluate the level of distraction induced by the use of the HMI, the duration of driver's eye glances to the interface(s) and their relative occurrence with respect to those directed to the road scene ahead, are often used [32].

Of course the energy actually spent is the ultimate measure of effectiveness of the system as a whole [33]. The energy efficiency benefits of using an eco-driving system should be evaluated both in the short and in the long term, where the superiority over classical eco-driving trainings should be more apparent.

8.3.3 Automated Drive

By removing the burden from human drivers, automated drive systems are expected to significantly ease implementation of eco-driving functions. This could be as simple as issuing the eco-speed or -lane determined at the motion planning layer (Sect. 3.3.3) to the lower level feedforward and feedback motion controllers (Sect. 3.3.4) that actuate the throttle, the brake, and the steering control motors.

For instance in anticipative car following, an MPC planner issues its acceleration command to the low level controllers. In one possible implementation, pre-mapped feedforward controllers could map the commanded acceleration to throttle or brake pedal positions, while a feedback loop is in charge of velocity tracking. A potential challenge is mismatch between simplified models used at the planning layer and actual dynamics of the vehicle, which could cause degradation in tracking performance. One possible solution in an MPC framework is augmenting a (step) disturbance observer [34] to the simplified model thus introducing integrating modes that improve tracking performance. Computation and communication delays need to be accounted for and addressed or they may cause poor tracking or chattering.

There are challenges at the low-level control layers as well. Poorly designed low-level controllers can diminish the energy gains planned at higher level or could even cause an increase in energy use. For example, an over-aggressive low-level controller may cause chatter between throttling and braking in an effort to precisely follow the eco-speed. This could arise for instance in a vehicle with an internal combustion engine if the response time of the engine is not properly taken into account. The situation can be worse if the vehicle is having a turbocharger or an automatic transmission with torque converter, both of which add extra delays to the vehicle response. Controller de-tuning may be an option in scenarios such as eco-approach to traffic signals but may not be possible for instance in tight platooning. Electric vehicles may be easier to control in similar scenarios due to their more responsive electric motors and drivetrain structures.

Control and perception modules can operate at their own pace and communicate via an environment such as Robot Operating System (ROS) using publish and subscribe functions [35]. A highly sophisticated motion planner that incorporates macro traffic planning may run on a backend cloud. In this case, the communication between the motion planner and the vehicle modules can be achieved by communication networks such as dedicated short-range communications (DSRC) and cellular communications. In all these circumstances, global time-stamping of data would be preferable for synchronization. In addition, potential risks such as deadlocks, time delays, and data losses should also be considered in the system designs and implementations.

References

1. Dib Wissam, Chasse Alexandre, Moulin Philippe, Sciarretta Antonio, Corde Gilles (2014) Optimal energy management for an electric vehicle in eco-driving applications. Control Eng Pract 29:299–307
2. Thibault L, De Nunzio G, Sciarretta A (2018) A unified approach for electric vehicles range maximization via eco-routing, eco-driving, and energy consumption prediction. IEEE Trans Intell Veh 3(4):463–475
3. Savitzky A, Golay MJE (1964) Smoothing and differentiation of data by simplified least squares procedures. Anal Chem 36(8):1627–1639

4. Ojeda LL, Chasse A, Goussault R (2017) Fuel consumption prediction for heavy-duty vehicles using digital maps. In: Proceedings of international conference on intelligent transportation systems (ITSC), pp 1–7. IEEE

5. Perantoni G, Limebeer DJN (2014) Optimal control for a formula one car with variable parameters. Veh Syst Dyn 52(5):653–678

6. Douglas DH, Peucker TK (1973) Algorithms for the reduction of the number of points required to represent a digitized line or its caricature. Cartogr: Int J Geogr Inf Geovisualization 10(2):112–122

7. Kamal MAS, Mukai M, Murata J, Kawabe T (2011) Ecological driving based on preceding vehicle prediction using MPC. IFAC Proc Vol 44(1):3843–3848

8. Kamal MAS, Mukai M, Kawabe T (2013) Model predictive control for ecological vehicle synchronized driving considering varying aerodynamic drag and road shape information. SICE J Control Meas Syst Integr 6(5):299–308

9. Zheng Y, Li SE, Li K, Borrelli F, Hedrick JK (2017) Distributed model predictive control for heterogeneous vehicle platoons under unidirectional topologies. IEEE Trans Control Syst Technol 25(3):899–910 May

10. McDonough K, Kolmanovsky I, Filev D, Szwabowski S, Yanakiev D, Michelini J (204) Stochastic fuel efficient optimal control of vehicle speed. In: Optimization and optimal control in automotive systems, pp 147–162. Springer

11. Zhang C, Vahidi A (2011) Predictive cruise control with probabilistic constraints for eco driving. In: Proceedings of dynamic systems and control conference and symposium on fluid power and motion control, pp 233–238. American Society of Mechanical Engineers

12. Wan N, Zhang C, Vahidi A (2019) Probabilistic anticipation and control in autonomous car following. IEEE Trans Control Syst Technol 27:30–38

13. Dollar RA, Vahidi A (2017) Quantifying the impact of limited information and control robustness on connected automated platoons. In: Proceedings of international conference on intelligent transportation systems (ITSC), pp 1–7. IEEE

14. Dollar RA, Vahidi A (2018) Efficient and collision-free anticipative cruise control in randomly mixed strings. IEEE Trans Intell Veh 3:439–452

15. Lin P-S, Fabregas A, Chen H, Rai S (2010) Impact of detection and communication degradations on traffic signal operations. In: Proceedings of annual meeting and exhibit, ITE

16. Mahler G, Vahidi A (2014) An optimal velocity-planning scheme for vehicle energy efficiency through probabilistic prediction of traffic-signal timing. IEEE Trans Intell Transp Syst 15(6):2516–2523

17. Alessio A, Bemporad A (2009) A survey on explicit model predictive control. In: Nonlinear model predictive control, pp 345–369. Springer

18. Zanon M, Frasch JV, Vukov M, Sager S, Diehl M (2014) Model predictive control of autonomous vehicles. In: Optimization and optimal control in automotive systems, pp 41–57. Springer

19. Mayne DQ, Rawlings JB, Rao CV, Scokaert POM (2000) Constrained model predictive control: stability and optimality. Automatica 36(6):789–814

20. Gould NIM, Toint PL (2000) A quadratic programming bibliography. Numer Anal Group Intern Rep 1:32

21. Betts JT (2010) Practical methods for optimal control and estimation using nonlinear programming, vol 19. Siam

22. Diehl M, Ferreau HJ, Haverbeke N (2000) Efficient numerical methods for nonlinear MPC and moving horizon estimation. In: Nonlinear model predictive control, pp 391–417. Springer

23. Bae S, Kim Y, Guanetti J, Borrelli F, Moura S (2018) Design and implementation of ecological adaptive cruise control for autonomous driving with communication to traffic lights (2018). arXiv:1810.12442

24. Shaobing X, Peng H (2018) Design and comparison of fuel-saving speed planning algorithms for automated vehicles. IEEE Access 6:9070–9080

25. Kaijiang Y, Yang J, Yamaguchi D (2015) Model predictive control for hybrid vehicle ecological driving using traffic signal and road slope information. Control Theory Technol 13(1):17–28

26. Guillemin F, grondin O, Moulin P, Dib W (2015) An innovative eco-driving coaching solution: Geco. In: Proceedings of ITS world congress
27. Toffetti A, Iviglia A, Arduino C, Soldati M (2014) "ecodriver" HMI feedback solutions. In: Proceedings of the human factors and ergonomics society Europe chapter 2013 annual conference
28. Seewald P, Kroon L, Brouwer S et al (2015) Ecodriver. d13. 2: results of HMI and feedback solutions evaluations. Technical report, European Union
29. Hibberd DL, Jamson AH, Jamson SL (2015) The design of an in-vehicle assistance system to support eco-driving. Transp Res Part C: Emerg Technol 58:732–748
30. Vaezipour A, Rakotonirainy A, Haworth N (2016) Design of a gamified interface to improve fuel efficiency and safe driving. In: Proceedings of international conference of design, user experience, and usability, pp 322–332. Springer
31. IFP Energies Nouvelles (2018) Premier challenge national écoconduite interentreprises 2018: essai transformé! https://www.ifpenergiesnouvelles.com/node/616
32. Birrell SA, Fowkes M (2014) Glance behaviours when using an in-vehicle smart driving aid: a real-world, on-road driving study. Transp Res Part F: Traffic Psychol Behav 22:113–125
33. Birrell SA, Fowkes M, Jennings PA (2014) Effect of using an in-vehicle smart driving aid on real-world driver performance. IEEE Trans Intell Transp Syst 15(4):1801–1810
34. Borrelli F, Morari M (2007) Offset free model predictive control. In: Proceedings of conference on decision and control, pp 1245–1250. IEEE
35. Thrun S, Montemerlo M, Dahlkamp H, Stavens D, Aron A, Diebel J, Fong P, Gale J, Halpenny M, Hoffmann G et al (2006) Stanley: the robot that won the darpa grand challenge. J Field Robot 23(9):661–692

Chapter 9
Detailed Case Studies

In this chapter we present a few case studies that demonstrate and possibly expand the concepts presented earlier in this book in practical applications. Each case study is often a compilation of several research papers led by the authors and is meant to highlight system integration and practical considerations.

9.1 Eco-Approach to Signalized Intersections

In Sect. 1.3.2 an overview of published results on eco-approach to signalized inter-sections was presented. Section 4.1.3 provided a basic overview of signalized inter-sections and Sect. 7.7 presented more insight about potential energy saving via a fundamental numerical and analytical treatment.

In this section we provide a case study summarizing published results of a decade of work on the topic at Clemson University. More specifically deterministic and probabilistic planning of eco-approach to traffic lights [1, 2], impact on mixed traffic via traffic microsimulation analysis [3], and real-world experimental implementation [4] are discussed.

9.1.1 Numerical Approach

The goal is to find a velocity profile which reduces the energy consumption during a trip based on full or partial SPAT information. This problem can be formulated as an energy (fuel) minimization problem as was treated numerically in Sect. 7.7.1 and analytically in Sect. 7.7.2. Energy minimization requires inclusion of dynamic models of a specific vehicle and its propulsion system (ICEV, EV, HEV) to relate

© Springer Nature Switzerland AG 2020
A. Sciarretta and A. Vahidi, *Energy-Efficient Driving of Road Vehicles*,
Lecture Notes in Intelligent Transportation and Infrastructure,
https://doi.org/10.1007/978-3-030-24127-8_9

energy use to the velocity profile. To avoid the ensuing computational complexity and to decouple the choice of optimal speed from a vehicle's make and model, a simpler cost function can be used that penalizes a weighted sum of the total trip time and the acceleration and deceleration, instead of total energy use. The underlying assumptions in this choice are that idling at a traffic light and excessive acceleration that lead to braking both cost energy with little to no benefit to the driver. Other factors such as motion constraints imposed by red intervals, road speed limits, and the fact that very low velocities will be unacceptable to consumers, can be accounted for by constraining the solution space. In this case study, the fuel economy for a specific vehicle model is evaluated a-posteriori, by feeding the optimal velocity profile to a high-fidelity dynamic model of the vehicle.

We first describe, in Sect. 9.1.1.1, the scenario when deterministic and accurate SPAT information over the entire prediction horizon is available. When the phase and timing of upcoming signals are uncertain, a probabilistic term can be added to the cost function, as described in Sect. 9.1.1.2.

9.1.1.1   Planning with Deterministic SPAT Information

When only a single traffic light is on the horizon, one can use deterministic knowledge of green and red intervals to plan within the allowable velocity limits a timely arrival at a green as shown in Fig. 9.1.

To obtain a best achievable energy efficiency baseline, the optimal control problem is first solved assuming full and deterministic knowledge of signals' phase and timing over the planning horizon. Instead of the cost function in (8.2), a simpler and heuristic cost function is used that can be later transformed to a quadratic program for efficient numerical solution. The following cost function was chosen,

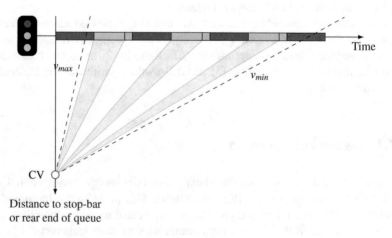

Fig. 9.1 Feasible velocity intervals in order to avoid stopping at red, if possible

$$J = \sum_{i=k}^{k+N-1} \left[w_1 \frac{\Delta t_i}{\Delta t_{min}} + w_2 \left| \frac{a_i}{a_{max}} \right| + c(s_i, t_i) \frac{1}{\varepsilon} \right] , \tag{9.1}$$

where J is the total cost and is indexed over position s with index i that starts at the current position step k and ends N steps later at step $k + N - 1$ with N being the length of the prediction horizon. Here $\Delta t_i = t_{i+1} - t_i$ is the time required for a vehicle to cover the fixed distance $\Delta s = s_{i+1} - s_i$ between steps s_i and s_{i+1} given the velocity at s_i and the acceleration a_i; Δt_{min} is the minimum time to complete the step if starting and ending at the maximum velocity and is used as a scaling factor, a_i is the constant acceleration assumed during step i, and a_{max} is the maximum allowed acceleration. The constants w_1 and w_2 are weighting terms. Motion constraints imposed by a red interval are imposed as a soft constraint by inclusion of the term $c(s_i, t_i) \frac{1}{\varepsilon}$ in the cost function. The value of $c(s_i, t_i)$ is zero except for spatiotemporal intervals when a light is red in which case its value is set to one, and ε is a very small constant (for example 10^{-6}), such that idling at red is discouraged.

The vehicle kinematics, realized by the following two-state equations, are imposed as equality constraints. Here s is the independent variable, velocity v and time t are the two states, and acceleration a is the input:

$$\frac{dv(s)}{ds} = \frac{a(s)}{v(s)} , \tag{9.2}$$

$$\frac{dt(s)}{ds} = \frac{1}{v(s)} , \tag{9.3}$$

Discretizing the above equations with a constant sampling interval of Δs and with a zero-order hold on acceleration, we obtain:

$$v_{i+1} = \sqrt{(v_i)^2 + 2a_i \Delta s} , \tag{9.4}$$

$$t_{i+1} = t_i + \frac{2\Delta s}{v_i + \sqrt{(v_i)^2 + 2a_i \Delta s}} , \tag{9.5}$$

The hard inequality constraints: $v_{min} \le v_i \le v_{max}$ and $a_{min} \le a_i \le a_{max}$ are also enforced. Here v_{min} and v_{max} are the road speed limits and can also include lowest speed acceptable to a driver; a_{min} and a_{max} are the feasible bounds for deceleration and acceleration.

The above optimal control problem can be solved numerically using Dynamic Programming (DP) and based on the discretization on position, time, and velocity as schematically shown in Fig. 9.2. The solution is calculated in one backwards sweep along the position axis taking advantage of Bellman's principal of optimality. The outline of the DP algorithm was described in Algorithm 1 of Sect. 6.2.2.2.

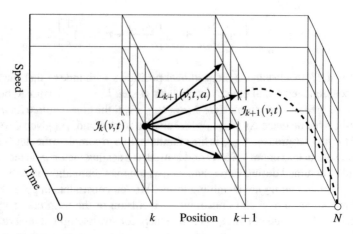

Fig. 9.2 Schematic of the DP grid

9.1.1.2 Planning with Probabilistic SPAT Information

Because perfect full-horizon SPAT information is not always available as explained in Sect. 8.2.4, here we consider the scenario where SPAT is known only probabilistically for instance based on historical data. The cost function in (9.1) is modified to the following to take into account the probabilistic nature of SPAT information,

$$ J = \sum_{i=k}^{k+N-1} \left[w_1 \frac{\Delta t_i}{\Delta t_{min}} + w_2 \left| \frac{a_i}{a_{max}} \right| + c(s_i, t_i) | \ln \left(p(s_i, t_i) \right) | \right] . \qquad (9.6) $$

All parameters and variables in (9.6) are the same as those described for (9.1); the only new variable is $p(s_i, t_i)$ which represents probability of green at time t_i for a light situated at position s_i. Therefore higher costs are assigned to solutions that pass through time intervals where probability of green is lower. At the limit when probability of green at s_i and t_i is zero, $\ln \left(p(s_i, t_i) \right) = \infty$ and passing through a red would be discarded. Where $p(s_i, t_i) = 1$, this term of the cost function drops to zero and increases the likelihood that the corresponding velocity will be selected. The probability of green for each light can be generated based on real-time and/or historical information as described in Sect. 8.2.4. Minimization of the cost function (9.6) with the equality and inequality constraints described in the previous subsection, remains a deterministic optimal control problem. The problem is solved using DP but in a receding horizon manner; as new information becomes available, the DP is re-solved taking into account the updated information over the remaining trip horizon. Note that an alternative approach will be a Stochastic Dynamic Programming (SDP) formulation where the expected value of the cost in (9.1) is penalized.

9.1.2 Simulation Results

9.1.2.1 Single Vehicle Simulation

A vehicle driving down a 800 m long street with three traffic signals was simulated in three scenarios. In the first scenario it is assumed that the vehicle has no information about the future state of the traffic light, in the second scenario real-time probabilistic SPAT information as described in Sect. 8.2.4 is assumed, and the third scenario full a-priori knowledge of SPAT information is assumed. Each scenario was run 1000 times in a Monte Carlo type experiment in which the start of red phases were randomized within a window of sufficient length for the vehicle to complete the route. The total cycle length, and length of each red were kept constant. Also the proportion of red to green times across all simulations were constrained to be the same. The start of each red of a traffic signal was chosen independently of the start of red of the next traffic signal. In all simulations, the penalty weights in the cost functions (9.1) and (9.1) are set to $w_1 = 1/8$, $w_2 = 1/8$. The value of ε is set at 10^{-6}. To solve the DP, the solution space is discretized to distances of 20 m, time increments of 1 s, and velocity steps of 1 m/s as schematically shown in Fig. 9.2. In this discretization grid choice, the computational time and memory requirements were reasonable for implementation on a PC.

In calculating the fuel economy, it was not computationally feasible to run all cases through a simulation cycle with a full high-fidelity vehicle model. Therefore a simplified vehicle model was developed using efficiency maps taken from AUTONOMIE [5] and a simplified gear shifting logic. For example, the effects of engine start and stop transients on fuel economy were not modeled in the simplified fuel economy calculations. The simulated vehicle is a two-wheel-drive, automatic transmission, conventional-engine vehicle. This vehicle had a total mass of 1580 kg, an engine producing a peak of 115 kW, and a constant electrical load of 200 W. The velocity profiles generated by the dynamic program were fed to this model to calculate the fuel economy for each case. This considerably simplified model provides a significant reduction in computational time when calculating the fuel economies for large numbers of simulation cases.

The Monte-Carlo simulation results found in Table 9.1 indicate that, for the road conditions described and with only real-time information and the probabilistic models, an average of 16% increase in fuel economy could be expected, representing approximately 62% of the benefit of full and exact traffic signal timing information.

With Monte-Carlo simulations indicating positive results for fixed time traffic lights, next an example of traffic signals with adaptive timing is presented. Twenty four hours of recorded traffic signal timing data from a series of signals in an urban corridor in Northern California is used. Figure 9.3 shows the lengths of the green phases for four different movements of one of these traffic signals.

A vehicle was simulated driving through the three traffic signals every 10 min over the 24 h yielding a total of 144 simulated drives per level of information. The real-world distance between the signals is preserved in the simulation, such that the

Table 9.1 Monte-Carlo simulation results reflect the positive influence of information, on average, on fuel economy

	Mean (MPG)	Standard deviation (MPG)
No information	25.9	5.0
Real time information	29.9	3.7
Full information	32.5	3.0

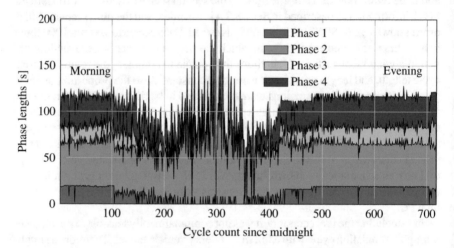

Fig. 9.3 Histories of green phases of four different movements of a traffic light on the chosen real-world route, for every cycle over a 24 h period (midnight to midnight)

Table 9.2 Fuel economy results from recorded real-world traffic signal timings with simulated vehicles moving between the lights reflect the positive influence of information

	Mean(MPG)	Standard deviation (MPG)
No information	31.7	3.1
Real time information	33.7	3.0
Full information	34.5	3.6

simulated vehicle has to cover the same distance using the same traffic signal timing offsets as a real driver would encounter. The total simulation distance is 1320 m. The lights occur at 520, 800, and 1200 m mark from the start. The DP resolution is the same as those set before. No other vehicles are considered to be on the road. In the case of real-time information the probability of green is calculated using Eq. (8.16) using a 24 h average of red and green lengths as t_r and t_g. If more relevant averages (for example a short-term average, a time of day average, or other statistical means) are available, they may continue to improve the performance of this real-time information case.

The simulation indicates that drivers with access to real-time probabilistic information were able to improve fuel economy over drivers with no information by approximately 6% (Table 9.2). This accounts for roughly 70% of the potential gains available through access to full and exact future knowledge of traffic signal timing. To deal with unexpected traffic, pedestrians crossing out of cross-walks, and other disturbances, the DP can be simulated frequently reproducing its cost-to-go map and optimal policy when necessary.

9.1.2.2   Multi Vehicle Microsimulation

While eco-approach to traffic signals could improve the energy efficiency of the ego vehicle, its impact on energy efficiency of upstream traffic deserves further investigation. One such microsimulation study was presented in [3]. Simulations were conducted in Quadstone Paramics [6] and custom code was developed to simulate vehicles with the eco-approach functionality. The equipped vehicles receive the timing of next upcoming traffic light in advance and adjust their speed for a timely arrival at green based on an analytical solution to the optimal control problem. The simulation is run in an urban corridor network. The 2.4 km (1.5 mile) path contains four signalized intersections with fixed timings. The speed limit of each link is 80 km/h (50 mph). Conventional vehicles do not have prior access to traffic signal information and always try to reach the maximum road speed limit unless affected by nearby vehicles or traffic signals. Three different traffic demand levels (300, 600, and 900 vehicles per hour per lane) and seven different penetration levels of equipped vehicles (100%, 90%, 70%, 50%, 30%, 10%, 0) are considered; therefore 21 simulations were conducted. Fuel consumption was estimated using a simplified model adopted from [7] which relates the fuel consumption rate to vehicle's velocity and acceleration. The parameters of the model are those in [7] and obtained by a third order polynomial fit to experimental data. It was further assumed that the engine was idling during negative acceleration consuming a constant idling fuel rate.

Figure 9.4 summarizes the energy consumption results for equipped vehicles as well as conventional vehicles. As shown in this figure, equipped vehicles (three bottom curves) consume much less fuel than conventional vehicles. This is due to fewer stops and closer-to-optimal operation of the engine. Another very interesting trend seen in Fig. 9.4 is that with the increment of the percentage of equipped vehicles, conventional vehicles consume less fuel. In other words, equipped vehicles have a positive impact on the energy efficiency of the entire mix of vehicles. With the increment of equipped vehicles, other conventional vehicles are more likely to follow them and benefit indirectly. However the energy efficiency of equipped vehicles generally decreases as their penetration increases. This could be due to slow-down of some equipped vehicles hindering procession of those behind them through a green.

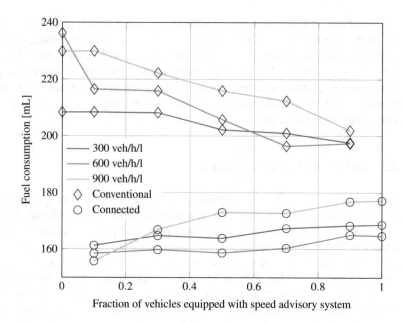

Fig. 9.4 Fuel consumption of vehicles with and without speed advisory system under different traffic demand levels and different penetration levels of equipped vehicles

9.1.2.3   Experimental Verification

Real-world implementation of traffic signal eco-approach or advisory has been reported in several recent publications. In [4] the concept is tested in the city of San Jose, California where the authors had real-time information of around 800 traffic lights. The connected vehicle identifies the next relevant traffic light and subscribes to it to receive updates about the state of the light using the cellular network via User Datagram Protocol (UDP) messaging. Because the vehicle only subscribed to the next upcoming light, a rule-based algorithm was employed to determine the feasible speed range for green arrival and use of dynamic programming was not needed. A user interface was created and recommended the appropriate speed range to pass through the next upcoming traffic signal during the green phase. The appropriate speed recommendation was displayed to the driver as green zones on the speedometer, as seen in Fig. 9.5.

A BMW 5 Series vehicle was used in these street experiments. Four drivers were asked to follow the speed recommendation shown on the dash display as long as safety was not jeopardized. The drivers were then asked to repeat the test, this time with the velocity advisory system deactivated. The tests were conducted in four different days and in real mixed traffic conditions. The fuel consumption of each driver was recorded for approximately one hour sessions and the results can be found in Table 9.3. In addition, the mean non-zero velocity, mean positive non-zero acceleration, and standard deviation of the same are all reported for evaluation of the

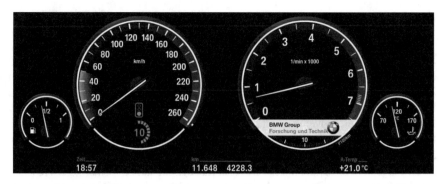

Fig. 9.5 Driver's dash display, including speed recommendation and countdown. A similar interface has been used in [8, 9]

Table 9.3 Field testing in San Jose, CA. Drivers were aware of the velocity advisory system and were specifically asked to follow the dash-display recommendations

Driver #	System inactive				System active			
	MPG	Mean velocity (m/s)	Mean accel. (m/s^2)	σ accel.	MPG	Mean velocity (m/s)	Mean accel. (m/s^2)	σ accel.
Driver 1	13.48	7.3	1.4	0.6	14.44	5.9	1.2	0.5
Driver 2	12.71	6.6	1.2	0.5	13.55	6.5	1.2	0.5
Driver 3	13.16	7.4	1.5	0.7	15.91	6.2	1.2	0.4
Driver 4	10.91	7.5	1.5	0.7	11.22	7.7	1.5	0.8

system. The fuel consumption measurements in Table 9.3 show that on average a 9.5% decrease in fuel usage is possible if drivers follow the displayed recommendations as closely as possible.

9.2 Cooperative Intersection Control

Results in previous section demonstrate that individual vehicles potentially save energy when they adjust their speed for a timely arrival at a green light. One could expect even higher efficiency with cooperative intersections in an all-autonomous vehicle environment as described in Sect. 1.4.3.

In recent years a few groups have proposed methods for such cooperative intersection control concepts. Here we summarize the results in [10, 11] in which the arrival time assignment was formulated as a Mixed Integer Linear Program and was implemented in a Vehicle-In-the-Loop (VIL) experiment.

9.2.1 Formulation as an Optimization Problem

Figure 9.6 schematically shows a shaded two-way intersection and the goal is to schedule the two conflicting directions of movements. This is done by defining a larger square centered at the intersection denoted by access area. Each vehicle i, is assigned an access time t_i, to the edge of this square when it is empty of vehicles from the opposing movement. The access area is sized large enough and based on the speed limit, to ensure there is enough time to react safely to a vehicle that violates its access time in the opposing movement.

The objective of increasing intersection throughput can be formalized here as an optimization problem. If n vehicles are subscribed at each instant to an intersection, minimizing the maximum assigned access times to these vehicles will push more vehicles through the intersection in a given time span. But the optimization objective could also consider the desired speed of each vehicle. One choice for the objective function can then be a weighted sum of both objectives:

$$J = w_1 \max(\{t_1, \ldots, t_n\}) + w_2 \sum_{i=1}^{n} |t_i - t_{des,i}| , \qquad (9.7)$$

where t_i and $t_{des,i}$ are assigned and desired access times for vehicle i respectively and w_1 and w_2 are penalty weights. This optimization is expected to not only improve intersection flow but can reduce energy consumption due to reduced number of stops.

Given the speed limit v_{max} and the maximum acceleration constraint a_{max}, the earliest possible access time for vehicle i, denoted by $t_{min,i} = t_0 + \Delta t_1 + \Delta t_2$, can be calculated as illustrated in Fig. 9.7:

$$t_{min,i} = t_0 + \min\left(\frac{v_{max} - v_i}{a_i}, \frac{\sqrt{v_i^2 + 2a_i d_i} - v_i}{a_i}\right) + \max\left(\frac{d_i}{v_{max}} - \frac{v_{max}^2 - v_i^2}{2a_i v_{max}}, 0\right) ,$$
$$(9.8)$$

Fig. 9.6 Schematic of the proposed collaborative intersection control system. Gray denotes intersection area while white denotes access area

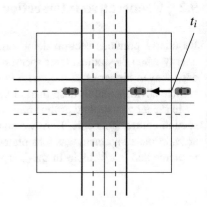

Fig. 9.7 The earliest possible access time based on speed limit and maximum accelerations

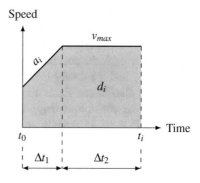

where t_0 is the current time, d_i is the distance of vehicle i to the intersection, v_i is the current speed of the vehicle i, and a_i is the maximum feasible acceleration for vehicle i thus yielding the minimum travel time. The minimum access time $t_{min,i}$ serves as a lower bound to the assigned access time for each vehicle:

$$t_i \geq t_{min,i} \, . \tag{9.9}$$

Two consecutive vehicles that are traveling on the same movement should be separated by a time headway of t_{gap_1} as they enter the intersection area. If vehicle j is the immediate follower of vehicle k in the same movement, the headway constraint can be expressed as

$$t_j - t_k \geq t_{gap_1} \, . \tag{9.10}$$

Two vehicles traveling on different phases (conflicting movements) need to be separated by a larger time headway to ensure that a vehicle can only enter the access area after all conflicting vehicles have left the intersection. For each two vehicles j and k that are on different phases of an intersection, the following OR constraint needs to be enforced:

$$t_j - t_k \geq t_{gap2} \quad \vee \quad t_k - t_j \geq t_{gap_2} \, , \tag{9.11}$$

where $\vee$ is the OR operator. The time headway between access times t_{gap_2} can be determined based on the dimensions of intersection and speed limits to allow enough time for a vehicle to come to a stop in the event that the vehicle in the opposing movement violates its assigned access time.

9.2.2 Numerical Solution

The optimization problem presented in Sect. 9.2.1 can be converted to a Mixed Integer Linear Program (MILP) using standard techniques. More specifically, in the cost function (9.7) the term $\max(\{t_1, \ldots, t_n\})$ can be replaced by a new slack variable t_{max} and by imposing new constraints $t_i \leq t_{max}$. In the same cost function, the terms $|t_i - t_{des,i}|$ can be replaced by new slack variables δt_i and imposing $\delta t_i \geq \pm(t_i - t_{des,i})$. As a result the cost function (9.7) will be a linear function of the optimization variables t_i and the newly introduced slack variables. The constraints in (9.9) and (9.10) are also linear. Each disjunctive OR constraint in (9.11) can be converted to two linear constraints using the big M method and by introducing new binary variables $B_l \in \{0, 1\}$ and a large constant M as follows:

$$
\begin{aligned}
t_j - t_k + M B_l &\geq t_{gap_2} , \\
t_k - t_j + M(1 - B_l) &\geq t_{gap_2} ,
\end{aligned}
\tag{9.12}
$$

where $0 \leq l \leq m$, m is the number of disjunctive constraints, and j and k correspond to any two vehicles on different phases of an intersection. When $B_l = 0$, the first constraint in (9.12) indicates $t_j - t_k \geq t_{gap_2}$ and the second constraint in (9.12) is automatically satisfied given that M is sufficiently large. When $B_l = 1$ then $t_k - t_j \geq t_{gap_2}$ the first constraint in (9.12) is automatically satisfied and the second constraint is active.

The above optimization problem can then be written in a canonical linear programming form. That is to minimize $c^T x$ subject to $Ax \leq b$ and $x \geq 0$. Here $x = (t_1, \ldots, t_n, \delta t_1, \ldots, \delta t_n, t_{max}, B_1, \ldots, B_m)$ is the vector of optimizing variables, n is the number of all subscribed vehicles, and m is the number of artificial binary variables of Eq. (9.12). By including B_l for each disjunctive (OR) constraint among optimization variables, we ensure that the most favorable of the two OR constraints is chosen.

The above MILP problem was solved using IBM's CPLEX optimization package on an Intel Core i5@2.5 GHz Windows 7 laptop with 8 GB of RAM. For 50 subscribed vehicles, the average intersection controller execution time was 120 ms but varied between 28 ms and 2400 ms. These times include the MILP solver execution time plus the time needed for pre-processing the probe vehicle data and expressing the problem in canonical form. The MILP problem was solved once every 4 s to adapt to unmodeled effects and to deviation of vehicles from their assigned access times.

9.2.3 Simulation Results

Here we summarize some of the results detailed in [10, 12]. Figure 9.8 shows 9 vehicles approaching an intersection with X and O representing vehicles on conflicting movements. The distance of each to the intersection at the time of their subscription is shown on the vertical axis in the two plots of Fig. 9.9. The assigned access times are shown on top horizontal axis. Two extreme choices of penalty weights w_1 and w_2 are shown in the two plots. It can be seen that the MILP objective of reducing the

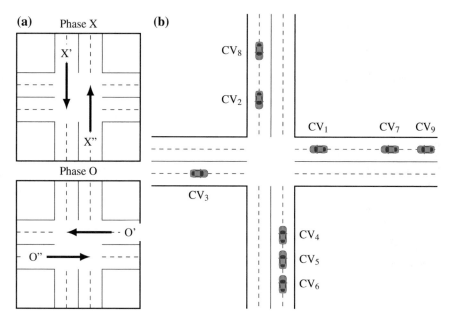

Fig. 9.8 Nine vehicles at a simplified two phase intersection with four movements

maximum assigned access time along with collision avoidance constraints groups together the vehicles of the same movement, when possible.

The performance of the proposed intersection control concept was also compared to a signalized intersection with a pre-timed traffic light in a one hour microsimulation with many vehicles. The intersection had four 500 m legs. The signal phase and timing for the benchmark pre-timed traffic signals were optimized off-line using Synchro signal optimization [13] resulting in a cycle time, green split, and yellow interval of 100, 44.5, and 3.5 s respectively. The vehicles were sampled from a negative exponential distribution [14] at 750 vehicles per hour for all four legs of the intersection. The vehicles' arrival pattern was recorded and replayed in all simulations. The average and maximum speeds were set to $v_{avg} = 15.6$ m/s (35 mph) and $v_{max} = 20$ m/s (45 mph). The pre-timed case was simulated twice: (i) with no speed advisory in which vehicles did not receive SPaT preview and (ii) with speed advisory to vehicles, similar to that of Sect. 9.1, in which vehicles received SPaT preview when they were within a 500 m range of the intersection. The penalty weights in the MILP objective function (9.7) were set to $w_1 = 50\%$ and $w_2 = 50\%$.

Table 9.4 summarizes some of the performance metrics for the three simulations that were conducted. It can be seen that number of stops has an almost 100-fold reduction with respect to a pre-timed signalized intersection. Average idle time for stopped vehicles is also cut in half. The average travel time shows considerable improvement as well. We expect the reduced idling and travel time positively impact energy efficiency. The experimental results presented next confirm this hypothesis.

Fig. 9.9 Examples of MILP solution with 9 subscribed vehicles (vertical axis at time zero: remaining distance of each individual vehicle to the access area, horizontal axis at distance zero: access time assigned to each individual vehicle, solid lines: minimum access times $t_{min,i}$, dashed lines: desired access times $t_{des,i}$, $i \in [1, 9]$, colors and marks refer to the four directions of Fig. 9.8); **a** all weight given to intersection throughput improvement, **b** all weight given to satisfying the desired speeds of all vehicles [10]

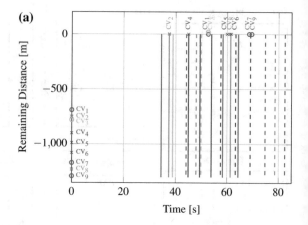

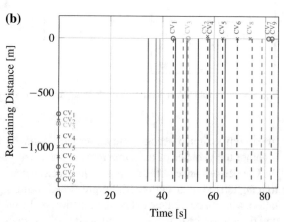

Table 9.4 Microsimulation results comparing MILP controlled intersection concept with pre-timed signalized intersections with and without speed advisory

Performance metric	Pre-timed	Pre-timed+advisory	MILP
Intersection traversals	2900	2900	2900
Simulation time [min]	61	61	61
Number of intersection stops	1171	872	13
Total intersection idling Delay [min]	3640	1843	2
Avg. idle per stopped vehicle [s]	20	15	9
Avg. travel time per vehicle [s]	50	51	36

9.2.4   Experimental Results

In order to investigate the energy efficiency potential of the proposed approach, a
Vehicle-In-the-Loop (VIL) test concept was proposed in [11] in which a real vehicle
approaching a signal-less intersection on a test track interacts with hundreds of simulated vehicles approaching a simulated version of that intersection. Simulated and
real vehicles all subscribe and communicate similarly to the intersection controller
and are treated equally. The position of the real vehicle is injected in the microsimulation and therefore is easily visualized. The proposed approach is more realistic
than a simulation-only environment, while also ensuring a safer environment for test
vehicles because conflicting movements (and potential crashes) occur in a simulated
environment. The VIL concept is shown in Fig. 9.10.

The test vehicle was a human driven Honda Accord LX with a 2.4 L 4-Cylinder
SI gasoline engine. The vehicle was driven on an isolated straightaway located at
International Transportation Innovation Center (ITIC) test track in Greenville, South
Carolina. A custom-coded user interface on a mobile phone allowed speed control
by the driver for a timely arrival at assigned access times. A cellular network was
used for communication between the in-vehicle mobile phone and remote intersection
controller. The mobile phone sent the vehicle position and velocity to the intersection
controller every 5 s, received the assigned access time, calculated appropriate speed
to meet this access time, and visualized the calculated speed by a narrow green arc
on a circular speedometer. The traffic microsimulator node ran inside the test vehicle
as shown in Fig. 9.10c for real-time monitoring of the simulations.

The fuel rate was estimated from data logged in real-time from the vehicle On-
Board Diagnostics (OBD-II) port with details reported in [11]. Extra care was taken
in calibrating the fuel rate estimator so that it matched the actual fuel consumption
measured from a gasoline tank fill-up.

Within the same VIL framework described above 3 sets of tests were run: a pre-
timed intersection baseline, pre-timed with speed advisory baseline similar to that of
Sect. 9.1, and our proposed MILP controlled intersection. Each set consisted of 12
laps around the test track with wide U-turns at both ends of the track. The start time

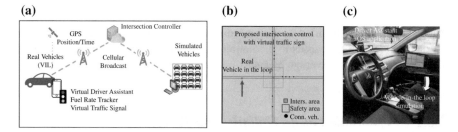

Fig. 9.10 Vehicle-in-Loop experimental setup in [11] showing **a** interactions between a real vehicle
and microsimulation environment via 4G network, **b** Java microsimulation interface, and **c** in-vehicle
setup

Table 9.5 VIL experiment results for real vehicle: Comparing MILP controlled intersection concept with pretimed baselines

Performance metric	Pre-timed	Pre-timed+advisory	MILP
Intersection traversals	12	12	12
Simulation time [min]	57.5	55	51
Number of intersection stops	10	0	0
Total intersection idling delay [min]	4.3	0	0
Avg. idle per stopped vehicle [s]	26	0	0
Avg. travel time per vehicle [s]	108	99	79
Fuel consumption [L]	1.13	1.11	0.91

at the beginning of each lap was randomized using a random number generator to prevent unintended bias due to cyclic runs.

Table 9.5 summarizes some of the performance metrics for the test vehicle. The test vehicle passed the MILP-based imaginary intersection 12 times without stopping. This resulted in 19.5 and 18.0% reduction in fuel consumption compared respectively with the two pretimed benchmarks.

9.3   Anticipative Car Following

In Sect. 1.3.3 we highlighted the potentials offered by CAVs for proactive and anticipative car following to lower energy use compared to current reactive car following practices. This approach is the eco-ACC introduced in Sect. 8.1.3. An analytic treatment of this scenario was presented in Sect. 7.8.

Here we expand on what was presented before in the book and present a detailed case study based on a compilation of the approach and results in [15–18]. We start by formulating car following as an optimal control problem. The unknown "disturbance" is the future position of the preceding vehicle which motivates methods for its deterministic or probabilistic prediction. We show via microsimulation analysis that one could gain on average by such anticipative car following measures.

9.3.1   Formulation as an Optimization Problem

In anticipative car following the goal is to reduce energy-consuming braking or stop and go events by judiciously adjusting the following distance between the two vehicles as a buffer. This desire can be cast as an optimization with a cost function that balances the car following distance against acceleration command by penalizing a weighted sum of both as described in Sect. 8.1.3. For instance, in a model predictive approach the following quadratic cost minimization can be performed at the beginning of each receding horizon similar to that shown in [16],

$$\min_{u(i)} J = w_s \|s_p(N) - s(N) - T\dot{s}(N) - L_{min}\|^2 +$$

$$+ \sum_{i=0}^{N-1} \left(w_s \|s_p(i) - s(i) - T\dot{s}(i) - L_{min}\|^2 + w_u \|u(i)\|^2 \right) , \quad (9.13)$$

where N is the number of time steps in a prediction horizon, progression of steps along the horizon is indexed by i, $u(i)$ are the acceleration commands and optimization variables, $\|.\|$ denotes the two norm and w_s and w_u are penalty weights. Here s_p and s are the position of the preceding and ego vehicles respectively, and L_{min} is the minimum desired gap between them when the ego vehicle is stopped. The distance headway $T\dot{s}$ is the product of time headway T and velocity of the ego vehicle $\dot{s}$ and is meant to induce larger gaps at higher ego vehicle speeds.

The vehicle longitudinal kinematics along with a first order lag between the acceleration command input u and the vehicle's acceleration a can be enforced as equality constraints and obtained by discretizing the following continuous time equations:

$$\dot{s}(t) = v(t) , \quad (9.14)$$

$$\dot{v}(t) = a(t) , \quad (9.15)$$

$$\dot{a}(t) = -\frac{1}{\tau}a(t) + \frac{1}{\tau}u(t) , \quad (9.16)$$

where τ approximates the time constant from acceleration command to actual acceleration. After discretization, the continuous-time t is indexed by i as the independent variable.

Hard constraints on vehicles states and on the following distance must be enforced at each step in time. An important safety constraint is a lower bound on car following distance. An upper bound can also be optionally enforced to avoid leaving large gaps that could negatively impact traffic flow or encourage cut-ins. In summary:

$$L_{min} \leq s_p(i) - s(i) - T\dot{s}(i) \leq L_{max} \quad i = 1, \ldots, N . \quad (9.17)$$

Minimum and maximum speed limits should also be enforced,

$$v_{min} \leq v(i) \leq v_{max} \quad i = 1, \ldots, N . \quad (9.18)$$

As shown in [17] and illustrated in Fig. 9.11, the powertrains maximum acceleration capacity depends strongly on velocity as seen in combined engine-transmission maps. The velocity dependent acceleration constraint can be approximated as piece-

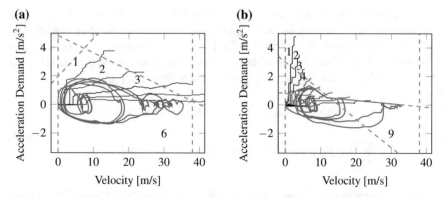

Fig. 9.11 Velocity dependent acceleration constraints (dashed orange) for **a** a passenger vehicle where conjunctive maximum acceleration constraints yield a convex velocity-acceleration admissible set and **b** a heavy duty vehicle where disjunctive maximum acceleration constraints yield a non-convex velocity-acceleration admissible set. In both scenarios, the blue phase portrait trajectories are sample operating point trace of vehicles running MPC planning under US06 drive cycle [17]

wise linear combinations of velocity and acceleration as detailed in [17] and illustrated in Fig. 9.11. Depending upon the convexity of the acceleration-velocity constraint-admissible set, these piecewise linear constraints may be applied conjunctively or disjunctively. As shown in [17] disjunctive OR constraints can be converted to conjunctive AND constraints by introducing new integer optimization variables in the big M method described in Sect. 9.2.2.

In [18] a terminal constraint is also imposed on velocity and position of the ego vehicle to prevent a collision post-prediction horizon. This terminal constraint on velocity and position of ego vehicle is constructed using kinematic relationships and assuming that the preceding vehicle will apply maximal braking post-prediction horizon (worst case scenario). While the resulting terminal constraint is nonlinear, a linear approximation to it could be used.

The above linear constraints along with the quadratic cost function in (9.13) form a quadratic program over each horizon. Efficient QP solvers exist that could solve this problem in real-time. Even when integer variables are introduced to handle disjunctive linear constraints shown in Fig. 9.11b, the resulting Mixed Integer Quadratic Program (MIQP) can still be solved relatively fast as documented in [17].

The main challenge is the uncertainty about the position of the preceding vehicle s_p over the optimization horizon. Note that s_p appears in the cost function (9.13) as well as in the constraint (9.17). The optimization could be solved under the worst case scenario assuming the preceding vehicle comes to a sudden emergency stop at each step of the horizon. Such worst case assumptions could induce very conservative and perhaps unnecessarily large headways between vehicles. Here we employ the methods described in Sect. 8.2.3 for predicting the motion of preceding vehicle.

With a probability distribution for $s_p(i)$, the gap constraints can be enforced probabilistically as a so-called chance constraint. For instance a minimum gap constraint in Eq. (9.17) can be instead written as:

$$P(s(i) + T\dot{s}(i) \leq s_p(i) - L_{min}) \leq 1 - \alpha \quad i = 1, \ldots, N , \qquad (9.19)$$

which means that the chance of violating the constraint should be less than $1 - \alpha$. Note that at any current step $i = 0$, s, $\dot{s}$, and s_p are all deterministic rather than probabilistic, and MPC finds solutions that do not violate constraints. The probabilistic constraint will be converted to a deterministic constraint using probability distribution of $s_p(i)$. If we denote $R_{1-\alpha}$ as the position where the cumulative distribution function of $s_p(i)$ is equal to $1 - \alpha$, then the equivalent deterministic constraint is

$$s(i) + T\dot{s}(i) \leq R_{1-\alpha} - L_{min} . \qquad (9.20)$$

Similarly, the maximum distance constraint can be enforced probabilistically. With transformation of the probabilistic constraints to deterministic ones, we end up with a standard MPC problem. This is the approach employed in [15, 16].

9.3.2 Numerical Solution

In [18] a parameter optimization was performed to find optimal values of the penalty weights and the prediction horizon length. Once the parameters were fixed, the Gurobi optimization package [19] was used to solve the QP MPC problem for a passenger vehicle. For a heavy truck the maximum acceleration constraint is disjunctive as was illustrated in Fig. 9.11b which resulted in a Mixed Integer Quadratic Program (MIQP) formulation for the MPC problem. For both cases, passenger cars and heavy truck, two scenarios were considered: (i) When a CAV follows another CAV in which future intentions of the preceding CAV were available to the following CAV for the duration of prediction horizon, and (ii) when a CAV follows a conventional vehicle in which a probabilistic model similar to those described in Sect. 8.2.3 was used to estimate the intentions of the preceding vehicle.

Table 9.6 shows the computation times for both QP and MIQP MPCs. One MPC vehicle was simulated following an open-loop vehicle. The optimization was solved on a laptop PC equipped with 16.0 GB RAM and a 2.70 GHz CPU. In Table 9.6, Optimization Time refers to the time required to solve the mathematical program (QP or MIQP) and Compt. Time refers to the total time required to run a single vehicle's control move determination, including both preview handling and optimization.

Table 9.6 Computation time for anticipative car following MPC. The MPC is converted to a QP for a passenger car and to a MIQP for a heavy truck. When a CAV follows another CAV the preview source is connectivity and full intentions over the horizon is communicated from the preceding CAV. On the other hand when the preceding vehicle is a conventional vehicle, preview source is a probabilistic model

Algorithm	Preview source	Worst case in-horizon con-straints	Worst case terminal con-straints	Mean comp. time [s]	Max comp. time [s]	Mean opt. time [s]	Max opt. time [s]
Car	Connectivity	No	Yes	0.0337	0.0561	0.0108	0.0444
Car	Probabilistic	Yes	Yes	0.0757	0.1134	0.0110	0.0892
Truck	Connectivity	No	Yes	0.0435	0.0789	0.0148	0.0504
Truck	Probabilistic	Yes	Yes	0.0571	0.0919	0.0069	0.0425

9.3.3 Simulation Results

Here we report the results of the implementation in [18] in which a mixed string of 8 vehicles following a target vehicle are simulated. The mix includes conventional cars with no connectivity that use a standard Intelligent Driver Model (IDM) for car following as described in Sect. 4.2.1.3. The IDM parameters were sampled from empirical data. More specifically, the desired time headway and maximum and minimum acceleration levels were sampled from log-normal distributions fit to the empirical data of [20]. Different penetrations of CAVs in the mix are explored. Each CAV solves a variant of the receding horizon car following approach described above. When a CAV is immediately preceded by another CAV it receives the intended position of the preceding vehicle but when it is following an IDM vehicle it uses a probabilistic prediction of the preceding car motion. Post horizon, collision avoidance is ensured by considering worst case hard constraints at the end of each prediction horizon when two CAVs follow each other. When a CAV follows an IDM vehicle, worst case collision constraints are enforced along the prediction horizon as well.

Both passenger vehicle and heavy duty CAVs are injected in the mix. Observing that the penetration of heavy trucks is 25% in some US highways, 0, 1, or 2 heavy duty trucks are injected in the mix of the 8 vehicles in the string. The receding horizon problem for the passenger vehicle is converted to a QP and that of a heavy duty CAV becomes a MIQP as described in Sect. 9.3.1. A quasi-random approach is used to create different placement of the vehicle types in the string leading to a total of 2224 scenarios and simulations as detailed in [17]. Figure 9.12 shows the cumulative fuel consumption results at different CAV penetrations and for 0, 1, and 2 heavy duty vehicles. The fuel consumption was estimated using quasi-static engine fuel maps. As expected the energy efficiency increases with the penetration of CAVs. A comparison between homogeneous human-like IDM strings and those composed

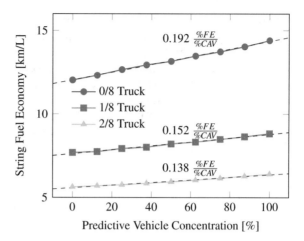

Fig. 9.12 Combined string fuel economy at various penetration levels of predictive and heavy vehicles

entirely of CAVs is available from the endpoints of each line in the plot. In absence of heavy duty vehicles a 1.9% improvement in fuel economy is shown for every 10% increase in CAV penetration.

9.4 Anticipative Lane Selection

In Chap. 4 mandatory and discretionary lane change behavior of human drivers were briefly discussed and the MOBIL lane change model was presented. Connected and automated vehicles can more judiciously make a lane change decision and execute it as explained in Sect. 1.3.4. Here we present a case study based on results in [21] in which lane selection is formulated as an optimization problem and in anticipation of neighboring vehicles' intentions.

9.4.1 Formulation as an Optimization Problem

A lane command u_l as well as a longitudinal acceleration command u_a are the high level control inputs. The longitudinal state space Eq. (9.16) relates u_a to vehicle longitudinal acceleration, velocity, and position. The lateral position of the vehicle $l(i)$ is expressed in lane width units with respect to a reference frame that aligns integer values of l with lane centers. For example, on a two-lane road $l = 1$ could coincide with the center of the right lane, $l = 2$ denotes the center of the left lane, and $l = 1.5$ is on the visible marking between the lanes. In [21] a critically damped second order lag is assumed between the lane command u_l and the actual lane l of the form

$$\ddot{l}(t) + 2\zeta\omega_n\dot{l}(t) + \omega_n^2 l(t) = \omega_n^2 u_l(t) . \qquad (9.21)$$

The damping ratio ζ is chosen to be unity and the natural frequency ω_n is chosen to be 1.1 rad/s resulting in a settling time of around four seconds to match a naturalistic lane change behavior. Binary "lane indicator" variables will be defined later that determine whether each lane is occupied as a function of l.

The state dynamics in (9.16) and (9.21) are discretized, transforming the continuous-time formulation with t as the independent variable into a discrete-time problem with i as the independent variable. The cost function (9.13) can be modified to include lane choices along with new constraints that arise in multi-lane traffic. Similar to the approach in Sect. 9.3.1, the following moving horizon cost function can be used,

$$
\min_{u_a(i),u_l(i)} J = \sum_{i=0}^{N-1} \left(w_v(v(i) - v_{ref}(i))^2 + w_a(u_a(i))^2 + w_l(l(i) - l_{ref}(i))^2 \right) +
$$
$$
w_v(v(N) - v_{ref}(N))^2 + w_l(l(N) - l_{ref}(N))^2 ,
$$
$$(9.22)$$

where $i = 0$ denotes the current time step and N is the prediction horizon. Here v_{ref} is the desired velocity and l_{ref} denotes the desired lane[1] that could be dictated from the vehicle navigation system. For instance a vehicle may prefer to stay in the right-most lane in anticipation of an imminent exit. The above cost function strikes a trade off between remaining in the desired lane and following the desired velocity while it penalizes the acceleration command to reduce unnecessary braking and energy loss. Minimizing the moving horizon cost could command changing to a faster lane to pass a slow moving vehicle instead of reducing speed in the desired lane. After passing the slow vehicle, return to the desired lane could be commanded due to the residual lane cost. This behavior is consistent with what a human driver would do.

The position of the ego vehicle s in its lane is constrained by other reachable vehicles in the same lane. The ego can be either behind or in front of another vehicle which is a non-convex OR constraint:

$$
s \leqslant s_{rear,p} - L_{ego} \quad \vee \quad s \geqslant s_{front,p} , \tag{9.23}
$$

where $\vee$ is the OR operator. Here $s_{rear,p}$ denotes the position of the rear bumper of the neighboring vehicle p and $s_{front,p}$ is the position of its front bumper. The length of the ego vehicle is denoted by L_{ego}. Similar to the approach in Sect. 9.2 the disjunctive OR constraint can be converted to an AND constraint using the big M method:

$$
s \leqslant s_{rear,p} - L_{ego} + M\beta_p \quad \wedge \quad s \geqslant s_{front,p} - M(1 - \beta_p) , \tag{9.24}
$$

where $\wedge$ is the AND operator. Here $\beta_p \in \{0, 1\}$ is a new binary variable defined for each reachable obstacle p in the ego vehicle lane and M is a large enough

---

[1] l_{ref} does not need to be an integer, in fact it is useful to offset it slightly to break symmetry in common situations.

constant. When $\beta_p = 0$, the first constraint in (9.24) may be active and indicates $s \leqslant s_{rear,p} - L_{ego}$ and the second constraint in (9.24) is trivially satisfied given that M is sufficiently large. When $\beta_p = 1$ then $s \geqslant s_{front,p}$ and the first constraint in (9.24) is automatically satisfied. The optimal solution chooses the value for β_p that results in minimum value of the cost function in (9.22).

Finally for each lane a new "lane indicator" binary variable $\mu_{nl} \in \{0, 1\}$ can be defined. When the ego is fully or partly in lane nl, $\mu_{nl} = 1$, otherwise if it does not occupy lane nl at all $\mu_{nl} = 0$. The inequality constraints in (9.24) are only relevant if the two vehicles occupy the same lane. So in multi-lane scenarios the constraint in (9.24) should be expressed as

$$s \leqslant s_{rear,p} - L_{ego} + M\beta_p + M(1 - \mu_{nl}) \quad \wedge \quad s \geqslant s_{front,p} - M(1 - \beta_p) - M(1 - \mu_{nl}) , \tag{9.25}$$

which should be imposed for every reachable vehicle in the occupied (or to be occupied) lanes.

The lane indicator binary variables μ_{nl} are determined as a function of ego vehicle lane state $l(i)$. For instance in a 2-lane scenario only two of these binary variables μ_1 and μ_2 are defined. Consider the variable δ to be the maximum deviation, in units of lane width, for a vehicle to remain wholly in a lane. If $1 + \delta \leq l(i) \leq 2 - \delta$ then the vehicle occupies both lanes and we set $\mu_1 = \mu_2 = 1$. If $l(i) \leq 1 + \delta$ then vehicle is in lane 1 and we set $\mu_1 = 1$ and $\mu_2 = 0$. If $l(i) \geq 2 - \delta$ then vehicle is in lane 2 and we set $\mu_1 = 0$ and $\mu_2 = 1$.

Setting these lane indicators must be handled by using inequality constraints as proposed in [21] since the MIQP solver demands a certain canonical form that does not accommodate if-then-else rules. For instance in a two lane scenario the following four constraints correctly set the values for μ_1 and μ_2,

$$-l(i) - M\mu_1 \leq -(2 - \delta) , \tag{9.26}$$

$$l(i) + M\mu_1 \leq M + 2 - \delta , \tag{9.27}$$

$$l(i) - M\mu_2 \leq 1 + \delta , \tag{9.28}$$

$$-l(i) + M\mu_2 \leq -1 - \delta + M , \tag{9.29}$$

where again M is a large enough number and $\mu_1, \mu_2 \in \{0, 1\}$ are binary optimization variables to be determined for each time step.

9.4.2 Numerical Solution

The above moving horizon optimization was solved using the Gurobi optimization package [19]. In order to reduce the computational effort of a mixed integer quadratic program which can be high for reasonable choices of prediction horizon, a move blocking approach is used in [21] to reduce the number of integer variables. The lane

Table 9.7 Simulation results of anticipativelane selection algorithm

	Time [min]	Fuel [L]
Rule-based	13.8	17.1
MPC	12.9	15.7
Free flow	12.7	15.3

change command u_l is held constant over every three steps while the acceleration command u_a can assume a different value at each step of the prediction horizon. With a sampling time of 0.4 s, a 10 s prediction horizon could be handled in real-time. With move blocking, longer horizons could be executed real time all on a laptop PC equipped with a 4 core, 2.7GHz CPU and 16GB RAM. More details can be found in [21].

9.4.3   Simulation Results

In [21] the simulated receding horizon approach was tested first in a two lane scenario in MATLAB with 4 MPC CAVs in their desired first lane encountering a slow moving vehicle. Each CAV has a different desired speed and a set of full factorial simulations was performed to include all possible placements of the CAVs with respect to each other. The MPC algorithm was compared to a reactive rule-based algorithm in which the CAVs used an IDM model for car following. They changed lane reactively when slowed down by a user defined threshold behind the slow moving vehicle provided that necessary space in the adjacent lane was available.

As shown in Table 9.7, the MPC algorithm reduced fuel consumption by 8.4% and travel time by 6.2% compared to the rule-based algorithm in full factorial simulations. The results were also compared to free flow traffic results. Excess travel time, defined as the increase in travel time over the congestion-free value, decreased by 79% relative to the reactive algorithm. Correspondingly, excess fuel consumption was reduced by 80% compared to a baseline of 18.1 mL for the average vehicle.

9.5   Eco-Routing and Eco-Coaching

Eco-routing has been extensively described in Chap. 5. Eco-coaching, an implementation variant of eco-driving is the subject of Sect. 8.1.1. This section is mainly based on the publication [22], which collects various results of the European Commission funded project OPTEMUS (2015–2019).[2] The scope is to discuss the energy efficiency benefits of the eco-routing and eco-coaching functions as experimentally assessed on a demonstrator car.

---

[2] www.optemus.eu.

9.5.1 Experimental Setup

In order to evaluate the impact of the eco-routing and eco-coaching (Sect. 8.1.1) strategies in real-world conditions, an experimental setup as illustrated in Fig. 9.13 was implemented. The system consists of:

- A dedicated smartphone application, which serves as a Human-Machine Interface (HMI) and hosts the eco-driving algorithms ensuring high response time;
- A cloud computing server which communicates with a Geographical Information System (GIS) to retrieve real-time traffic data and hosts the routing and driving range algorithms which are both costly in terms of computation time;
- An On Board Diagnosis (OBD) dongle which monitors the battery state-of-charge. This device is optional and could be replaced by an observer.

For the pre-trip and in-trip assistance, the driver can obtain from the HMI the most energy-efficient route, as well as the energy driving range available with the current battery state-of-charge. Furthermore, to perform in-trip eco-driving assistance, the smartphone application computes the optimal speed on the previous sub-trip at each

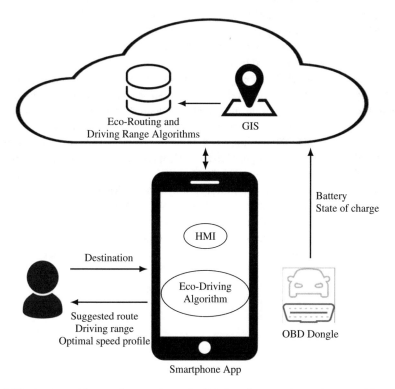

Fig. 9.13 System architecture for experimental validation of eco-routing and eco-coaching algorithms

speed breakpoint, as shown in Fig. 8.2. The application shows the optimal trajectory together with the actual speed profile in order to provide the drivers with visual feedback on their driving style. An energy consumption evaluation of the driving style on the sub-trip is also shown.

9.5.2 Experimental Results

In order to experimentally assess the energy benefits in using the proposed strategies, a series of field tests has been conducted. The tests have been conducted in the urban and sub-urban area of Turin, Italy, with a Fiat 500e (83 kW, 200 N electric motor) driven by a professional driver.

9.5.2.1 Energy Consumption Model

For the energy consumption model validation, a total of 35 trips were recorded, featuring an overall traveled distance of 434.6 km and a total travel time of 16.1 h (i.e. an average of about 12.4 km and 27.5 min per trip). The vehicle reference data for model validation were provided by the CAN-Bus data acquisition system. The topological data (i.e. road network, road signs, etc.) and the traffic information (i.e. average speeds) were provided by HERE [23].

The presented models for energy consumption and travel time, in Eqs. (4.41) and (4.31) respectively, were compared to standard state-of-the-art approaches. In particular, the energy consumption model was compared to a simple model reducing Eq. (4.28) to the average speed V_i for the entire road link ("NOacc"), and yet a simpler version neglecting also the auxiliary power term ("NOacc+NOaux"). Analogously, the travel time model was compared to a simple model obtained by considering only average velocity, $\tau_i = \ell_i / V_i$ in (4.31) ("NOacc"). The results of the experimental validation in Table 9.8 are expressed in terms of symmetric Mean Absolute Percentage Error [24] (sMAPE[3]) with respect to the CAN-bus measured energy consumption and travel time. The energy consumption estimation performed by the different models for one specific case is shown in Fig. 9.14. The proposed model largely outperforms the state-of-the-art approaches. Accuracy of the prediction models is crucial for a reliable navigation strategy.

---

[3] sMAPE is an accuracy measure based on percentage errors. It is used to solve the issue of heavier penalty on negative errors than on positive errors.

Table 9.8 Experimental validation results of the energy consumption model

	Energy			Time	
	Proposed model	NOacc	NOacc+ NOaux	Proposed model	NOacc
sMAPE	8.5%	17.4%	30.4%	7.9%	13.4%

Fig. 9.14 Energy consumption estimations over one trip compared to the CAN-bus reference measurement

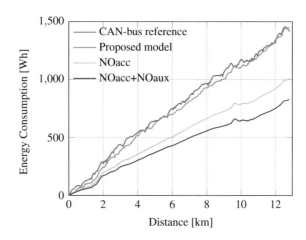

9.5.2.2 Eco-Routing

The aim of the conducted experiments for eco-routing validation is threefold. The first goal, achieved in simulation, is to show that a bi-objective eco-routing is highly effective in discarding those energy-optimal routes that penalize travel time. This feature could increase drivers' compliance to the route planning assistance. The second goal, achieved in simulation, is to prove that the shortest route is likely to be out of the Pareto efficient solutions in terms of energy consumption and travel time. The third goal, achieved both in simulation and experimentally for a selected origin/destination (O/D) pair, is to show that the eco-route is actually more energy-efficient than the shortest and the fastest route.

The energy and time weights w_k (5.7) and τ_k (5.15) were computed by using average traffic speed information for a regular working day at 09:00. The non-dominated points in the objective space calculated by the proposed algorithm are shown in Fig. 9.15. The eco-route and the fastest route correspond to the two solutions for $\lambda = 1$ and $\lambda = 0$ in formulation (5.17). A route corresponding to one of the Pareto-optimal trade-offs is labeled as multi-route. The shortest route is away from the Pareto front of the non-dominated points, and therefore not interesting either in terms of energy consumption nor in terms of travel time. The four routes are displayed on a map in Fig. 9.16.

For the experimental routing validation, the professional driver was instructed to drive on the eco, the shortest, and the fastest routes previously identified in simulation.

Fig. 9.15 Routing
simulation results.
Non-dominated points in the
criteria space calculated by
the proposed algorithm. The
performance of the four
routes is also displayed

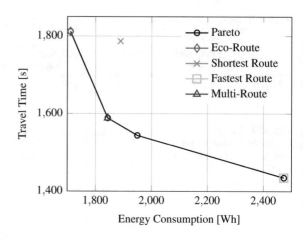

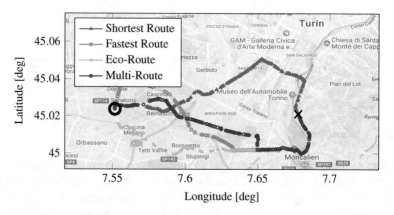

Fig. 9.16   The four routes obtained in simulation for the selected origin/destination pair

The driver performed three repetitions for each route, starting the experiments always
at the same hour of the day, therefore over several working days. As summarized in
Table 9.9, the experimental results showed that the eco-route for the identified O/D
pair is actually the most energy-efficient among the three alternatives. In particular,
the eco-route shows on average an energy gain of 4.5% with respect to the shortest
route, and 12.4% with respect to the fastest one. In terms of energy prediction accuracy over the three repetitions of each route, the sMAPE (between the measurement
and the estimation) ranged from 4.5% for the eco-route to 9.3% for the fastest route.
In terms of travel time prediction, the sMAPE ranged from 3.5% for the eco-route
to 12.7% for the fastest.

Table 9.9 Experimental eco-routing results

		Energy			Travel time	
		Mean	sMAPE	Gain	Mean	sMAPE
Eco	CAN	1784	4.5%	/	1637	3.5%
	Prediction	1759.5			1696	
Short	CAN	1868	5.9%	4.5%	2236	10.6%
	Prediction	1854			2230	
Fast	CAN	2041	9.3%	12.4%	1457	12.7%
	Prediction	2305			1655	

9.5.2.3 Driving Range Estimation

As discussed in Sect. 5.2 typical strategies for the estimation of electric vehicles driving range make assumptions on the average energy consumption per kilometer. Such an average energy consumption, often corresponding to the worst-case consumption for conservative estimation, is then used to calculate the driving range in terms of distance.

In Fig. 9.17, the proposed strategy for the calculation of the driving range is compared to a typical approach based on an average energy consumption and the corresponding driving range in terms of distance. In the experiment, which can be conducted only in simulation, a conservative average energy consumption of 0.18 kWh/km was chosen (the value is consistent with a worst-case energy consumption observed during the experimental campaign). The available energy capacity was set to 1 kWh. The chosen average energy consumption translates into a radius of 5.5 km, which corresponds to a quite symmetric driving range as shown in blue in Fig. 9.17. However this approach neglects important factors such as road grade, traffic conditions, type of employed route. The proposed strategy is able to take into account all these aspects, and every destination inside the driving range may be reached by following an eco-route. The driving range (in green in Fig. 9.17) is asymmetric about the origin due to the presence of hilly terrain in the road network and different consumption patterns. In this example, the energy driving range varies from a minimum of about 5 km to a maximum of about 10 km.

Furthermore, it may happen that the region is not simply connected, as discussed in [25], meaning that some destinations in the driving range are unreachable with the current battery state of charge, even by following an eco-route. Such critical destinations may be shown to the driver for more precise assistance, and they are shown with orange dots in Fig. 9.17. In this case, the unreachable destinations correspond either to points close to the driving range boundary or to particularly energy-expensive roads, such as motorways.

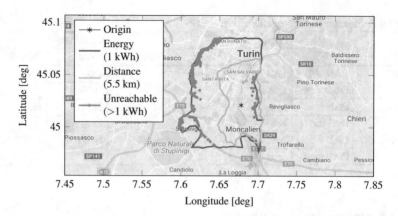

Fig. 9.17 Comparison of the energy driving range based on the prediction of energy consumption and the standard distance driving range

Fig. 9.18 Itinerary for the eco-coaching experimental campaign. The itinerary is about 16 km long, with an estimated travel time of 40 min

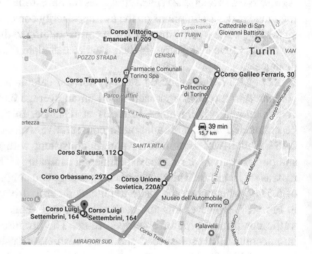

9.5.2.4 Eco-Coaching

The eco-coaching strategy was tested in the city center of Turin. The following test procedure was followed: (i) three repetitions of the itinerary given in Fig. 9.18 without eco-driving assistance (i.e. "pre-eco"), (ii) the driver was then introduced to the eco-driving smartphone application, and (iii) three repetitions of the itinerary were then performed with the eco-driving assistance (i.e. "eco").

An example of the corresponding speed profiles is given in Fig. 9.19. Table 9.10 shows the energy consumption measured from the CAN bus with and without the eco-driving assistance. It shows that with the eco-coaching assistance the energy consumption was reduced by 9% on average, while travel time was reduced by 3%. It is therefore possible to improve the energy efficiency of the trips without driving

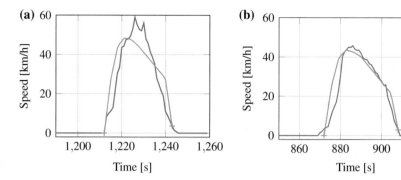

Fig. 9.19 Example of measured (blue) and optimal (orange) vehicle speed profiles for the same road segment without (**a**) and with the eco-coaching assistant (**b**)

Table 9.10 Experimental eco-coaching results

	Energy [Wh]	Time [s]	Speed[km/h]
Average of pre-eco trips	2246	2623	19.9
Average of eco trips	2041	2543	21.3
Variation	−9%	−3%	+3%

more slowly since the average speed is not decreased. One additional fact is that *each* eco-trip made with the eco-driving assistance has a lower energy consumption.

9.5.2.5 Overall Gains

Eco-routing and eco-coaching have been validated independently on a similar urban driving conditions with the same vehicle and driver. On average, the eco-coaching and eco-routing allow the driving range to increase by 9 and 12% respectively (as compared to the fastest route), by reducing the energy consumption. The driving range prediction strategy allows the driver to have a more precise knowledge of which destination is reachable. The state-of-the-art iso-distance approach is significantly less precise and is therefore necessarily tuned in a conservative way. By overcoming this limitation, the proposed strategy allows the driver to use the full potential of the available driving range.

References

1. Asadi Behrang, Vahidi Ardalan (2011) Predictive cruise control: Utilizing upcoming traffic signal information for improving fuel economy and reducing trip time. IEEE Trans Control Syst Technol 19(3):707–714
2. Mahler G, Vahidi A (2014) An optimal velocity-planning scheme for vehicle energy efficiency through probabilistic prediction of traffic-signal timing. IEEE Trans Intell Transp Syst 15(6):2516–2523
3. Wan N, Vahidi A, Luckow A (2016) Optimal speed advisory for connected vehicles in arterial roads and the impact on mixed traffic. Transp Res Part C: Emerg Technol 69:548–563
4. Mahler G, Winckler A, Fayazi SA, Vahidi A, Filusch M (2017) Cellular communication of traffic signal state to connected vehicles for eco-driving on arterial roads: system architecture and experimental results. In: Proceedings of intelligent transportation systems conference, pp 1–6. IEEE
5. Argonne National Lab. Autonomie vehicle system simulation tool (2019). https://www.autonomie.net/
6. Paramics Q (2009) The paramics manuals, version 6.6. 1. Quastone Paramics LTD, Edinburgh, Scotland, UK
7. Kamal MAS, Mukai M, Murata J, Kawabe T (2011) Ecological driving based on preceding vehicle prediction using MPC. IFAC Proc Vol 44(1):3843–3848
8. Xia H, Boriboonsomsin K, Schweizer F, Winckler A, Zhou K, Zhang W-B, Barth M (2012) Field operational testing of eco-approach technology at a fixed-time signalized intersection. In: Proceedings of international conference on intelligent transportation systems (ITSC), pp 188–193. IEEE
9. Andreas W, Andreas W (2013) Advanced traffic signal control algorithms, appendix A: Exploratory advanced research project. Technical report, BMW final report
10. Fayazi SA, Vahidi A (2018) Mixed-integer linear programming for optimal scheduling of autonomous vehicle intersection crossing. IEEE Trans Intell Veh 3(3):287–299
11. Fayazi SA, Vahidi A (2017) Vehicle-In-the-Loop (VIL) verification of a smart city intersection control scheme for autonomous vehicles. In: Proceedings of conference on control technology and applications (CCTA), pp 1575–1580. IEEE
12. Fayazi SA, Vahidi A, Luckow A (2017) Optimal scheduling of autonomous vehicle arrivals at intelligent intersections via MILP. In: Proceedings of American control conference (ACC), pp 4920–4925. IEEE
13. Trafficware. Synchro studio (2019). https://www.trafficware.com/synchro.html
14. Mathew TV (2014) Transportation systems engineering. Cell Transmission Models, IIT Bombay
15. Zhang C, Vahidi A (2011) Predictive cruise control with probabilistic constraints for eco driving. In: Proceedings of dynamic systems and control conference and symposium on fluid power and motion control, pp. 233–238. American Society of Mechanical Engineers
16. Wan N, Zhang C, Vahidi A (2019) Probabilistic anticipation and control in autonomous car following. IEEE Trans Control Syst Technol 27:30–38
17. Dollar RA, Vahidi A (2017) Quantifying the impact of limited information and control robustness on connected automated platoons. In: Proceedings of international conference on intelligent transportation systems (ITSC), pp 1–7. IEEE
18. Dollar RA, Vahidi A (2018) Efficient and collision-free anticipative cruise control in randomly mixed strings. IEEE Trans Intell Veh 3:439–452
19. Gurobi Optimization. Gurobi optimizer 5.0 (2013). http://www.gurobi.com
20. Pourabdollah M, Bjärkvik E, Fürer F, Lindenberg B, Burgdorf K (2017). Calibration and evaluation of car following models using real-world driving data. In: Proceedings of international conference on intelligent transportation systems (ITSC), pp 1–6. IEEE
21. Dollar RA, Vahidi A (2018) Predictively coordinated vehicle acceleration and lane selection using mixed integer programming. In: Proceedings of dynamic systems and control conference, pp V001T09A006–V001T09A006. American Society of Mechanical Engineers

22. De Nunzio G, Sciarretta A, Gharbia IB, Ojeda LL (2018) A constrained eco-routing strategy for hybrid electric vehicles based on semi-analytical energy management. In: Proceedings of international conference on intelligent transportation systems (ITSC), pp 355–361. IEEE
23. HERE. HERE APIs (2019). https://developer.here.com/develop/rest-apis
24. Armstrong JS (1985) Long-range forecasting. Wiley, New York ETC
25. De Nunzio G, Thibault L (2017) Energy-optimal driving range prediction for electric vehicles. In: Proceedings of intelligent vehicles symposium (IV), pp 1608–1613. IEEE

Appendix A
Parametric Optimization Method
for Eco-Driving of ICEVs

In Sect. 6.4.3, the eco-driving of an ICEV has been formulated as an ED-OCP resulting in a TPBVP. Solving this problem can be reduced to finding the switching times) between the modes that can be possibly part of the optimal solution.

The parametric optimization method requires the preliminary solution of the law of motion (2.1)–(2.2) for the modes A $(T_e = T_{e,max}(v), F_b = 0)$, C $(T_e = T_{e,min}(v) = 0$,[1] $F_b = 0)$, and B $(T_e = T_{e,min}(v) = 0, F_b = F_{b,max})$, with $\gamma(t)$ variable according to the law (2.17). Consequently, speed trajectories $v_A(\tau)$, $v_C(\tau)$, and $v_B(\tau)$ are calculated such that $v_k(0) = 0$, together with related trajectories of position $s_k(\tau) = \int_0^\tau v_k(\tau)d\tau$ and fuel consumed $E_{f,k}$, for each $k \in \{A, C, B\}$.

For a three-phase sequence, for instance, A-S-A,[2] the optimal speed profile is defined by six parameters: $\tau_{\{1,...,4\}}, v_\sigma, \Delta t$, which are related to the switching times by the relationships $t_1 = \tau_2 - \tau_1$ and $t_2 = t_1 + \Delta t$, see Fig. A.1. With the four sequence-depending boundary conditions $v_A(\tau_1) = v_i$, $v_A(\tau_4) = v_f$, $s_A(\tau_4) - s_A(\tau_3) + s_A(\tau_2) - s_A(\tau_1) + v_\sigma \Delta t = s_f, \tau_4 - \tau_3 + \tau_2 - \tau_1 + \Delta t = t_f$ and the two additional conditions $v_A(\tau_2) = v_A(\tau_3) = v_\sigma$, the six parameters can be calculated which determine the optimal speed profile.

For a four-phase sequence, for instance, A-S-C-B, the optimal speed profile is defined now by eight parameters: $\{\tau_{\{1,...,6\}}, v_\sigma, \Delta t\}$, which are related to the switching times by the relationships $t_1 = \tau_2 - \tau_1$, $t_2 = t_1 + \Delta t$, and $t_3 = t_2 + \tau_4 - \tau_3$ see Fig. A.2. The four boundary conditions $v_A(\tau_1) = v_i$, $v_B(\tau_6) = v_f$, $s_B(\tau_6) - s_B(\tau_5) + s_C(\tau_4) - s_C(\tau_3) + s_A(\tau_2) - s_A(\tau_1) + v_\sigma \Delta t = s_f, \tau_6 - \tau_5 + \tau_4 - \tau_3 + \tau_2 - \tau_1 + \Delta t = t_f$ and the additional conditions $v_B(\tau_5) = v_C(\tau_4), v_A(\tau_2) = v_\sigma = v_D(\tau_3)$ add up to seven conditions. Hence, a degree of freedom is remaining. The latter is chosen in such a way to minimize the function

---

[1]In Sect. 6.4.3 the assumption $u_{p,min} = 0$ was used. Removing it, this mode is generalized by the override operation (engine brake).

[2]Here S does not represent a "true" singular arc with constant speed but the PnG mode, which yields a speed that is constant on average. The latter has been obtained in Sect. 6.4.3 under the assumption of fuel cutoff while coasting.

© Springer Nature Switzerland AG 2020
A. Sciarretta and A. Vahidi, *Energy-Efficient Driving of Road Vehicles*,
Lecture Notes in Intelligent Transportation and Infrastructure,
https://doi.org/10.1007/978-3-030-24127-8

Fig. A.1 Parametric optimization method in the case of a three-phase optimal trajectory: speed trajectory, switching times t_1, t_2, and PnG arc around v_σ (**a**); trajectories $v_A(\tau)$ with the characteristic times $\tau_1, \ldots, \tau_4$

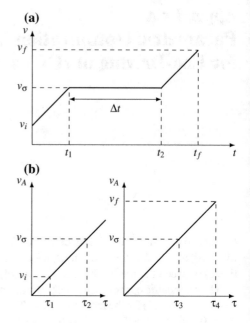

Fig. A.2 Parametric optimization method in the case of a four-phase optimal trajectory: speed trajectory, switching times $t_1, \ldots, t_3$, and PnG arc around the speed v_σ (**a**); trajectories $v_A(\tau)$, $v_C(\tau)$, and $v_B(\tau)$ with the characteristic times $\tau_1, \ldots, \tau_6$

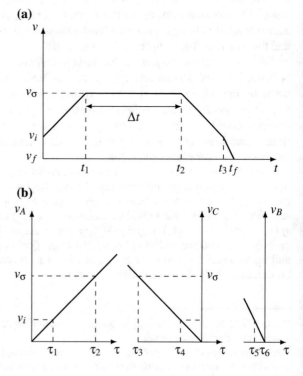

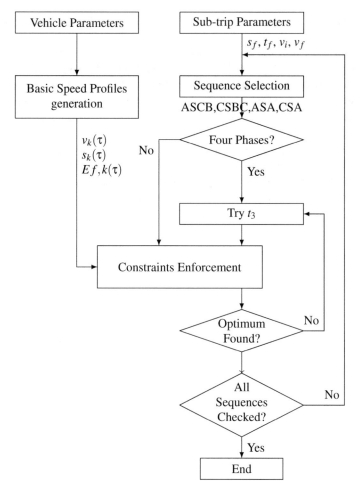

Fig. A.3 Flowchart of the parametric optimization method to solve the TPBVP resulting from the ED-OCP for ICE vehicles

$$E_f = E_{f,A}(\tau_2) - E_{f,A}(\tau_1) + E_{f,C}(\tau_3) - E_{f,C}(\tau_4) + E_{f,B}(\tau_5) - E_{f,B}(\tau_6) + \Delta t P_{f,\sigma} \,,$$
$$(A.1)$$

subject to the constraints above. The quantity $P_{f,\sigma}$ is the fuel power for the engine operating point corresponding to the PnG operation around v_σ.

To summarize, once use of optimal control has determined the modes of operation, the ED-OCP is reduced to a one-dimensional parametric optimization for the complete four-mode sequence. For speed profiles of three-mode basic sequences, no further optimization is necessary. In this case, the velocity trajectory is given by the mission constraints only.

The online algorithm used in the eco-coaching system presented in the main text operates as follows, see Fig. A.3. In a first step, the vehicle parameters are set. From such information, the speed, position, and fuel trajectories for the three modes A, C, and B are preliminarily calculated as described above.

During online use, the identification of the boundary conditions of the actual segment (v_i, v_f, t_f, s_f) allow a reduction of the number of admissible mode sequences. When several sequences are still possible, the rest of the algorithm is repeated and the results are compared to find the optimal mode sequence. For three-mode sequences, constraints are enforced and all switching times are directly calculated. For four-mode sequences, the only degree of freedom, chosen to be the switching time t_3 (or τ_5), is found by finding the minimum of the function (A.1). The search limits are $t_{3,min}$ such that $v_B(\tau_{5,min}) = v_\sigma$ and $t_{3,max} = t_f$. Within these boundaries, the function $E_f(t_3)$ is monotone in most practical situations and thus the *internal halving algorithm*, a direct search method, can be used to find its minimum.

Appendix B
Domain of Feasibility of the Analytical Optimal Speed Profiles for EVs

In this appendix we derive the feasibility domain—in terms of the boundary conditions t_f, s_f—of the solution of the ED-OCP for the simplified EV modeling assumptions of Sect. 6.5 (parabolic speed profile). We start the discussion with the unconstrained speed profile, then we introduce speed and position constraints.

B.1 Unconstrained Case

Under the aforementioned assumptions, the parabolic speed profile

$$v(t) = v_i - \frac{4v_i t}{t_f} - \frac{2v_f t}{t_f} - \frac{6s_f t^2}{t_f^3} + \frac{6s_f t}{t_f^2} + \frac{3v_i t^2}{t_f^2} + \frac{3v_f t^2}{t_f^2}, \tag{B.1}$$

is the optimal speed profile to cover a distance s_f in t_f units of time starting at v_i and ending at v_f, and without constraints. Its derivative (acceleration) and second derivative are evaluated as

$$\dot{v}(t) = a(t) = -\frac{4v_i}{t_f} - \frac{2v_f}{t_f} - \frac{12s_f t}{t_f^3} + \frac{6s_f}{t_f^2} + \frac{6v_i t}{t_f^2} + \frac{6v_f t}{t_f^2} \tag{B.2}$$

$$\ddot{v}(t) = \dot{a}(t) = -\frac{12s_f}{t_f^3} + \frac{6v_i}{t_f^2} + \frac{6v_f}{t_f^2}. \tag{B.3}$$

The conditions that we impose on these profiles are treated below one by one.

The Function $v(t)$ is Always Positive

To impose that, we find its extremum by equating $\dot{v}$ to zero, yielding

A. Sciarretta and A. Vahidi, *Energy-Efficient Driving of Road Vehicles*,
Lecture Notes in Intelligent Transportation and Infrastructure,
https://doi.org/10.1007/978-3-030-24127-8

$$\hat{\tau} \triangleq \frac{\hat{t}}{t_f} = \frac{3\bar{v} - v_f - 2v_i}{6\bar{v} - 3v_f - 3v_i} , \tag{B.4}$$

where $\bar{v} = s_f/t_f$. The extremal value of speed is obtained as

$$\hat{v} = v(\hat{\tau}) = \frac{9\bar{v}^2 - 6\bar{v}(v_i + v_f) + (v_f^2 + v_iv_f + v_i^2)}{6\bar{v} - 3(v_i + v_f)} . \tag{B.5}$$

These two functions are represented in Fig. B.1 as a function of $\bar{v}$. The function $\hat{\tau}$ has a horizontal asymptote at $\hat{\tau} = 1/2$, a zero for

$$\bar{v} = \frac{2v_i + v_f}{3} \triangleq \bar{v}_1 \tag{B.6}$$

and a vertical asymptote for

$$\bar{v} = \frac{v_i + v_f}{2} \triangleq \bar{v}_2 , \tag{B.7}$$

and thus can be rewritten as

$$\hat{\tau} = \frac{1}{2} \cdot \frac{\bar{v} - \bar{v}_1}{\bar{v} - \bar{v}_2} . \tag{B.8}$$

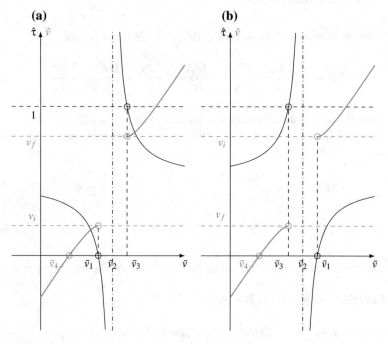

Fig. B.1 Extremal speed and time $\hat{v}$ and $\hat{\tau}$ as a function of $\bar{v} \triangleq s_f/t_f$, for $v_i \leq v_f$ (left) and $v_f \leq v_i$ (right)

Moreover, $\hat{\tau} = 1$ for

$$\bar{v} = \frac{2v_f + v_i}{3} \triangleq \bar{v}_3 . \tag{B.9}$$

The function $\hat{v}$ can be rewritten as

$$\hat{v} = \frac{3}{2} \cdot \frac{(\bar{v} - \bar{v}_{4+})(\bar{v} - \bar{v}_{4-})}{\bar{v} - \bar{v}_2} , \tag{B.10}$$

with

$$\bar{v}_{4\pm} \triangleq \frac{v_i + v_f}{3} \pm \frac{\sqrt{v_i v_f}}{3} . \tag{B.11}$$

The function $\hat{v}$ has its own extrema that are found by equating the derivative $d\hat{v}/d\bar{v}$ to zero. These extrema are found at $\bar{v}_1$ and $\bar{v}_3$. It is easy to verify that

$$\hat{v}(\bar{v}_3) = v_f \quad \text{and} \quad \hat{v}(\bar{v}_1) = v_i . \tag{B.12}$$

It is also easy to verify that either $\bar{v}_3 \leq \bar{v}_2 \leq \bar{v}_1$ (for $v_f \leq v_i$) or $\bar{v}_1 \leq \bar{v}_2 \leq \bar{v}_3$ (for $v_i \leq v_f$). Moreover, $\min(\bar{v}_1, \bar{v}_3) \leq \bar{v}_{4+} \leq \bar{v}_2$ and $\bar{v}_{4-} \leq \min(\bar{v}_1, \bar{v}_3)$ always hold. Therefore, the domain of $\bar{v}$ is divided in four regions:

- $\bar{v} \leq \bar{v}_1$. Here $0 \leq \tau \leq 1$ and $\hat{v} \leq \min(v_i, v_f)$. The speed profile has a minimum.
- $\bar{v}_1 \leq \bar{v} \leq \bar{v}_2$. Here $\tau \leq 0$. The speed profile is monotone.
- $\bar{v}_2 \leq \bar{v} \leq \bar{v}_3$. Here $\tau \geq 1$. The speed profile is monotone.
- $\bar{v} \geq \bar{v}_3$. Here $0 \leq \tau \leq 1$ and $\hat{v} \geq v_f$. The speed profile has a maximum.

All of the latter three regions are feasible, while in the first region we must require that $\hat{v} \geq 0$, that is,

$$\bar{v} \geq \bar{v}_{4-} . \tag{B.13}$$

In summary, the feasibility of the speed profile imposes the constraint

$$F_{UB1}(t_f, s_f) = \frac{s_f}{t_f} - \frac{v_i + v_f}{3} + \frac{\sqrt{v_i v_f}}{3} \geq 0 . \tag{B.14}$$

For $v_i = 0$ or $v_f = 0$, conditions (7.3), resp. (7.14) are retrieved.

The Speed is Bounded by a Maximum Value v_{max}

To impose that, we consider only the fourth subdomain of $\bar{v}$, where we require that

$$\hat{v} = \frac{3}{2} \cdot \frac{(\bar{v} - \bar{v}_{4+})(\bar{v} - \bar{v}_{4-})}{\bar{v} - \bar{v}_2} \leq v_{max} . \tag{B.15}$$

By developing all the factors, we obtain that the profile is feasible if

$$F_{LB1}(s_f, t_f) = \frac{v_i + v_f + v_{max} + \sqrt{v_i v_f + v_{max}^2 - v_i v_{max} - v_f v_{max}}}{3} - \frac{s_f}{t_f} \geq 0 . \tag{B.16}$$

Again, for $v_i = 0$ or $v_f = 0$, conditions (7.4) and (7.15) are retrieved, respectively.

The Acceleration is Bounded by a Maximum Value a_{max}

From the fact that the acceleration derivative (B.3) is a constant, it is clear that the acceleration is either always increasing or decreasing. The top value is thus either at $\tau = 0$ or $\tau = 1$. We evaluate these two values as

$$a(0) = -\frac{4v_i}{t_f} - \frac{2v_f}{t_f} + \frac{6\bar{v}}{t_f} \tag{B.17}$$

and

$$a(t_f) = \frac{2v_i}{t_f} + \frac{4v_f}{t_f} - \frac{6\bar{v}}{t_f}. \tag{B.18}$$

The two aforementioned cases are discriminated by $\bar{v}$ being less or greater than $\bar{v}_2$ defined above. If $\bar{v} \leq \bar{v}_2$, then $a(0) \leq a(t_f)$ and the speed profile is feasible if

$$F_{UB2a}(t_f, s_f) = 6s_f - 2v_i t_f - 4v_f t_f + t_f^2 a_{max} \geq 0. \tag{B.19}$$

If, on the contrary, $\bar{v} \geq \bar{v}_2$, then $a(0) \geq a(t_f)$ and the speed profile is feasible if

$$F_{LB2a}(t_f, s_f) = -6s_f + 2v_f t_f + 4v_i t_f + t_f^2 a_{max} \geq 0. \tag{B.20}$$

Note that, for $a_{max} = 0$, the two conditions above become $\bar{v}_3 \leq \bar{v} \leq \bar{v}_1$, which is only possible in the case $v_i \geq v_f$.

The Acceleration is Bounded by a Minimum Value $a_{min} < 0$

The maximum deceleration is obtained at $t = 0$ if $\bar{v} \leq \bar{v}_2$ or at $t = t_f$ if $\bar{v} \geq \bar{v}_2$. By applying the limiting condition in both cases, we obtain the two inequalities

$$F_{UB2d}(t_f, s_f) = 6s_f - 2v_f t_f - 4v_i t_f - t_f^2 a_{min} \geq 0. \tag{B.21}$$

$$F_{LB2d}(t_f, s_f) = -6s_f + 2v_i t_f + 4v_f t_f - t_f^2 a_{min} \geq 0. \tag{B.22}$$

Clearly inequalities (B.21)–(B.22), given the symmetry of the parabolic speed profile, are the same as the (B.19)–(B.20), except that the roles of v_i and v_f are interchanged, as well as a_{max} is replaced by $|a_{min}|$.

In the case where $a_{min} = -a_{max}$ (a reasonable case recalling the fact that the underlying model is grounded on EV characteristics), the (B.21)–(B.22) can be lumped with (B.19)–(B.20) as follows:

$$F_{UB2}(t_f, s_f) = 6s_f - 2\min(v_i, v_f)t_f - 4\max(v_i, v_f)t_f + t_f^2 a_{max} \geq 0, \tag{B.23}$$

$$F_{LB2}(t_f, s_f) = -6s_f + 2\max(v_i, v_f)t_f + 4\min(v_i, v_f)t_f + t_f^2 a_{max} \geq 0. \tag{B.24}$$

Fig. B.2 Domain of
feasibility (shaded gray area)
of the parabolic speed profile
in the plane s_f–t_f. The
curves shown are: F_{UB1}
(orange), F_{UB2} (green), F_{LB1}
(purple), and F_{LB2} (blue)

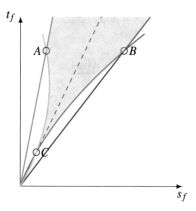

Overall

By putting together (B.14)–(B.24), we can find the domain of feasibility of the
parabolic speed profile. For a given s_f, conditions (B.24) and (B.16) are both
lower bounds for t_f, while (B.14) and (B.23) are upper bounds, whence comes
their labeling. Let us find the intersections between the various curves, which are
depicted in Fig. B.2. The two curves (B.23)–(B.24) intersect at the origin of the axes
($t_f = 0$, $s_f = 0$) and at

$$t_f = \frac{|v_f - v_i|}{a_{max}} \triangleq t_C, \quad s_f = \frac{1}{2} \cdot \frac{|v_f^2 - v_i^2|}{a_{max}} \triangleq s_C . \tag{B.25}$$

It is clear that the values of t_f less than t_C are infeasible.

Curves (B.14) and (B.23) intersect at

$$t_f = \frac{2}{a_{max}}(\max(v_i, v_f) + \sqrt{v_i v_f}) \triangleq t_A, \quad s_f = t_A \bar{v}_{4-} \triangleq s_A . \tag{B.26}$$

Curves (B.16) and (B.24) intersect at

$$t_f = \frac{2(v_{max} - \min(v_i, v_f) + \sqrt{v_i v_f + v_{max}^2 - v_i v_{max} - v_f v_{max}})}{a_{max}} \triangleq t_B . \tag{B.27}$$

Clearly, these quantities are always larger than zero.

B.2 Constrained Case

Speed Constraint

In the presence of a constant speed constraint, the optimal speed profile is calculated
from considerations similar to those in Sect. 7.5.2 and reads

$$v(t) = \begin{cases} v_i + \dfrac{2(v_{max} - v_i)}{t_1}t - \dfrac{(v_{max} - v_i)}{t_1^2}t^2, & t \in [0, t_1) \\[3mm] v_{max}, & t \in [t_1, t_2] \\[3mm] v_f + \dfrac{2(v_{max} - v_f)}{t_f - t_2}(t_f - t) - \dfrac{(v_{max} - v_f)}{(t_f - t_2)^2}(t_f - t)^2, & t \in (t_2, t_f] \end{cases} \quad \text{(B.28)}$$

where t_1 and t_2 are the entry and exit times of the boundary interval, respectively.

The time t_2 is found by imposing the equality of the control input time derivative (second derivative of speed) in the two unconstrained phases, resulting in

$$t_2 = t_f - t_1 \sqrt{\frac{v_{max} - v_f}{v_{max} - v_i}} , \quad \text{(B.29)}$$

from which, the special case of symmetric speed profile is retrieved for $v_i = v_f = 0$. Finally, the time t_1 is found by imposing the overall distance, which results in

$$t_1 = \frac{3(v_{max}t_f - s_f)\sqrt{v_{max} - v_i}}{(v_{max} - v_f)^{3/2} + (v_{max} - v_i)^{3/2}} , \quad \text{(B.30)}$$

from which, the special case (7.37) is retrieved for $v_i = v_f = 0$.

A first obvious constraint is that

$$v_{max} - \frac{s_f}{t_f} \geq 0 . \quad \text{(B.31)}$$

On the other hand, we might want to limit the acceleration of the constrained speed profile to a maximum value a_{max}. When the speed constraint is active, the top value of the acceleration is necessarily at $t = 0$ and its value is

$$a(0) = 2(v_{max} - v_i)/t_1 . \quad \text{(B.32)}$$

The constrained speed profile is thus feasible if

$$F_{LB2'a}(t_f, s_f) \triangleq v_{max}t_f - s_f +$$
$$- \frac{2\left((v_{max} - v_i)^2 + (v_{max} - v_f)\sqrt{(v_{max} - v_i)(v_{max} - v_f)}\right)}{3a_{max}} \geq 0 , \quad \text{(B.33)}$$

which is clearly more restrictive than (B.31). Therefore the latter will not be considered further.

In addition, we might want to limit the deceleration of the constrained speed profile to a maximum value $|a_{min}|$, where $a_{min} < 0$. The maximum deceleration is necessarily obtained at $t = t_f$ and its value is

$$a(t_f) = -\frac{2\sqrt{(v_{max} - v_f)(v_{max} - t_i)}}{t_1} . \quad \text{(B.34)}$$

Fig. B.3 Domain of feasibility of the parabolic speed profile (shaded gray area) and the speed-constrained speed profile (dark gray area) in the plane s_f–t_f. The curves shown are: F_{UB1} (orange), F_{UB2} (green), F_{LB1} (purple), F_{LB2} (blue), and $F_{LB2'}$ (black)

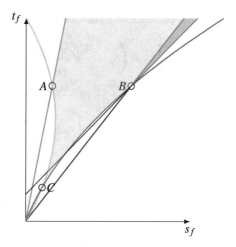

The constrained speed profile is thus feasible if

$$F_{LB2'd}(t_f, s_f) \triangleq v_{max}t_f - s_f - \frac{2\left((v_{max} - v_f)^2 + (v_{max} - v_i)\sqrt{(v_{max} - v_i)(v_{max} - v_f)}\right)}{3|a_{min}|} \geq 0 . \quad (B.35)$$

Contrarily to the unconstrained case, the maximum deceleration condition can be more restrictive than the maximum acceleration condition, as it is easily seen by inspection and comparison of (B.33) and (B.35).

In the common case when $a_{min} = -a_{max}$, (B.33)–(B.35) can be lumped together as

$$F_{LB2'}(t_f, s_f) \triangleq v_{max}t_f - s_f - \frac{2\left((v_{max} - \min(v_i, v_f))^2 + (v_{max} - \max(v_i, v_f))\sqrt{(v_{max} - v_i)(v_{max} - v_f)}\right)}{3a_{max}} \geq 0 .$$
$$(B.36)$$

Curves F_{LB1} (B.16), F_{LB2} (B.24), and $F_{LB2'}$ (B.36) intersect each other at t_B. For $t_f \geq t_B$, the domain of feasibility is thus wider than that allowed by the unconstrained solution only, as shown in Fig. B.3.

Position Constraint

The leader motion is given by $s_p(t) = s_{p,0} + v_{p,0}t + \frac{a_p t^2}{2}$, provided that that its speed $v_p(t) = v_{p,0} + a_p t$ remains positive. When the leader acceleration a_p is negative, it is possible that the leader stops at a time $v_{p,0}/|a_p| < t_f$, at which the position reached is $s_{p,0} + v_{p,0}^2/(2|a_p|)$.

In the presence of a position constraint, the optimal speed profile is calculated as

$$
v(t) = \begin{cases}
\begin{aligned}
& v_i + \left(a_p + \frac{4(v_{p,0} - v_i)}{t_1} + \frac{6s_{p,0}}{t_1^2} \right) t - \\
& \quad - \left(\frac{6s_{p,0}}{t_1^3} + \frac{3(v_{p,0} - v_i)}{t_1^2} \right) t^2,
\end{aligned} & t \in [0, t_1) \\[2em]
\begin{aligned}
& v_{p,0} + a_p t_1 + \left(a_p - \frac{6s_{p,0}}{t_1^2} - \frac{2(v_{p,0} - v_i)}{t_1} \right) (t - t_1) + \\
& \quad + \left(v_f - 3v_{p,0} + 2v_i - 6\frac{s_{p,0}}{t_1} - a_p t_f + 6s_{p,0}\frac{t_f}{t_1^2} + \right. \\
& \quad \left. + 2(v_{p,0} - v_i)\frac{t_f}{t_1} \right) \frac{(t - t_1)^2}{(t_f - t_1)^2},
\end{aligned} & t \in (t_1, t_f]
\end{cases}
$$

$$(B.37)$$

The contact time t_1 is found by imposing the overall distance, which results in a cubic equation

$$
(v_i - v_f + a_p t_f)t_1^3 + (4v_{p,0}t_f + v_f t_f - 2v_i t_f + a_p t_f^2/2 - 3s_f)t_1^2 +
$$
$$
+ (6s_{p,0}t_f + v_i t_f^2 - v_{p,0}t_f^2)t_1 - (3s_{p,0}t_f^2) = 0 \quad (B.38)
$$

from which, the special case (7.37) is retrieved for $v_i = v_f = s_{p,0} = a_p = 0$.

We do not consider in this section speed constraints ($v_{max} \to \infty$). Thus the condition F_{LB1} given by (B.16) becomes the trivial condition $t_f \geq 0$. In turn, the limit case for the unconstrained profile is that the cubic equation $s^*(t) = s_p(t)$, where s^* is the unconstrained position profile, has just one negative real root. That reduces to imposing the sign on the discriminant of the cubic equation, i.e.,

$$
F_{LB3}(t_f, v_f) \triangleq - \left(18abcd - 4b^3d + b^2c^2 - 4ac^3 - 27a^2d^2 \right) \geq 0, \quad (B.39)
$$

where

$$
a = \frac{v_f}{t_f^2} + \frac{v_i}{t_f^2} - \frac{2s_f}{t_f^3}, \tag{B.40}
$$

$$
b = \frac{3s_f}{t_f^2} - \frac{v_f}{t_f} - \frac{2v_i}{t_f} - \frac{a_p}{2}, \tag{B.41}
$$

$$
c = v_i - v_{p,0}, \tag{B.42}
$$

$$
d = -s_{p,0}, \tag{B.43}
$$

which yields (7.71) in the aforementioned special case.

This lower bound can be exceeded by the position-constrained speed profile. However, we must obviously require that the final position does not exceed the final leader position, that is,

$$F_{LB3''}(t_f, s_f) \triangleq s_p(t_f) - s_f \geq 0 , \tag{B.44}$$

where

$$s_p(t_f) = \begin{cases} s_{p,0} + v_{p,0}t_f + \frac{a_p}{2}t_f^2, & \text{if } (-\frac{v_{p,0}}{a_p} \geq t_f) \\ s_{p,0} - \frac{v_{p,0}^2}{2a_p}, & \text{otherwise} \end{cases} . \tag{B.45}$$

Condition (B.44) reduces to (7.72) in the special case treated in that section.

In the domain of feasibility of the position-constrained profile, the lower bound F_{LB2} (B.24) is replaced by the conditions that the maximum acceleration of the position-constrained profile is lower than a_{max} and the maximum deceleration is lower than $|a_{min}|$. These two conditions respectively yield

$$F_{LB2''a}(t_f, s_f) \triangleq a_{max} - a_p - \frac{4(v_{p,0} - v_i)}{t_1(t_f, s_f)} - \frac{6s_{p,0}}{t_1(t_f, s_f)^2} \geq 0 \tag{B.46}$$

and

$$F_{LB2''d} \triangleq a_{max} - a_p \frac{t_f + t_1}{t_f - t_1} + \frac{6s_{p,0}}{t_1^2} + v_{p,0} \frac{2t_f - 4t_1}{t_1(t_f - t_1)} - v_i \frac{2}{t_1} + v_f \frac{2}{t_f - t_1} \geq 0 . \tag{B.47}$$

Condition (B.46) is always satisfied in the domain of the constrained profile. Curves F_{LB2} (B.24), F_{LB3} (B.39), and $F_{LB2''d}$ (B.47) all intersect at the same point (s_D, t_D).

Example feasibility domains are sketched in Fig. B.4 for both accelerating and decelerating leader vehicle.

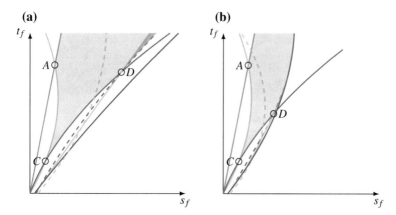

Fig. B.4 Domain of feasibility of the parabolic speed profile (shaded gray area) and the position-constrained speed profile (dark gray area) in the plane s_f–t_f. The curves shown are: F_{UB1} (orange), F_{UB2} (green), F_{LB2} (blue), $F_{LB2''a}$ (dashed green), $F_{LB2''d}$ (dashed blue), F_{LB3} (yellow), and $F_{LB3''}$ (red), for $a_p \geq 0$ (left) and $a_p \leq 0$ (right)

Index

© Springer Nature Switzerland AG 2020
A. Sciarretta and A. Vahidi, *Energy-Efficient Driving of Road Vehicles*,
Lecture Notes in Intelligent Transportation and Infrastructure,
https://doi.org/10.1007/978-3-030-24127-8

Printed in the United States
By Bookmasters